Outdoor Recreation Management

The ability to manage natural resources within an outdoor recreation context is becoming increasingly important as greater participation in recreational activities is placing pressure on the environment.

The second edition of this text provides a comprehensive, non-specialised introduction to outdoor recreation management as an area of study and of real world significance. The book:

- clarifies the links between leisure, recreation, tourism and resource management
- reviews contemporary outdoor recreation and resource management concepts and issues
- critically examines approaches to outdoor recreation planning and management in diverse settings
- considers the future of outdoor recreation management.

This comprehensively revised new edition has many sections rewritten and expanded to reflect contemporary development in leisure and outdoor recreation management in countries such as Australia, Canada, the UK, the US and New Zealand. With extensive use of figures, plates, tables and boxed case studies highlighting theoretical and applied developments, the second edition of *Outdoor Recreation Management* is accessible and student-friendly, offering a truly international focus.

John J. Pigram is Adjunct Professor attached to the Centre for Ecological Economics and Water Policy Research, the University of New England, Australia. He has long-standing research and teaching interests in outdoor recreation and tourism and is co-author of several books including the *Encyclopedia of Leisure and Outdoor Recreation* (also by Routledge).

John M. Jenkins is Associate Professor of Leisure and Tourism Studies and Research Associate at the Centre of Full Employment and Equity, The University of Newcastle, Australia. John is co-author of several books, including the *Encyclopedia of Leisure and Outdoor Recreation* and *Tourism and Public Policy*. He has written more than 50 book chapters and journal articles on issues concerning leisure, outdoor recreation in protected areas, and tourism policy and planning.

Outdoor Recreation Management

Second edition

**John J. Pigram and
John M. Jenkins**

LONDON AND NEW YORK

Second edition published 2006
by Routledge
2 Park Square, Milton Park, Abingdon, Oxon OX14 4RN

Simultaneously published in the USA and Canada
by Routledge
270 Madison Ave, New York, NY 10016

First published 1999
Reprinted 2002, 2003 (twice)

Routledge is an imprint of the Taylor & Francis Group

© 2006 John J. Pigram and John M. Jenkins

Typeset in Times by
HWA Text and Data Management, Tunbridge Wells
Printed and bound in Great Britain by
TJ International Ltd, Padstow, Cornwall

British Library Cataloguing in Publication Data
A catalogue record for this book is available from the British Library

Library of Congress Cataloging in Publication Data
Pigram J.J.
    Outdoor recreation management / John J. Pigram and John M.
    Jenkins.– 2nd ed.
       p.   cm.
    Includes bibliographical references and index.
    1. Outdoor recreation–Management. 2. Outdoor recreation–Social
    aspects. 3. Leisure–Social aspects. I. Jenkins, John M. (John
    Michael), 1961–. II. Title.
    GV191.66.P5 2005 2006
    790'.06'9–dc22   2005013883

ISBN10: 0–415–36540–6 (hbk)
ISBN10: 0–415–36541–4 (pbk)

ISBN13: 9–78–0–415–36540–6 (hbk)
ISBN13: 9–78–0–415–36541–3 (pbk)

Contents

Illustrations

Plates

Tables

Preface

Much has happened in the world and in the field of outdoor recreation since the first edition of *Outdoor Recreation Management* was published in 1999.

The new century has seen regimes and governments come and go in the international scene, coupled sometimes with appalling and tragic instances of inhumanity and devastation. Against this background, it may seem of little concern to some to refocus on outdoor recreation and its management. Yet, the need remains, and indeed, is reinforced in times of stress, for societies to strive to ensure even greater opportunities to bring balance and fulfilment to people's lives through the recreative use of leisure.

The first edition of this book was international in scope. This second edition builds on that focus, exploring in more detail concepts and case studies drawn from a broad sampling of settings, experiences and evolving issues in outdoor recreation management. The past five years have seen significant developments in areas such as: planning and management of urban parks, national parks and natural areas; legislation affecting leisure and recreation; research into the biophysical and socio-economic impacts of outdoor recreation and tourism; new concepts and methodologies employed in studies of recreation decision-making behaviour and satisfaction; access agreements for public and private lands; provision for the leisure pursuits of people with a disability; and more creative public policy in developed and developing countries.

The important considerations identified in the Preface to the earlier edition have, if anything, been reinforced in their relevance in the ensuing period. Contributions to the literature in leisure, outdoor recreation and tourism have grown impressively, complemented by the ever-widening access to materials on the internet. The authors' commitment to an integrated approach to the management of outdoor recreation opportunities, in tandem with resource conservation, remains. The extent and nature of accessibility to these opportunities, and the associated conflicts which can occur, are of enduring interest. Furthermore, the reaction of governments and agencies to the challenge of providing accessible and satisfying leisure experiences to rapidly ageing populations will call for innovative strategies, public–private partnerships and local entrepreneurship. Unfortunately, impediments posed by over-conservative planning, inter-agency rivalry, entrenched organisa-tional cultures and political priorities, continue to thwart progress in meeting the recreation needs and demands of an increasingly diverse and discriminating clientele.

Industrialised societies remain dominated by policy and planning ideologies which promote the role of market forces in determining resource allocation and distribution. The dehumanising process of automation in search of greater economies and a lower bottom line encourages a philosophy where 'everything is costed – nothing is valued' (quoted in www.nicholasdattner.com). Regrettably, this is also true of approaches to the supply and management of opportunities for outdoor recreation. However, there will always be some segments of the population for which the user-pays principle has little relevance. Unemployment and under-employment, redundancy and early retirement mean that an increasing proportion of society is looking for rewarding outlets for use of time at their disposal. Rather than being seen as a positive trend, the non-voluntary acquisition of greater amounts of leisure time is now recognised as a major social problem. Coping with an unstructured existence can be a difficult and frustrating process. Recreation has the potential to fill this void if the human conditioning, which sees a programmed lifestyle based around regular employment as critical, can be offset. This is particularly so when work hours or the journey-to-work are long in duration. In short, people from all walks of life and levels of society should benefit from access to rewarding recreational opportunities. This book seeks to make a contribution to meeting that challenge.

Acknowledgements

A number of people helped in the preparation of this book. Paul Stolk at The University of Newcastle has been a valuable research assistant on various projects. He also scanned and edited photos and figures. Rudi Boskovic did cartography work for the first edition, some of which carried over into this edition.

Reviewers of the first edition and the proposal for this second edition made very valuable comments which we carefully considered. Others, such as Tony Veal and Stephen Wearing, also gave us feedback, while C. Michael Hall and Stephen J. Page kindly gave the first edition a nice 'rap' in their *Geography of Tourism and Recreation: Environment, Place and Space*.

Andrew Mould and Zoe Kruze of Routledge showed great interest in this edition. Zoe also went to great lengths to help us keep on track and pull everything together as a coherent package of text, plates, figures, tables of contents, references, permissions and so on.

John Hodgson, HWA Text and Data Management, organised a speedy and highly competent job in working through the editing and production of page proofs.

John Pigram is grateful for support provided by the University of New England, where he is Adjunct Professor with the Centre for Ecological Economic and Water Policy Research.

John Jenkins would like to thank The University of Newcastle for valuable resources needed to complete the manuscript. In the time that has elapsed since the publication of the first edition, John Pigram retired from the University of New England. I have been incredibly fortunate to have worked with John Pigram for about 15 years, though we first met when I was an undergraduate student in his Geography of Leisure and Recreation class in 1985. John has been a fantastic and generous mentor to me, a strong and inspirational academic whom I have long admired. Richard Butler, Ralf Buckley, C. Michael Hall, Julie Hodges, Les Killion, Kevin Lyons, Kevin Markwell, Dave Mercer, Rob Schaap, Tony Sorensen, Tony Veal, Jim Walmsley and Stephen Wearing have also provided great inspiration and ideas in various ways over the years.

1 Introduction

Concepts, issues and themes

Outdoor recreation issues may be relatively neglected in our national political discourse, but they are not trivial and never will be on our shrunken planet.

(Carroll 1990: xvii)

This book presents a comprehensive, non-specialised introduction to outdoor recreation management, as an area both of academic study and of real-world significance. Its underlying principle is the potential of recreation to contribute to pleasurable, satisfying use of leisure. Outdoor recreation is recognised as an important form of resource use, and much attention is given to how resources can be managed to provide a quality environment for sustained and satisfying recreational use.

Our decision to focus on outdoor recreation management, and to do so in an international context, was influenced by the wide-ranging and somewhat fragmented research in the field, the restricted geographical focus of many recreation texts, and the need to provide an overview of past and present understandings of outdoor recreation research and activity spanning mainly the developed world. Consequently, this book:

- clarifies the links between leisure, recreation, tourism and resource management;
- reviews contemporary outdoor recreation and resource management concepts and issues;
- critically examines approaches to outdoor recreation planning and management in diverse recreational settings; and
- considers the future of outdoor recreation and the potential influences of economic, social, political and technological developments.

Leisure and outdoor recreation are widely recognised as important elements in people's lives, and are receiving increasing academic attention and respectability (e.g. Mercer 1980a; Chubb and Chubb 1981; Patmore 1983; Van Lier and Taylor 1993; Lynch and Veal 1996; Walmsley and Jenkins 2003). They are vital social issues (e.g. see Owen 1984; Veal 2002) and rewarding forms of human experience, constituting 'a major aspect of economic development and government responsibility' (Kraus 1984: 3). Outdoor recreation brings joy and pleasure to many people, with the provision of appropriate recreational opportunities 'critical to the satisfaction of an individual's need for cognitive and aesthetic stimulation, one of six needs identified by Maslow (1954) as basic to human well-being' (Faulkner 1978, in Walmsley and Jenkins 1994: 89). Put simply, 'in the framework of our civilisation, tourism and recreation have moved from the relatively unimportant margins to a very salient position' (Mieczkowski 1990: 347). People will continue to treasure the outdoors, as they have throughout human history – tracts of land will continue to be set aside for recreation; the beauty of nature will continue to be expressed in art and the development of formal gardens, as it was during the Renaissance; and consumptive activities such as hunting and fishing (whether or not people support such activities), and more passive activities in wilderness areas, such as bushwalking, will afford some the opportunities to experience 'a closer affinity between primitive and modern concepts of outdoor recreation' (Jensen 1977: 15). That said, outdoor recreation is now highly commodified and an expanding 'big business' phenomenon.

This chapter places outdoor recreation in its broader societal context. It defines relevant terms, clarifies related concepts, and discusses the significance of leisure and outdoor recreation in industrialised nations. Approaches to the study of outdoor recreation and the focus of the present book are outlined.

Key definitions and concepts: leisure, recreation and tourism

'The word leisure originates from the Latin *licere*, meaning to be permitted. From *licere* came the French word *loisir*, which means free time, and such English words as license, meaning permission' (Kando 1975: 22). Leisure is important and means different things to different people. For some people, leisure is just as important as work and discrete periods of time are given to leisure each and every day. For others, leisure time is hard to find amidst work (including the journey to work) and the pressures of day-to-day life. The relaxation people experience during leisure may be central to reducing stress in daily living. Indeed, leisure is important to personal development, and viewed holistically, it 'brings a degree of balance to spirit, mind, and body...' (Walmsley and Jenkins 2003: 279).

There are many definitions or conceptualisations of leisure in contemporary society (e.g. see Pieper 1952; DeGrazia 1962; Parker 1971; Kaplan 1975; Godbey and Parker 1976; Patmore 1983; Lynch and Veal 1996). However, three main aspects are commonly noted. First, leisure equates with the enjoyment and satisfaction derived from free-time activities. Second, leisure represents a spiritual condition or state of mind, with the emphasis on self-expression and subjectively perceived freedom (Neulinger 1982). Third, leisure, in one or more of the above contexts, may be associated with activity and is a concept which appears to fit more comfortably with a Western world perspective than with traditional cultures. According to some observers of indigenous Australians,

> there is no evidence in accounts of traditional lifestyle of a separate category of activity that resembled what would now be called leisure or recreation.

Nor do there appear to be any Aboriginal words that equate to these concepts ... We are led to the conclusion that there was no need for such a category of behaviour as leisure (which allows refreshment, escape, personal development and the pursuit of pleasure) as there appeared to be little activity that was considered as drudgery (from which escape was needed).

(Veal and Lynch 2001: 33)

Leitner and Leitner (2004a: 10) defined leisure based on the notion of discretionary or unobligated time. They noted, however, that 'there is considerable debate over whether leisure can or should be viewed as a category of time expenditure'. They then identified several contrasting views of leisure:

- *Classical or traditional leisure* (Kraus 1984), which sees leisure as a highly desirable state of mind or state of being, arising from activities that are intrinsically motivated;
- *Antiutilitarian* view of leisure (Neulinger 1981), which argues that leisure needs no justification and is a state of mind;
- *Social instrument* view, where leisure promotes self or personal growth and serves a useful purpose in so far as people help others;
- *Social class symbol*, a view in which free time symbolises wealth, and hence leisure is a symbol of social class. This view was well documented more than a century ago by Thorstein Veblen in *The Theory of the Leisure Class* (1899). In contemporary society it is evidenced by ownership of yachts and expensive recreational equipment, travel, gastronomy, and membership of exclusive clubs;
- *Activity*, a view in which leisure is closely linked to recreation in that recreation is activity undertaken during free time/leisure time;
- *Casual and serious leisure*, a distinction of Stebbins (1982, 1997), who argued that serious leisure has long-term commitment, may be career-oriented and involves substantial personal involvement and commitment. It 'is systematic pursuit of an amateur, hobbyist, or volunteer activity that participants find so substantial and interesting that, in the typical case, they launch themselves on a career centred on acquiring and expressing its special skills, knowledge and experience' (Stebbins 2003b: 452). Casual leisure, on the other hand, is intrinsically rewarding and may not require special training or skills. 'It is fundamentally hedonic, pursued for its significant level of pure enjoyment or pleasure' (Stebbins 2003a: 46);
- Leisure and work often cannot be separated; leisure can be found in almost any human endeavour (e.g. see Beatty and Torbert 2003). Perhaps this is a more *holistic* view of leisure.

Aristotle viewed leisure as the state of being *free* from the necessity to labour. Freedom is generally considered the key element of leisure. Thus, many definitions link the notion of leisure with free time – periods which are relatively free of economic, social or physical constraints or obligations. In these terms, leisure is a residual component – discretionary time over and beyond that needed for existence

(Clawson and Knetsch 1966). There are several problems with this point of view in that it assumes the dominance of a work rather than a leisure ethic, and it does not give due recognition to the difficulty in distinguishing obligated time from free time. In particular:

- leisure can be experienced within the context of primary role obligations – leisure and work can become indistinguishable. Professional athletes, writers or, more generally, people who derive relaxation and revitalisation from their work, blur the divisions between work and leisure. Perhaps some professional athletes do not consider the financial or other tangible rewards from their pursuits as payment for work itself, but rather as rewards for being skilled and highly competitive at their chosen recreational activity;
- the pursuit of leisure can be influenced significantly by personal associations, values and choices. The leisure of parents or guardians may be constrained or eroded if they feel obliged to commit time to the amusement of their dependants; this 'obligated' time, however, may be one of their few means of escape from the work place. If the differentiating factors, then, are freedom of choice and freedom from necessity to fulfil occupational and family duties and expectations or other obligations (Farina 1980), then the leisure-work dichotomy is tenuous; work can acquire some characteristics of leisure (Jamrozik 1986), and much leisure may take a form which is not the preferred choice of an individual;
- leisure is a fundamental and essential part of people's lives – leisure time is needed for psychological and physical well-being, perhaps as much as work. One of the primary needs of people is leisure that affords psychological strength and refreshment (Perez de Cuellar 1987). Indeed, leisure is well recognised as a means of coping with stress and unexpected life events. Leisure helps people cope with life (e.g. Coleman and Iso-Ahola 1993; Iwasaki and Mannell 2000; Hutchison *et al*. 2003).

The concept of leisure clearly implies more than the antithesis of the necessity to labour or work. Lack of employment does not necessarily equate with leisure. Unemployed people do not always make a conscious choice between work and non-work/leisure. For them, an abundance of time free from work is often dictated by their ability to secure employment. Frustration and anti-social behaviour can occur because of the difficulty of occupying time out of work with meaningful, fulfilling or 're-creative' activities. Enforced 'idleness' or 'free time', as a result of unemployment, underemployment, disability, redundancy or early retirement, is a fact of life in many countries; so much so that the work ethic, which has typified Protestant society for generations, may no longer be relevant to many people. However, imposing a leisure ethic in its place can only be appropriate if this new-found leisure is free of guilt, discomfort and anxiety about survival (Bannon 1976). Such an outcome seems unlikely, and is certainly not a prominent aim in the public policies and programmes of modern industrialised societies.

The industrial system has long held out one rather striking promise to its participants. That is, the eventual opportunity for a great deal more leisure... To urge more leisure is a feckless exercise so long as the industrial system has the capacity to persuade people that goods are more important. Men will value leisure over work only as they find the uses of leisure more interesting or rewarding than those of work, or as they win emancipation from the management of their wants, or both. Leisure is not wanted *per se* but only as these prerequisites are provided.

(Galbraith 1972: 357–9)

Leisure might take place in time free from work, but it is becoming increasingly commodified. Leisure requires that people have money to purchase 'time', recreational access and supporting resources. Kando's (1975) remarks more than a generation ago remain relevant:

the coming of automation, cybernation and affluence would logically seem to produce the leisure society. However, this does not occur because of two developments. First, the society's value system is such that the new status hierarchy places an increasing premium on work; second, the society's economic structure – corporate capitalism – demands costly mass consumption and spectacular mass recreation rather than freedom in leisure.

(Kando 1975: 15)

The view that modern technology would create widespread leisure has been challenged (see below). If anything, modern technology has led to technocratic consumption and much regimented and institutionalised recreation, and costly mass spectacles (Kando 1975: 16) (e.g. the internationalisation of sporting culture such as the American Football 'Superbowl', the America's Cup, the Commonwealth and Olympic Games, Formula 1 Grand Prix, and the Rugby Union and Soccer World Cups via the media and the marketing of associated clothing and equipment). We have entered a broad phase of big-business 'spectatorism', often at home, 'oriented towards high profile sporting events, live theatre extravaganzas, concerts, festivals or the like' (Mercer 1994a: 20). In this context, recreation serves two functions. To paraphrase Kando (1975: 15), from the standpoint of the individual, it restores a person's energy to work. From society's standpoint, it fulfils a major functional prerequisite, namely that of sustaining the economic system. A similar viewpoint was expressed by Braverman:

the atrophy of community and the sharp division from the natural environment leaves a void when it comes to the 'free' hours ... the filling of the time away from the job also becomes dependent on the market, which develops to an enormous degree those passive amusements, entertainments, and spectacles that suit the restricted circumstances of the city and are offered as substitutes for life itself. Since they become the means of filling all the hours of 'free' time, they flow profusely from corporate institutions which have transformed every means of entertainment and 'sport' into a production process for the enlargement of capital. By their very profusion, they cannot help but tend to a standard of mediocrity and vulgarity which debases popular taste, a result which is further guaranteed by the fact that the mass market has a powerful lowest-common-denominator effect because of the search for maximum profit. So enterprising is capital that even where the effort is made by one or another section of the population to find a way to nature, sport, or art through personal activity and amateur or 'underground' innovation, these activities are rapidly incorporated into the market so far as is possible.

(Braverman 1975: 278–9)

At the individual level, perceptions of leisure depend very much on a person's subjective, individual and social/political circumstances, and on their view of the world (e.g. see Parker 1983). The sharp distinction implied between discretionary time and time needed for existence is in fact blurred, because leisure often tends to overlap with other uses of time such as work, family obligations and eating (see Figure 1.1). Despite these qualifications, it is probably true to say that for most people, leisure remains closely associated with uncommitted time. In the first edition of this book, the following definition of leisure was endorsed:

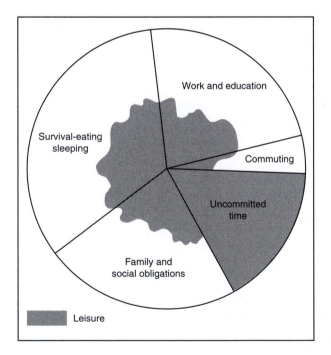

Figure 1.1 The diffusion of leisure time

Source: Department of Environment, Housing and Community Development (1977: 1).

Leisure consists of relatively self-determined activity–experience that falls into one's economically free time roles, that is seen as leisure by participants, that is psychologically pleasant in anticipation and recollection, that potentially covers the whole range of commitment and intensity, that contains characteristic norms and constraints, and that provides opportunities for recreation, personal growth and service to others.

(Kaplan 1975: 26)

Godbey's (1985: 9) definition, however, gives a more philosophic view. He argues 'Leisure is living in relative freedom from the external compulsive forces of one's culture and physical environment so as to be able to act from internally compelling love in ways which are personally pleasing, intuitively worthwhile, and provide a basis for faith'. So, 'True leisure transcends the realm of function or justification by objective criteria' (Goodale and Godbey 1988: 240). We endorse these two definitions with an important caveat. In modern settings, Kaplan's definition would be more compelling if there were not the requirement that leisure falls into one's economically free time roles. This issue is highlighted in the following discussion.

Confusion also arises over the indiscriminate use of the terms 'leisure' and 'recreation', which are closely related and often used interchangeably. The simplest distinction identifies leisure with time and recreation with activity. Recreation is activity voluntarily undertaken, primarily for pleasure and satisfaction, during leisure time, but it 'can also be seen as a social institution, socially organised for social purposes' (Cushman and Laidler 1990: 2). Whereas it is possible to conceive some jobs as having a recreative element, the definition normally requires that no obligation, compulsion or economic incentive be attached to the activity. Recreation, therefore, contrasts with work, the mechanics of life and other activities to which people are normally highly committed. Certain activities are often thought of as inherently recreational. However, in a similar way to leisure, the distinguishing characteristic is not the activity or experience itself, but the attitude with which it is undertaken. For many professional golfers, for instance, perhaps golf is, or becomes, merely an occupation (though it could also be recreation); for the weekend golfer, presumably golf is looked upon as recreation and sport, even though at times it can involve much physical effort and frustration.

According to *The Macquarie Dictionary* (1987), to 'recreate' is 'to refresh by means of relaxation and enjoyment, as after work … to restore or refresh physically or mentally'. Mercer (2003: 412) states that recreation 'refers to activities, either active or passive, enjoyed either outdoors or indoors, which take place during leisure – as opposed to non-work – time'. This gives us a pretty clear picture of what recreation might entail and gives a good insight into the overlap between leisure and recreation. Curiously, the second edition of *The Dictionary of Human Geography* (1986: 391) defined recreation as 'leisure-time activity undertaken away from home', arguing 'some degree of movement is inherent in all recreational activity … such that recreation involves trips of less than day length whilst tourism involves overnight stays away from home'. This definition is difficult to endorse.

The concept of recreation, like that of leisure, is personal and subjective. Thus, value judgements as to the worth or 'moral soundness' of a particular activity often are inappropriate (Godbey 1981). Generally, recreation implies revitalisation of the individual, although purists would argue that *recreation* is, or should be, the culmination of recreational activity – 'the activity is the medium: it is not the message' (Gray and Pelegrino 1973: 6). If this argument were to be accepted, recreation could only be defined in terms of end-results, and potentially recreative activities which, for whatever reason, fail to 'revitalise' the participant, would be excluded. Rather than attempt to split ends from means, it would seem more useful to identify leisure as a process and recreation as a response. As Owen (1984: 157) puts it: 'Leisure has now come to be viewed as a process (Kaplan 1975) and recreation as an experience (Driver and Tocher 1974), which is goal oriented', with participation expected to yield satisfactions (London *et al.* 1977), and therefore physical and emotional rewards (also see Shivers 1967).

> In the specific case of outdoor recreation that element of reward may be stronger since participation will usually [and so not always] require the physical removal of the participant from the home or the workplace in order to engage in the activity in question. There is thus an additional cost in effort, time and/ or money which must be part of the decision to participate.
>
> (Williams 1995: 6)

The term 'outdoor recreation' is more familiar in certain cultural contexts than others, but we do not agree with Mercer's (1994a: 4) view that, 'insisting on the distinction between "indoor" and "outdoor" recreation is as futile as emphasising the contrast between "urban" and "rural" leisure provision'. Regardless of indoor recreation developments, adaptations of such activities as cricket, soccer, tennis, athletics and rock-climbing, or of whether facilities have adjustable roofs, outdoor recreation is just what the category 'outdoor recreation' portrays – recreation that occurs outdoors in urban and rural environments. In this context, then, outdoor recreation raises significant resource management issues which indoor recreation activities do not.

Finally, discussion of tourism in the context of leisure and outdoor recreation is sensible. Much tourism is recreational, in that a good proportion of tourist activity takes place during leisure time, often outdoors, for the purpose of personal pleasure and satisfaction. Outdoor recreation overlaps with tourism in the distinctive characteristics and behaviour associated with each; tourism and outdoor recreation activity involve both travel and interaction with other people, and with the environ-ment, in its widest meaning (see Chapters 5 and 11). Some observers assign an emphasis on economic aspects and profit-making to tourism, while linking outdoor recreation primarily with noncommercial objectives (e.g. Gunn 1979). Unfortu-nately, others make a fundamental distinction between tourism and recreational travel (Britton 1979; Boniface and Cooper 1987). However, these distinctions create and foster an artificial gulf between tourism and outdoor recreation in applied and theoretical terms, leading to unnecessary obstacles to, among other things, understanding people's recreational motivations, choices, behaviour and experiences. In this book, tourism is considered within an essentially recreational framework, and 'may be thought of as the relationships and phenomena arising out of the journeys and temporary stays of people traveling primarily for leisure or recreational purposes' (Pearce 1987: 1).

Tourism receives some special consideration in this book (Chapter 11). Attractions, facilities and services, developed for tourists in industrialised societies, are often used by local residents (and vice versa), and therefore will, in almost any case, impact upon the physical environment and ecosystems, local resident perceptions of, and attitudes to, a range of recreational facilities and services and, of course, tourists themselves.

The significance of leisure and outdoor recreation

> Three great waves have broken across the face of Britain since 1800. First, the sudden growth of dark industrial towns. Second, the thrusting movement along far flung railways. Third, the sprawl of car-based suburbs. Now we see, under the guise of a modest word, the surge of a fourth wave which could be more powerful than all the others. The modest word is leisure.
>
> (Dower 1965: 5)

Leisure

Leisure was once the privilege of the élite. Recently, it has been argued that leisure has become largely the prerogative of the masses. People's historical preoccupation with work as a means of livelihood appears to have been tempered by priorities geared, in part, towards the acquisition of more leisure. Developed countries are faced with the problems of adjusting to, and providing for, a society orientated perhaps as much towards leisure and recreation as it is towards work.

The dimensions of the leisure problem were discussed by Dower (1965), who described the leisure phenomenon as a 'fourth wave', comparable with the advent of industrialisation, the railway age, and urban sprawl.

> The leisure phenomenon can be measured not only in terms of time availability but also in terms of activities engaged in, that is, how people spend their leisure time. It can also be measured in terms of consumer expenditure, that is, the extent to which people spend their money on leisure goods and services. The trends are unmistakable. Growth in participation in virtually all leisure activities since the Second World War has been dramatic ... leisure has become a highly significant element of people's lives and of the economies of advanced industrial nations.
>
> (Veal 1987: 2–3)

The growth of leisure, however, does not only bring benefits to individuals and society. The disadvantages of a leisured existence were foreseen (perhaps somewhat cynically) by George Bernard Shaw, who is reputed to have described a perpetual holiday as 'a good working definition of hell' (Gray and Pelegrino 1973: 3). This might well be applicable to unemployed people and retirees, to adolescents living in remote rural areas, or to homeless people in inner city areas. Shaw's assertion reflects the apparent psychological inability of people to cope with the monotony and burden of a non-structured existence. Many people who are working attempt to occupy time, which might have been utilised for leisure, with additional employment. These situations give some substance to the notion that, for whatever reason, a life of leisure may not be the course of gratification it should be, or is not as accessible as many anticipated.

Work satisfaction does fulfil many human needs. However, for the majority, a reduction in work commitments must seem a highly desirable goal. At the same time, it is being realised that the fundamental consideration is not the overall amount of increased leisure gained, so much as the spatial and sectoral distributions of disposable time. Of practical importance in determining the recreational response, is whether this time is concentrated or dispersed. In Australia, for example, progressive reductions in contractual working hours have been introduced by way of a nineteen-day month, a nine-day fortnight or flexible work arrangements, in part, to improve recreational opportunities. Although the compression of leisure into standard packages probably suits the convenience of both employers and employees, concentrated periods of use place great pressures on the recreation environment, and, in particular, fragile areas (e.g. alpine and coastal areas). Therefore, it would seem desirable, in the interests of recreation resource management, to devise a system of more flexible work patterns incorporating extended, but staggered blocks of disposable time; a desirable goal to be sure, but complicated by social patterns and processes (religious beliefs; designated public holidays often incorporating long weekends; Christmas and New Year festivities; Easter; and school vacations).

Despite growth in participation in outdoor recreation activities, there remain inequities. Many writers have noted that, notwithstanding individual differences, the extent and nature of leisure and recreational activities people engage in are related to their position in the socioeconomic stratification of society and to the class structure (Jamrozik 1986: 189).

> In Britain and in most West European countries, leisure participation generally, and sports participation particularly, is dominated by men, young people, white people, car owners and those in white-collar occupations. The participation rates of women, older people, ethnic minorities, and those in blue-collar occupations are generally lower.
>
> (Glyptis 1993: 6)

Jamrozik (1986) made similar, though more general comments:

> The 'promise' of increased prosperity, and a life of leisure through technological innovation has not, so far, eventuated as expected. Material prosperity

has become a reality but it is not shared, either within countries or across countries. On the contrary, inequalities in income and wealth are increasing and now extend to access to employment and consequently to consumption, including leisure consumption.

(Jamrozik 1986: 204)

Little has changed. Not all sections of society in the so called 'developed' or 'less developed' worlds enjoy adequate access to leisure. In many developing countries, leisure and recreation facilities are limited, especially in those countries where people are starving. Leisure hardly seems possible for those suffering from malnutrition or HIV. At the same time, obesity is a well documented problem among specific groups such as school children in countries such as the US and Australia, and among urban women in parts of Asia and the Pacific Islands.

The reasons are manifold, starting with our genetic make-up and including global economic development, higher incomes, shifting diets, and a range of changes in the nature of work and leisure. Together it is what scientists refer to as the 'nutrition transition', a process that is accelerating rapidly. People living in urban areas throughout the developing world are much further along in the transition than their rural counterparts. Comparatively sedentary jobs demand less physical energy than rural labor. And as more and more women work away from home, traditional patterns of food preparation are changing. Cities also offer a greater range of aggressively marketed calorie-rich food choices. The fact that more people are moving to cities makes the problem [of obesity] more pervasive.

(International Food Policy Research Unit 2004: 9)

Women experience unequal access to, and participation in, leisure as an inevitable consequence of, among other things, 'sexist' policies in society. The problem appears to relate, at least in part, to an unequal incidence of leisure time. While the recent picture is somewhat mixed, it has been a widely held view that men have more leisure time than women, especially women in the workforce (Cushman *et al.* 1996a). The burdens of domestic and child-care responsibilities fall inequitably on women, who make up an increasing percentage of the workforce in Western countries. These factors, together with economic and cultural constraints, might explain why women tend to be more active in home-based recreation activities than men. Even weekends 'become more a matter of overtime work for married women and more a matter of recreation for men' (Rapoport and Rapoport 1975: 13). Furthermore, men are perceived to be more active in sports, while women tend to be more active in arts and cultural activities (Cushman *et al.* 1996a).

For the workforce in general, the same technological progress and social advances, which have permitted reductions in working hours, have imposed pressures on the way leisure is used, and on the extent and nature of recreational participation. LaPage (1970) suggested that non-work discretionary time, ostensibly available for recreation, is constantly eroded by the time necessarily spent in commuting to work and in travel to and from sites for social purposes. More than 40 years ago, Wilensky (1961: 136) deplored the fact that leisure was spent '... commuting and waiting – hanging on the phone, standing in line, cruising for parking space'; very little has changed. Urban sprawl and the concentration of much work in central business districts have, despite innovations and extensive public and private sector investment in transport technologies, resulted in lengthier commuting times for many workers in the Western industrialised world. The development of communication technologies, such as the internet and the mobile phone, means it is more difficult for people to remove themselves from the work place, which for many was once confined to an office and to particular hours. The picture has been further complicated by the effect of evolving social mores and changing lifestyles in urban and rural areas. Couples, where both partners are working, for instance, require more ancillary time in the home for necessary chores and maintenance, so that hours set free from work are taken up with domestic tasks.

Moreover, in a materialistic and seemingly sophisticated society, leisure without affluence often seems of little relevance. Preoccupation with material possessions can divert values away from the acceptance and simple enjoyment of leisure; so, the benefits of improved working conditions are often translated into money terms. Economic circumstances or personal inclination force a trade-off between more free time and increased disposable income, and can lead to the filling up of leisure hours with overtime or a second, or even third, job. This, in turn, curtails the opportunity for recreative use of leisure.

> One of the paradoxes of leisure is that while time and money are *complementary* in the production of leisure activities, they are *competitive* in terms of the resources available to the individual. Some leisure time and some money to buy leisure goods and services are *both* needed before most leisure activities can be pursued.
>
> (Martin and Mason 1976: 62)

Bearing in mind the continuing emphasis on material possessions in modern industrialised nations, there is little reason to expect a reversal of this trend towards acquisition of leisure durables at the expense of leisure time. However, the pursuit of affluence is self-limiting to the extent that, ultimately, time is needed to make use of the possessions acquired, and that, beyond a certain level, marginal tax rates usually ensure that additional income becomes an 'inferior good' compared with disposable time.

Sociocultural factors, too, can have a bearing on the appreciation of leisure and its use for recreation. Contrasting attitudes and value systems mean that some individuals and societies continue to equate leisure with frivolity and wasted time. For others, a hedonistic orientation, which clearly ranks free time more highly than work, appears to welcome the emergence of a leisured society without any sense of guilt. Pearson (1977) relates this tendency to institutional arrangements biased towards greater amounts of disposable time, and to an environment with significant leisure potential. Certainly, the efforts of labour unions, and enlightened social reforms and legislators have contributed to a greater leisured existence for a wider cross-section of the population. Yet, even in Australia, there is evidence of the persistence of a puritan work ethic. In the past it has been attributed to those new settlers from a different cultural background, who have migrated to that country since the Second World War. Most recently it is evident in the Australian federal (Coalition) government's forecasting of an economic downturn with the ageing of the Australian population. The government is now disseminating ideas about people working beyond the traditional retirement age and well into their seventies because the lower proportion of younger workers will not be able to sustain a horde of retired baby boomers.

Such anti-leisure sentiment, displayed by apparently compulsive workers and policy-makers, is, of course, not restricted to Australia. Indeed, it could arise in any situation as a function of deficiencies in the leisure environment, rather than from a conviction of the necessity and desirability of work. These same deficiencies can inhibit recreative use of leisure time. Individuals slumped in front of the television set may be there, in part, because of their socioeconomic circumstances and inclination, and/or because of their lack of a more constructive outlet for leisure. The physical and mental demands of work in an increasingly automated society emphasise the importance of, and need for, challenging and satisfying leisure pursuits. In addition, the existence and apparent acceptance of a persistent core of permanently unemployed and people working very few hours (e.g. in casual and seasonal employment), for whom the provision of satisfying recreational outlets is an urgent task, must be acknowledged. Yet, the leisure environment very often cannot provide the opportunities needed for more positive use of disposable time, whether voluntarily acquired or enforced. Identification and remedy of such deficiencies are necessary for a fuller realisation of what leisure has to offer people and society generally.

The arrival of television coincided with a change to greater in-home leisure generally. Television is only one of several leisure goods that made the home a centre of leisure activities in the second half of the twentieth century. Hi-fi equipment and recorded music, videos, computers and computer games, the internet and other factors, singularly or in combination, have provided a variety of new products and activities in the home; products and activities which are being produced and distributed on a worldwide scale (Cushman *et al.* 1996a). Simultaneously, housing conditions (e.g. household design and technologies – insulation; air conditioning) have improved considerably so that there has been less incentive to get away from the house for entertainment and enjoyment.

Although the home has become much more important in leisure, other developments in technology (e.g. air transport; snow mobiles and off-road vehicles; trail bikes; walking boots) have either widened the scope of outdoor recreation activities or at least made it easier, more comfortable and speedier for people to venture further afield than ever before in the search for leisure experiences. There have also been important changes in the structure of national and regional economies. The leisure and tourism sectors have increased in significance as areas of personal expenditure and employment, and as aspects of public policy (e.g. see Carroll

1990; Henry 1993; Veal 1994). Clearly, the position of leisure in society is cemented in social, psychological, political and economic factors.

Outdoor recreation

Devlin (1992) notes that:

> People's recreational use of leisure time will almost inevitably at some stage include outdoor recreation. This is currently true for 90 per cent of those who live in Western countries, and, for many of these participants, it is a form, this form of recreation, which represents a very important part of their lives.
>
> (Devlin 1992: 5, in Mercer 1994a: 4)

The 'leisure explosion' in the developed world has been paralleled by a striking upsurge in all levels of recreation activity. Institutional, technological and socioeconomic factors have been influential in this upsurge. Much leisure activity, as noted above, is of course home-centred, perhaps also home-technology-centred (e.g. television, computers, videos), a feature which is being reinforced and cultivated by capitalist society. Nevertheless, participation in outdoor recreation in Australia, Canada, New Zealand, the United Kingdom (UK), the United States (US) and other industrialised nations has grown rapidly since the Second World War, and particularly since the 1960s, though there have been periods of small declines or stagnation, while participation in organised sports has seen some dramatic changes (see Cushman *et al.* 1996b).

Despite its unquestionable scale and significance in social and economic terms, sport remains a minority participatory activity – many more people actually watch sport than participate in it. The leisure activities with mass appeal are still those that are more informal, social and passive.

Expenditure on recreation is increasing as a proportion of household expenditures in countries such as Canada and Australia. Local governments are spending more on recreation and culture. In Australia, in the twelve months to June 1999, average household expenditure on recreation was $89 per week, while 'Spending on recreation and culture has grown to around 20 per cent of local government outlays' (Montgomery 2004).

Surveys for the Countryside Commission (1991 in Glyptis 1993) showed that in 1990, 76 per cent of the population had visited the English countryside for purposes of recreation, generating over 1,600 million trips and 12,400 million pounds of expenditure. The English countryside attracts not only a large number of people, but also a high frequency of participation. In 1990, as much as 19 per cent of the UK population had visited the countryside within the past week, and nearly half had done so within the past month. At the other extreme, 2 per cent had never been to the countryside (Glyptis 1993: 5–6). More recently, total leisure day visits in England fell from 5.2 billion trips in 1998 to 4.5 billion in 2002/3 (a 14 per cent reduction), while total expenditure decreased from approximately £69.9 billion in 1998 to £61.9 billion in 2002/3. The most popular activities were going out for a meal or drink (18 per cent), walking, hill walking, rambling (16 per cent), visiting friends and relatives (14 per cent) and shopping (12 per cent). In the 2002/3 survey, 62 per cent of adults in Great Britain had visited the countryside at least once in the year prior to the survey; 21 per cent of respondents had visited in the two weeks prior to the survey.

The Australian Bureau of Statistics (ABS) conducts the Environmental Attitudes and Practices Survey, a household survey which seeks to collect data on environmental topics, such as visits to national and state parks and World Heritage Areas. In the 2001 survey it was found that approximately 54 per cent of all adults visited such areas at least once in the 12 months prior to survey, but interestingly, visitation to these areas had declined between 1992 and 2001. The ABS (2002) also published data on participation in recreation and sports activities. These data highlight the importance of outdoor recreational activities, but they also give some brief insights into the very marked differences arising in recreation participation among men and women, a theme taken up in Chapters 2 and 3.

In the United States, a 1994/5 survey of outdoor recreation participation revealed that 94.5 per cent of respondents aged 16 years or older participated in at least one outdoor recreation activity. In the 1999–2000 NSRE (National Survey on Recreation and the Environment), 94.5 per cent of the US population aged 16 years or older participated in some kind of outdoor recreational activity. Walking, viewing natural scenery and family gatherings were the most popular terrestrial activities. Motor-boating, fresh and warm-water fishing, swimming in a lake, river or ocean, and visiting a beach were the most popular water-based activities (Cordell *et al.* 1999). Overall participation in outdoor recreation in the United States is growing as the population grows. The most popular activities overall appear to be walking, waterside and beach activities, gathering outdoors with family and friends, and sightseeing (Table 1.1).

Table 1.1 Outdoor recreation participation in the United States (people 16 years and older)

Type of outdoor activity	% of population 16 years and older
Any outdoor recreation activity	94.5
Observing nature	76.2
Walking/hiking	66.7
Sightseeing	56.6
Picnicking	50.1
Fishing	28.9
Bicycling	28.6
Camping	26.3
Canoeing/kayaking/white-water rafting	15.9
Downhill skiing/snowboarding	10.7
Hunting	9.3
Water-skiing	8.9
Horseback riding	7.1
Sailing	4.8
Snowmobiling	3.6
Cross-country skiing	2.7

Source: Adapted from Cordell *et al.* (1999, in Leitner and Leitner 2004: 373).

Common trends in modern Western societies can be noted. Land and water-based or related activities that have witnessed growth, include: golf; bicycle riding; walking/day hiking and backpacking; photography; nature study; horseback riding; orienteering, mountaineering, rock climbing and caving; off-road (four wheel) driving; rafting, wind surfing, water-skiing, tubing and jet-skiing; snow-skiing/ snowboarding, and cross-country skiing. Indeed, it is widely written that nature-based recreation and tourism are fast growing worldwide, and that viewing or observing nature and wildlife is an increasingly popular attraction. This certainly seems to be the case in many countries, especially less developed countries, but there are regional and local variations where participation may in fact be declining or at least stagnating. For example, the percentage of Australians visiting World Heritage Areas, national and state parks actually decreased between 1992 and 2001 (Australian Bureau of Statistics 2005a).

A number of interrelated events and social and political developments, arising from global, regional and local forces, have led to growth and increased diversity in outdoor recreation participation and tourist travel, and to the establishment of public and private (including voluntary) recreation organisations and programmes. The extent and nature of recreational participation and personal travel have been affected by many factors (see Box 1.1).

Participation in recreation activity is influenced by, among other things, socio-economic factors (see Chapter 3). Income and education, which are often reflected in occupation and correlate highly with car ownership, probably have the greatest impact on recreation. 'Men of substantial mental accomplishment have not usually lacked interesting ways of employing their time apart from toil. And it seems likely that they will be somewhat less susceptible to the management of demand' (Galbraith 1972: 359).

Demographic variables such as age, sex, family structure, immigration and concomitant cultural assimilation and diversity, are also important in explaining recreation patterns. Participation in recreational pursuits tends to decline progressively with age, although television watching, golf and bowls have higher participation rates among the older age groups than the young (see Cushman *et al.* 1996a). In short, the types of leisure pursuits and recreational activities undertaken, change throughout a person's life-cycle (see Chapter 2).

An important demographic aspect is the general ageing of Western societies, so that provision must be made for a less active, but growing segment of the population, with considerable leisure time. In Australia, life expectancy has increased from about 57 years on average to about 80 years, while the fertility rate in Australia is about 1.76 children per woman when the replacement rate is 2.1 children per woman. 'Women aren't just happily making their choices and getting what they want' (Gittins 2004: 36). Those that wish for a satisfying career and two or three kids are:

> having a lot of trouble pulling off that double because governments have done far too little to renovate a male-centred labour market to accommodate the new reality of highly educated women – and because, for the past eight

Box 1.1 **Factors affecting recreational participation and travel**

Factors affecting recreational participation and travel include:

- population growth (including immigration in many countries/regions);
- changes in population characteristics – improved health care and diets, longer life spans and ageing populations;
- changing family structures to broader sociocultural and political acceptance of non-traditional family units (e.g. policy and legislative developments);
- shorter working weeks. The regular working week has been reduced from an estimated seventy-hour, six-day week in the mid-nineteenth century, to around forty hours or less, spread over as little as four days, although overtime and second jobs are common, and more households are dual-income. There is a markedly different employment structure, with more people working in service industries and in casual jobs. Seasonal work in service industries is common;
- increased affluence and higher disposable incomes (though arguably becoming more concentrated in some countries), affected to some extent by growth in the number and proportion of dual-income households in several countries;
- increased holiday entitlements. The right to generous periods of paid annual leave has been established, with the addition, at least in countries such as Australia, of an additional holiday pay loading to enable workers to take better advantage of their vacations. Not only have work periods been reduced, but various peripheral activities such as travel time and lunch breaks may be incorporated into the paid working day, so that non-obligated time is increased;
- increased mobility (by way of the development and wider use of private motor vehicles, and the greater availability, speed and comfort of other forms of transportation, particularly long-haul);
- urbanisation and suburbanisation;
- the influence of commercial interests (public relations and marketing) and technological developments in recreational equipment and infrastructure;
- the promotion of high-risk recreational activities;
- greater educational attainment;
- increasing attention to health and fitness programmes;
- growth in environmental and cultural awareness and interests;
- the age of retirement receding to the point where close to 60 years of age is the accepted norm and even earlier retirement is commonplace;
- a growing focus on human services and increased recognition of the needs of special groups and new roles for girls and women; and
- tourism development (Kraus 1984; Murdock *et al.* 1991; Parker and Paddick 1990; Lynch and Veal 1996; Cushman *et al.* 1996b).

years, and under the cover of providing 'choice', our crazy Prime Minister has been trying to tax mothers back into the home.

(Gittins 2004: 36)

Institutional, technological and socioeconomic forces operating at local to global levels, in combination and separately, have clearly influenced recreation patterns in the developed world. Growth in outdoor recreation and tourism, and the resulting escalating pressures on resources have necessitated both closer examination of planning and management of the recreational and tourist resource bases of countries and regions, and innovations in policy and planning approaches. Furthermore, recreation and tourism are becoming increasingly important elements in the relationship between the economic, environmental and social dimensions of countries, regions, cities and towns (e.g. see Mercer 1970; Cloke and Park 1985).

Nevertheless, research on outdoor recreation is generally disjointed (e.g. longitudinal studies are lacking), and is relatively scant in such countries as Australia and New Zealand, as compared with North America and the UK. Indeed, we know very little about the spatial and sectoral allocation and distribution of the benefits and costs of outdoor recreation.

Research reported by Hendry (1993) in New Zealand and Hamilton-Smith (1990) in Australia, suggests that the most frequent users of local government recreation services also tend to be the most well-off in the community. Access and use by low income groups, ethnic minorities, Aborigines, the aged, persons with disabilities and women are more restricted.

(McIntyre 1993: 33)

For many in these categories, lack of status, money, mobility, ability and agility, access or awareness, can all inhibit the purposeful use of leisure and, therefore, knowledge of, access to, and participation in recreational activities. In short, the use of leisure, and the nature and extent of participation in outdoor recreational activities, vary spatially and temporally, and fluctuate, sometimes unpredictably, with changes in taste and fashion, and with other developments on the local, regional, national and global scenes. Clearly, an understanding of outdoor recreation patterns and processes requires an appreciation of such factors as:

- people's motivations, choices, participation and recreational satisfaction;
- the roles of the private, public and volunteer sectors in the supply of recreational opportunities; and
- planning and policy-making.

In most circumstances, it might be assumed that the availability of more hours free from work would be regarded as a significant social advance. Yet, as mentioned above, for large sections of society the acquisition of greater amounts of time for leisure, and therefore for recreation, is problematic, and is consequently emerging as a major social problem.

Leisure and recreational opportunity first became recognised as a cause for concern during the Great Depression of the 1930s. The concern continues, but has expanded. Conferences on the subjects of leisure and recreation have proliferated around the world. These gatherings have been organised and sponsored by a diverse set of organisations, ranging from academic bodies to professional administrators and marketing groups. Such meetings and conferences (e.g. world and national leisure and recreation congresses) are now commonplace, while associations facilitating research activity and dissemination (e.g. World Leisure – WLRA; the Australian and New Zealand Association for Leisure Studies – ANZALS) have been established. Journals devoted to leisure, recreation and tourism issues have increased in number. Disciplines such as geography and sociology frequently include conference themes relating to leisure, recreation and/or tourism. Courses in leisure, recreation and, in particular, tourism are widespread and expanding into the Asia-Pacific area and other regions. This book is, in part, an outcome of the need to synthesise an ever-increasing flow of ideas, approaches, conceptual insights and applied research concerning outdoor recreation management.

Approach and structure of the present book

For resource managers, in particular, the focus of interest on outdoor recreation is largely on active, informal types of recreation (i.e. those activities engaged in beyond the confines of a building, sporting arena or home). This is not meant to denigrate the use of free time for individual indoor pursuits such as reading and hobbies, or for formal, structured and institutionalised activities such as organised sports. It simply recognises that the really important resource issues arise with the allocation and use of extensive areas of land and water for outdoor recreation. This is where space consumption and spatial competition and conflict are most likely to occur; 'it is in this context that spatial organisation and spatial concerns become paramount' (Patmore 1983: 225). By considering outdoor recreation as a process in spatial organisation and interaction, the resource manager can focus on those aspects with spatial implications (e.g. imbalance or discordance between population-related demand and environmentally-related supply of recreation opportunities and facilities) (Wolfe 1964; Toyne 1974). Obviously, too, this is where the opportunities and the need for recreation resource management are greatest.

Outdoor recreation can be studied in many ways. It can encompass different disciplinary frameworks (e.g. economics, sociology, political economy, geography and law), and thus can incorporate a combination of theoretical and applied research approaches. This book does not claim to present any 'ideal' approach, but it does seek to fill a gap by bringing together many disparate and complementary ideas and studies concerning outdoor recreation.

Outdoor recreation is not the prerogative of all, as we may have been led to believe. People's accessibility to outdoor recreation opportunities is constrained by barriers linked with age, gender, class, income, race, a lack of facilities and opportunities (see Chapter 3), and inappropriate policy-making, planning and management. Outdoor recreation puts pressure on the physical environment, is an increasingly significant factor in the economic concerns of households, communities and regions, and is receiving higher priority in political arenas. Outdoor recreation presents great challenges to planners and policy-makers. In some respects, those challenges represent 'wicked tasks'; tasks which have no definitive right or wrong answer, so that any planning, management or political response is open to challenge from various (sometimes unexpected) interests.

Several books and other sources of information on outdoor recreation and related resource/environmental management issues have been produced. Some of these are now somewhat dated in terms of their concepts, theories and case studies, and therefore in terms of their applied usefulness, while others have stood the test of time. These contributions are complemented by a growing number of leisure, recreation and tourism journals (e.g. *Annals of Tourism Research*; *Annals of Leisure Research*; *Australian Parks and Leisure*; *Journal of Leisure Research*; *Journal of Park and Recreation Administration*; *Journal of Tourism Studies*; *Journal of Travel Research*; *Leisure Sciences*; *Parks and Recreation*; and *Recreation Research Review*), conference and workshop proceedings, and the publications of such innovative government agencies as the United States Department of Agriculture (USDA) Forest Service.

This book has a wide catchment area, so to speak, and the authors were selective in their chapter foci and sources. Material for the book, like the first edition, is drawn mainly from North America, Australia, Britain, Europe and Southeast Asia. Review questions and guides to further reading are included at the end of each chapter so that readers can explore issues in greater depth than discussed in the text. An extensive bibliography is provided, and many useful websites are also included.

Guide to further reading

- Definitions of leisure, recreation and tourism; the relationships between leisure and work, and between leisure, recreation and tourism: Kaplan (1975); Patmore (1983); Murphy (1985); Fedler (1987); Goodale and Godbey (1988); Mieczkowski (1990); Leiper (1995); Lynch and Veal (1996); Butler *et al.* (1998); Walmsley and Jenkins (2003); Leitner and Leitner (2004a).
- History: Veblen (1899, republished 1970); Armitage (1977); Bailey (1978); Cunningham (1980); Yeo and Yeo (1981); Rosenzweig (1983); Clark and Crichter (1985); Cross (1990); Kraus (2001); Veal and Lynch (2001).
- Leisure and recreation participation trends and issues: Jackson and Burton (1989); Murdock *et al.* (1991); Veal (1994); Cushman *et al.* (1996b); Cordell (2000); Leitner and Leitner (2004a). Students should consult the statistics compiled by respective national, state/provincial, regional or local governments, and other agencies.

Review questions

1   Compile, and then compare and contrast, several definitions of leisure, recreation and tourism. Can you identify (1) major flaws in such definitions, and (2) important temporal shifts in such definitions?
2   Discuss the significance of leisure and outdoor recreation in modern society.
3   What factors have influenced outdoor recreation in contemporary society?
4   Looking to the future, what do you see as being likely areas of expansion and contraction in outdoor recreation activity?
5   Why is it important to examine outdoor recreation in the context of resource management?
6   Reflect on leisure in your life. What do you regard as leisure? List the recreational activities that you consider to be most central to your life. Why are those activities important? Have your leisure and recreation been influenced by any particular individuals or groups?

2 Motivation, choice and behaviour

I was aware that fewer than half the expeditions to climb Everest ever put a single member – client or guide – on the summit. But I wanted to join an even more select circle, the fifty or so people who had completed the so-called Seven Summits Quest, scaling the highest peaks on all seven continents. I also knew that approximately 150 people had lost their lives on the mountain, most of them in avalanches … Everest mocks its dead … Common as sudden, dramatic death is among mountain climbers, no one actually expects to be killed at high altitude. I certainly didn't … I positively loved mountain climbing: the camaraderie, the adventure and danger, and – to a fault – the ego boost it gave me

(Wethers 2000: 4–5)

Ever since the morning of May 1953, when Tenzing Norgay and I became the first climbers to step onto the summit of Mount Everest, I've been called a great adventurer. The truth is, I'm just a rough old New Zealander who has enjoyed many challenges in his life. In fact, as I look back after 50 years, getting to the top of Everest seems less important, in many ways, than other steps I've taken along the way – steps to improve the lives of my Sherpa friends in Nepal and to protect the culture and beauty of the Himalayas.

(Sir Edmund Hilary 2003: 38)

As with other aspects of human decision-making, explanation of leisure and recreation behaviour is complex. The unfettered personal connotations of leisure and the discretionary nature of recreation were noted in the opening chapter. An underlying dimension common to both leisure and recreation is discretion – the exercise of choice. This discretionary element helps explain why observers find difficulty in explaining why people choose particular leisure and recreation settings and activities, and in accounting for recreation choice behaviour. It might be argued that the choice process is no more complex than that involved in, say, the selection of a new residence. After all, choice is subject to a range of influences and is not a completely random process. Nor is it unique to any individual or group. So, as Clawson and Knetsch (1963: 14) argued, 'We must undertake the explanation and analysis of the regularities and patterns of behavior among individuals in the use of all kinds of recreation areas'. However, the unbounded nature of leisure and the subjective, even capricious characteristics of recreation decisions, make generalisation and prediction more challenging.

Motivation

> Why do people choose to use discretionary time for recreation? What motivates climbers to risk their lives in high-risk recreational activities? Why do some city office workers devote much of their lunch breaks to intense physical pursuits? Why do some individuals find great satisfaction in the isolation of wilderness recreation? Why for many is leisure behaviour associated with a stimulating social environment with modern comforts so desirable, but the isolation of wilderness so abhorred?

The process by which a person is moved to engage in particular forms of behaviour has been the subject of speculation and research over a long period. Understanding leisure behaviour is no less complex. According to Iso-Ahola (1980), human actions are motivated by subjective, defined goals and rewards which can be either extrinsic or intrinsic. When an activity is engaged in to obtain a reward, it is said to be extrinsically motivated. When an activity is engaged in for its own sake, rather than as a means to an end, it is said to be intrinsically rewarding. Although the distinction is blurred and open to subjectivity, Iso-Ahola believes that leisure behaviour is chiefly motivated by intrinsic factors related to self-expression, competence and satisfaction, which, in turn, implies freedom of choice. Motivation for leisure and recreation is a highly subjective issue. In particular, the motivation to engage in tourism or pleasure travel is elusive and its dynamic, multi-faceted nature does not lend itself readily to generalisation. The perceived appeal of places as tourist destinations is obviously a major factor. However, the phenomenon of tourism cannot be satisfactorily explained on the basis of physical or cultural attractions alone. A conscious decision must be made to seek a tourism experience and the reasons for this decision and all its ancillary aspects can be as diverse as the tourist population itself.

The question – 'what makes tourists travel?' – is no more or less difficult to answer than any other aspect of recreational behaviour, or of consumer behaviour generally. A predisposition or propensity to travel has much to do with it and a lot of effort by market researchers is put into identifying target groups at which specific promotional material can be aimed. Undoubtedly, the notion of change and contrast is attached to much tourism behaviour and a person's mental state can mould positive or negative attitudes to travel. Dann (1981) demonstrates that the conditions prevailing in the potential tourist's home environment ultimately provide the predisposition to travel. He believes two twin 'push' factors underpin tourist decision making. First, the existence of 'anomie', or the felt need to break out of dull, meaningless surroundings and situations may act as a motivational push factor to persuade people to seek temporary respite in another environment seen to be less affected by such characteristics. The second factor is 'ego-enhancement', or the desire to be recognised, feel superior, or create envy as the result of undertaking a particular trip or vacation. People experiencing a lack of belonging, or a denial of status advancement in their home situation, may seek to visit other places, often developing countries, where they are treated with greater respect.

Underlying both factors is a strong fantasy component, so that tourism becomes a form of escape. Fantasy motivation forms an important element in demand for pleasure travel and indicates its individualistic nature. It is not so much the recreational or tourist experience that matters, as the act of getting away which counts. Dann (2000) suggests that it is possible to present a continuum ranging from predominantly anomic motives to ego-enhancement motives. This offers the prospect of collective minds, and identifying a profile of tourists which, to a greater or lesser extent, reflects these idealised constructs.

Identifying clearly the relationship between an individual's motivation and selection of a destination and tourist experience is a difficult task. Krippendorf (1987) suggested a number of motivating factors including: recuperation and regeneration; compensation and social integration; escape; communication; broadening the mind; freedom and self-determination; self-realisation; and happiness. Reference also needs to be made to the various social influences which condition the decision to travel. These could include the family or societal group, social class, surrounding culture and workplace. The working environment can be particularly significant in that it may be conducive to compensatory effects manifested in tourism. Boring, monotonous jobs may prompt a search for excitement; workplace stress may generate a need for relaxation; and regimented working conditions may encourage the pursuit of a freer, unbounded alternative. Alternatively, types of occupation, if rewarding and satisfying may motivate the tourist to seek outlets to pursue those interests further in a different setting. Of course, it is impossible to ignore the influence of advertising on the decision to travel. Creation of an attractive image of a place to visit and experience involves giving the destination or feature an ambience and easily recognised attributes which will act as 'pull' factors to complement the push factors noted above and reinforce the motivation for pleasure travel.

Freedom of choice and leisure behaviour, be it tourism or outdoor recreation, should not be seen as totally unrestricted. Whereas individual motivation instils a propensity towards certain activities in recreation and tourism, actual participation largely reflects the selection of the best alternative or compromise under the circumstances. Choice is bounded by any number of constraints including physical capability, affordability, awareness, time restrictions and family obligations. The existence and intensity of these constraints vary between individuals and across demographic and socioeconomic groups. Thus, recreation and tourism are sensitive and vulnerable to any number of real or perceived concerns about safety and health, security of property, or financial wellbeing. Recreation also shares with tourism attributes of voluntary, discretionary behaviour. People are free to become tourists and to decide location, timing, duration, mode of travel, activities and costs to be incurred in outdoor recreation. Any one of these attributes may be modified or dispensed with by unforeseen or uncontrollable factors. Moreover, motivation and the process of choice are imperceptibly influenced by incremental adjustments to lifestyle, social mores, traditions and culture. In a world marked by a multiplicity of change agents, motivation itself is subject to the dynamics of an uncertain geopolitical, technological and socioeconomic environment in which human behavioural decision making takes place.

Despite these uncertainties, social psychological studies continue into motivational research in recreation and pleasure travel. One focus is on concepts and terminology, and whether and how motivation differs from related concepts such as aspiration, intention, reason, purpose, satisfaction, aim, and goal (Dann 2000). Motivation is inherently personal and subjective, so that what may be reasonable and logical to one individual may be quite illogical and motivationally suspect to another. This is what makes research into motivation challenging, especially in areas of human behaviour like leisure, recreation and tourism, so clearly identified by their discretionary nature.

A further source of complexity in the explanation of leisure behaviour arises from confusion over the nature of recreation demand and its relationship with recreation participation. In particular, there is an apparent inability to distinguish between the concept of demand in the broad, generic sense and its use to refer to existing levels of recreation activity. The latter, as indicated by numbers of participants or tourist visitation rates, is not a true measure of demand because it relates to observed or actual participation and behaviour, which is only a component of overall aggregate demand.

Recreation demand

'Recreation demand' is generally equated with an individual's preferences or desires, whether or not the individual has the economic, physical and other resources necessary for their satisfaction (Driver and Brown 1978). Recreation demand, so defined, is at the preference-aspiration-desire level, before it is expressed in overt, observable behaviour or participation. In this sense, it is a propensity concept, reflecting potential or behavioural tendencies, and is detached from subsequent recreation activity. As one authority put it: 'Recreation demand is a conditional statement of the participation that would result … under a specific set of conditions and assumptions about an individual … and the availability of recreation resources …' (US Bureau of Outdoor Recreation 1975: 10–22).

This broad notion of demand is supply-independent. It assumes no constraints on recreation opportunities or access to them. In these terms, recreation demand depends only on the specific characteristics of the population (e.g. age, income, family structure, occupation and psychological parameters), and not on the relative location of user groups, or the quality and capacity of facilities, or the ease of access.

However, actual consumption or participation in recreation activities is very much a function of the supply of those opportunities. Observed levels of leisure

behaviour may conceal frustrated demand, which can only be satisfied by the creation of new recreation opportunities or by increasing the capacity of existing facilities (e.g. by management strategies encompassing land or water acquisition, hardening the landscape, and interpretation). If opportunities are less than ideal, people will actually participate less in recreation than their theoretical level of demand would indicate.

In the real world, recreation demand rarely equals participation. The difference between *aggregate demand* and *actual participation* (or 'expressed', 'effective', 'observed', 'revealed' demand) is referred to as *latent demand* or *latent participation* – the unsatisfied component of demand that would be converted to participation if conditions of supply of recreation opportunities were brought to ideal levels (Figure 2.1).

Lipscombe (1986, 2003) described eight possible conditions that at times have been called recreation demand. These included:

- *Effective demand* (i.e. existing, manifest, exhibited, existent, observed, revealed and expressed demand) – indicates what exists, and is a measure of use of any particular resource in a set period. This aspect of demand is also referred to as 'participation' or 'consumption'.
- *Latent demand* – demand which exists, but for some reason or another has been constrained.
- *Induced demand* – demand which has been stimulated by the provision of further facilities, thereby converting a latent demand into an effective or expressed demand.

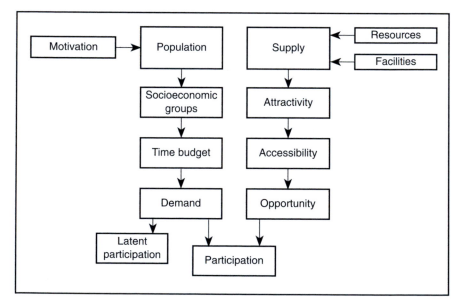

Figure 2.1 Recreation demand and participation

Source: Adapted from Kates, Peat, Marwick and Co (1970: 1.1).

- *Diverted demand* – arises when demand for a certain type of facility is diverted from one source of supply to another by the provision of a new supply.
- *Substitute demand* – refers to the shift in participation to other forms of recreational activity because of the provision of facilities for different purposes.

(Lipscombe 2003: 106–7)

The confusion between recreation demand and participation, and the possible implications of misinterpretation, have been noted by a number of authors (e.g. Knetsch 1972, 1974; Elson 1978; Lipscombe 1986). It is not enough simply to look at what people do and interpret this as reflecting what they want to do; it also reflects what they are able to do. Participation data are important, but they must be interpreted in terms of both supply and demand variables. Knetsch (1972) points out, for example, that if the participation rate in swimming in a given area is found to be very large, relative to that in some other area, it may be almost entirely due to greater availability of swimming opportunities. Using attendance figures as a measure of demand confuses manifest behaviour with recreation propensities and preferences.

Nor is the problem merely one of semantics; the planning implications are clear for adjusting the supply of recreation opportunities and estimating the probable effects of alternative policies and programmes. It is important for planners to have answers to questions concerning for whom, how much, what type and where, in regard to the introduction of new recreation facilities. Equating demand with existing consumption or participation rates, can lead to the assumption that people will want only increasing quantities of what they now have, thereby '… perpetuating the kind of facilities already existing in the areas already best served and further impoverishing already disadvantaged groups' (Knetsch 1974: 20).

Another problem in relying upon past (observed) participation to guide future decisions is that observed activity patterns reveal little regarding satisfaction or the quality of the recreation experience. As Stankey (1977) pointed out:

> … when opportunities are available, particularly at little or no cost, they will be used. But use should not lead us to automatically assume people are satisfied with existing opportunities or that alternative opportunities might not have been even more sought after.

(Stankey 1977: 156)

A deeper understanding of the true nature of recreation demand would throw light on the reasons for non-participation or under-participation in specific areas and activities, and reduce misallocation of resources (e.g. see Vining and Fishwick 1991). It should ensure also that any induced demand as a result of additional recreation investment is directed towards remedying these deficiencies. The supply of appropriate opportunities can release latent participation and translate it into effective demand, and can also be used to manipulate and redirect demand from one area or activity to another. Mercer (1980a) gives several examples of induced, substitute or diverted demand as the result of creating new resources and of improvements in access and technology. It should be noted, of course, that heightened levels of participation can just as readily be achieved by improvements in awareness, and by education, training and other triggers (see below).

Interest in studying recreation demand was generated in the 1960s as people gained more leisure time, and greater disposable income and mobility. Pressures on resources, particularly public lands, mounted. Interests in the costs and benefits of outdoor recreation grew, as instances of resource conflict, congestion and crowding, and environmental impacts, escalated. In the United States, public agencies took great interest in these matters and the extent and nature of recreation demand. Two commonly used non-market valuation tools since the 1960s are the Contingent Valuation Method (CVM) and the Travel Cost Model (TCM). CVM attempts to 'assess the value that individuals place on things which are generally provided free or at a subsidized price' (Veal 2003: 81). In CVM applications, people are generally asked about their willingness to pay for a resource (e.g. if the cost to you of developing and maintaining recreational trails in Algonquin Provincial Park was Can$5 per annum, would you support such a project?). TCM generally 'uncovers the consumer's surplus measure, the maximum willingness to pay' (Hanley *et al.* 2003a: 7). Preferences may be revealed by focusing not only on revealed behaviour but also people's preferences according to various attributes of a recreational resource. Applying the TCM to a study of mountaineering in the Highlands of Scotland, Hanley *et al.* (2003b) modelled the impacts of policy alternatives (parking fees or increasing the length of time it takes to walk in to an area by closing off car parks) to managing congestion and impacts stemming from increased use of such areas. Among other things, they found that 'a 5 pound car parking fee at the Cairngorms is predicted to reduce rock climbing trips to the Cairngorms by 31 per cent, which may be compared with a 44 per cent reduction achieved by a two-hour increase in approach time (or walk in time)' (p. 53).

Awareness of the factors generating recreation demand and the relationships between its various components are important in recreation planning and resource management. That said, it is obvious that most attention in the social sciences has been devoted to recreation behaviour *per se* (i.e. to actual participation or effective demand). It is in the spatial and temporal expression of demand and the use made of specific sites and facilities where many resource problems exist. Whereas these patterns of use are derived in part from underlying preferences, they reflect also the availability, quality and effective location of recreation opportunities. Explanation of revealed recreation behaviour, therefore, must be sought in terms of the interaction between recreationists and the resource base, and in terms of the processes by which outdoor recreation sites are chosen. With respect to the latter:

> human decision making cannot be understood by simply studying final decisions. The perceptual, emotional and cognitive processes which ultimately lead to the choice of a decision alternative must also be studied if we want to gain an adequate understanding of human decision making.
>
> (Svenson 1979, in Vining and Fishwick 1991: 114)

Recreation participation

A simplified representation of the factors which influence the decision to participate in recreation is set out in Figure 2.2. Once again, a broad distinction can be made between the potential demand or propensity for recreation and the supply of opportunities to realise these preferences or desires. The variables can be grouped into the demographic, socioeconomic and situational characteristics which generate a propensity to recreate, and those external factors which facilitate or constrain the decision and the choice of activity and site.

Demographic characteristics

The size, distribution and structure of the population are of crucial significance in explaining recreation patterns. Age, sex, marital status and family composition or diversity, have all been recognised as affecting recreation preference.

At the aggregate level, important demographic considerations are the overall size, structure and distribution of the population. Although population growth rates in Western countries remain low, significant shifts of population are taking place internally. One of the most widely publicised of these has been the migration from the Frost Belt to the Sun Belt States of North America. Whereas part of the attraction of the Sun Belt can be found in the outdoor recreation opportunities available, rapid, unplanned growth in these areas threatens the very qualities newcomers seek.

At the disaggregated, individual or family level, a good deal has been written on the effects of age and the progression of life from one phase to another through what is known as the life-cycle. It has been suggested that although sharp lines of

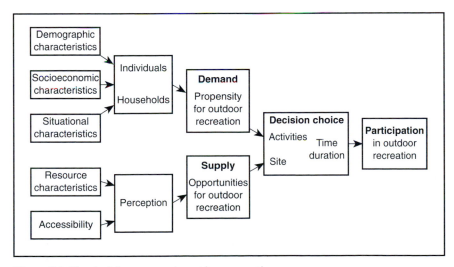

Figure 2.2 The decision process in outdoor recreation

Source: Pigram (1983).

division cannot be drawn, certain preoccupations and interests predominate at specific stages in the life-cycle. With regard to recreation, not only are preferences influenced by age, but also by an individual's physical, mental and social ability to participate. It is clear that the recreational importance of each phase is closely related to the family framework and to other 'life event' phases in an individual's 'life career' (Mercer 1981a). Apart from the family setting, these include the broader cultural background, government policies and the mass media. Mercer emphasises that the average life span subsumes and obscures major traumas such as illness, divorce, bankruptcy and the so-called 'mid-life crisis'. Moreover, during any life episode, recreation opportunities may be constrained by relative poverty, immobility and lack of time.

The implications of the family life-cycle approach are that recreation requirements can be expected to vary from individual to individual and between different people at different stages of the cycle, with important consequences for the planning and management of recreation space and resources. What is perhaps more important for current policy considerations, is that significant demographic changes are taking place within the family life-cycle, and that these, in turn, will generate altered priorities in recreation policy, planning and development.

In several countries of the Western world, the most dramatic demographic changes are shifts in age structure, stemming from the post-war 'baby boom' and the subsequent 'baby bust'. As these ripples move into maturity and beyond, their influence is reflected in recreation patterns so that resource managers need to be alert if a rapid and appropriate response is to be made. The changing status (some would say 'demise') of the family in modern society is another factor affecting individual and community participation in recreation. The prevalence of working couples and the freeing of women from many pre-existing constraints are gradually blurring sex-related differences in recreation participation. Childless couples, unmarried couples living together and greater numbers of elderly people living alone or in public and private community housing, all contribute to the growing complexity of 'family' life to which recreation planning must adapt.

Projected aged dependency ratios in OECD countries (see Figure 2.3) demonstrate the ageing profile of populations in Western industrialised countries (also see Chapter 1). Proportionally, a smaller and smaller workforce is going to be required to support a growing aged dependent component of the population. This ageing profile also underlines the need for greater provision of suitable recreation opportunities for older, active people. In Australia, for example, those aged 65 years and over are expected to make up over 20 per cent of the population by 2025, compared with approximately 12 per cent in the late 1980s (Borowski 1990), and to double by the middle of this century.

The emergence of a significant elderly and retired component in the population, for whom greater longevity, improved healthcare and better financial provision generate a new set of leisure opportunities and requirements, takes on greater significance because of the high concentrations of older people in particular areas. Older adults are participating in a wide range of outdoor recreation activities and organised sports, including Masters Games. The idea that as people get older they

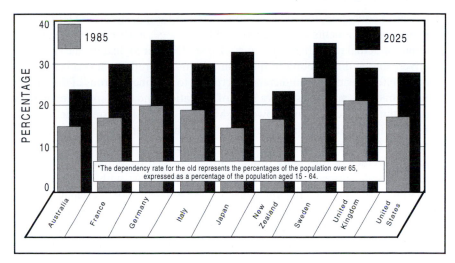

Figure 2.3 Projected aged dependency ratios in OECD countries

Source: Adapted fromWorld Population Prospects, UN; National Australia Bank, in Freeman (1992: 10).

necessarily do less and less physical activity is now misleading because increasing leisure participation by older people in Western economies is well documented (e.g. Vertinsky 1995; Grant 2001; Dionigi 2004). Indeed, according to MacPherson (1991: 337) people 60 years of age and over 'are reporting more frequent involvement in physical activity, involvement in more varied types of activity, and a greater range in the intensity of involvement, from passive flexibility exercises to marathon races' (also see Chapter 3).

Mercer (1980b) identified several localities in Australia, in particular the Gold Coast of southern Queensland, as geriatric colonies, with above average numbers of retired people. Retirement migration in Australia is becoming more pronounced in its regional effects in coastal areas in particular, as occurred on the South Coast of England, popularly known as 'Costa Geriatrica', or as in Florida, where the aged make up a significant percentage of the population. As Mercer (1980b) pointed out, such ageing of the population can occur very rapidly, and when accompanied by the departure of youth in search of employment or excitement, can give rise to a succession of strains and imbalances in the community.

Socioeconomic characteristics

Among the factors which influence the desires or inclinations of individuals for recreation, are social relationships and social structure, education, occupation and income. Recreation is a form of social interaction, and the way in which a society is organised affects recreation behaviour. For instance, interaction within and between families, peer groups and ethnic communities helps mould many facets of human behaviour, including goals and motivations for use of leisure.

Levels of education, too, whether considered in formal, structured terms or as incidental improvements in awareness and knowledge, must have a pronounced influence on actual recreation behaviour (also see Chapter 3). Indeed, the emphasis on advertising and marketing in the leisure industries reflects this relationship, while the efforts made by commercial enterprises to convince patrons of the quality of their attractions are themselves a form of education, and are also a facet of mass consumption and recreation. However, Mercer (1977) questions whether this correlation is causal when it comes to determining underlying propensities for recreation. The fact that the more highly educated person is likely to be more recreationally active may only reflect further correlation with a higher status occupation and reinforce income and class differences. As with so many of the factors impinging upon recreation demand, there is a degree of overlap, both with other influential factors and with the process of expression of demand through participation. Education contributes to knowledge, awareness and the development of attitudes and values, which, in turn, may generate aspirations and desires for recreation. At the same time, the acquisition of recreational skills through education can enhance opportunities for participation and for gaining satisfaction from recreation.

A similar problem occurs with income and occupation, each already highly correlated with the other. Undoubtedly, the amount of discretionary income available to an individual or family is a major factor affecting recreation participation, but does it help structure underlying recreation preferences? Do well-to-do people really prefer active outdoor recreation activities, or do their wealth and associated possessions merely open doors that are closed to the less affluent? Again, the former sharp distinction in attitudes to work and leisure between high and lower status occupations is becoming blurred. No longer can it be said with certainty that upper-class occupational groups show a preference for a more serious range of leisure pursuits, or view with disdain the thought of more mundane forms of recreation. Increased concern for conservation and environmental issues, especially among the 'baby boom' generation, has contributed to increased participation by a broad cross-section of the population in outdoor recreation, nature-based activities and use of national parks (Lacey 1996).

Moreover, the 38 or 40 hour working week is a myth; at least in some countries. In Australia, full time working males are working around the same number of hours as working males in the 1940s. In 1980 approximately 21 per cent of all full-time workers were working more than 49 hours per week. By 2000, this was 33 per cent. About one-quarter of all Australian workers work less than 30 hours a week. According to the International Labour Organisation (Messenger 2004), approximately 15 per cent of all Australian employees worked 50 or more hours per week in 1987. By 1990, this had leapt to about 20 per cent. The only countries where a greater proportion of people worked longer hours were Japan (28 per cent) and New Zealand (21 per cent). These figures contrast markedly with European countries, such as France, Germany and Spain. In these countries only 6 per cent of employees worked this length of time. In The Netherlands only 1.4 per cent of the workforce works 50 hours per week or more (Messenger 2004).

Generally, there are many people working longer hours, but preferring to work fewer, and many working few hours (or not at all), but preferring to work longer hours. There is a marked difference in leisure, work and wealth in societies around the world. In many countries, income is not equitably distributed and in fact is becoming increasingly concentrated among the already rich, while a good proportion of any increases in income earnings growth goes to executives.

Situational characteristics

The third group of factors which impinge upon recreational choice is linked to some of those previously discussed and shows similar ambivalence. Under the category of situational or environmental factors could be placed:

- *Residence* – which incorporates such aspects as location, type, lot size and existence of a garden or pool, and which, to some extent, is a function of income and occupation. At a larger scale, the place of residence can influence recreation patterns. Obvious examples are coastal locations, winter sports areas, and large urban centres.
- *Time* – which also frequently reflects occupation, although this is changing with innovations in working conditions and the high incidence of unemployment. It is not merely the amount of time which is important, but its incidence in terms of usable 'blocks' at convenient periods (e.g. weekends). In general, self-employed persons have greater control over their time budgets and are, or should be, in a position to allocate more time to leisure. This, in turn, has the potential to widen the dimensions of recreation participation.
- *Mobility* – which, for most people, freely translates to car ownership or access to a motor vehicle. If a vehicle is not available, a person's recreation action space (see Box 2.1) is obviously limited in terms of choice of site, journey, timing and duration of trip. Presumably, also, possession of a car generates a desire, or at least permits a propensity, for forms of recreation which otherwise could not be considered.

External factors

As noted above, some of the variables which are considered important in determining an underlying proclivity for recreation, can also be influential in the actual decision to participate. Several of the socioeconomic and situational factors, for example, appear to operate at various stages of the decision-making process. Furthermore, the role of resource-related characteristics is indicated in Figure 2.2. These characteristics have direct relevance to choice of recreation site, activities and travel, and are concerned with the opportunity to recreate (i.e. to activate latent participation).

Recreational opportunity depends upon the inter-related features of availability and accessibility of recreation resources or sites. The nature of recreation resources and their availability in functional terms depends upon such things as quality,

---

Box 2.1 Action space

Time and space are inherent properties for all phenomena. All human actions occur in time and space. An individual's action space is the full set of locations about which an individual has information and the subjective utility or preference the individual associates with these locations. An individual's activity (or 'movement') space is that part of the action space, the places and things, with which the individual has contact on a daily (or other time period) basis (e.g. see Walmsley and Lewis 1984).

The concepts of activity space and action space have been utilised in studies in behavioural geography. Horton and Reynolds (1971), for example, applied the concepts of action space and activity space in their study on the effects of urban spatial structure on individual behaviour in two compact residential areas in Cedar Rapids, Iowa. Their conceptualisation of the inputs to the formation of an individual's action space recognised the interactions of activity space, travel preferences, socioeconomic attributes, home location, length of residence at location, cognitive image of urban spatial structure, and the objective spatial structure of the urban environment.

The concepts of action space and activity space have had limited application in leisure and outdoor recreation studies, but are becoming more common through the use (and wider acceptance) of time diaries and large surveys of groups such as women and youth.

Source: Jenkins (2003: 5).

---

degree of development, carrying capacity, ownership, distribution and access. These, in turn, reflect economic, behavioural and political factors, which help shape public and private decision making for recreation provision.

Accessibility to recreation opportunities is a key influence on participation, and its several facets are examined in ensuing chapters. The importance of accessibility as the final deciding factor in determining the 'what' and 'where' of recreation participation, is stressed by Chubb and Chubb (1981: 153): 'If all other external and personal factors favour people taking part in an activity but problems with access to the necessary recreation resources make participation impossible, the favourable external and personal factors are of no consequence.' Accessibility also helps explain the contribution of the travel phases to the overall recreation experience

Recreation travel behaviour

Almost by definition, outdoor recreation implies that space, distance, and therefore time, separate recreationists from the sites and activities to which they wish to relate. A process in spatial interaction is stimulated as efforts are made to reduce spatial imbalance in recreational opportunities. The ease or difficulty of movement and communications are basic to the explanation of spatial interaction. Mobility and information diffusion thus become key elements in the spatial relationship between recreationists at the origin (i.e. place of residence) and the destination (i.e. the recreation site).

The friction of distance is important in all forms of recreation travel. For most movements, a distance-decay effect can be recognised, so that the strength of interaction declines as distance increases. Put simply, this means that recreation sites at a greater distance, or for which the journey is perceived as involving more time, effort or cost, are typically patronised less (a distance-decay effect). However, the effect of the friction of distance varies spatially, and with modes of movement and types of recreation activity. It can also change dramatically over time and space, with innovations in communication and transportation, and with advertising and promotion.

For some forms of recreation travel, the distance-decay effect may be heightened, manifesting itself in inertia or the reluctance to move at all. Alternatively, the reaction to distance may be in marginal terms. In most cases, the effect of distance will be negative, in that, beyond some point, further travel becomes less desirable; each kilometre offers more resistance or impedance than the last. Conversely, the effect may be positive, where the friction of distance is reversed; for some people and some occasions (e.g. ocean cruises), travel becomes so stimulating as an integral part of the recreation experience, that the further the distance, the greater the desire to prolong it.

The effect of travel and its key role in the satisfaction gained from the total recreation experience are important influences on recreation behaviour. The 'journey to play' can make or break the outing, and it is often the individual's perception of what is involved in the travel phases which is the crucial factor in the decision to participate or stay at home.

Recreation travel, in common with all aspects of recreation, is discretionary in nature in that it lacks the orderliness and monotony of, for instance, the journey to work. Yet, certain regularities can be discerned in recreation movement patterns in response to time–distance, connection, and network bias.

Time–distance bias, where the intensity of movement is an inverse function of travel time and distance, reveals itself in the distance–decay effect referred to earlier. Distance is constrained (or 'biased') by the time available and the type of recreation envisaged. Distance is also the basis for determining the extent of urban recreation hinterlands. In terms of travel distance, it is possible to conceptualise recreation traffic movements as a series of concentric rings progressively distant from the city to distinguish between day-trips, weekend trips and vacations. There is clearly scope for overlap between zones and such an arrangement may represent an oversimplification in an era of more sophisticated and efficient transportation systems.

Connectivity, and conversely barriers to movement, is another important aspect of transferability affecting the means or ease of spatial interaction. The presence or absence of interaction and the intensity of recreation travel are related to the existence and capacity of connecting channels of traffic flow. Recreational trip-making will respond positively or negatively to alterations in connectivity between origin and destination. An additional traffic facility such as a motor bypass, bridge or tunnel, can transform locational relationships by providing new or improved connections between places. Removal of linkages (e.g. destruction of a bridge) or impairment of capacity will lead to drastic alteration in patterns of recreation movement, and the resulting redistribution of traffic pressure can generate severe adjustments in dependent services and enterprises. There are many examples where new communities and recreation facilities have sprung up and established sites have gone into decline because of alterations to pre-existing routes and modes of movement. Closure of railway lines, relocation of river crossings, construction of highway-motorway bypasses, even the conversion of streets to one-way traffic, can all have dramatic effects on recreation travel behaviour.

Part of the explanation for regularities in recreation movements can be found in the characteristics of existing communication networks. Recreational travel is more likely where networks relate to shared information channels, a common transport system or the same sociocultural, national, political or even religious grouping. The huge volume of tourist flows based on group tours is but one example of the influence of network bias in promoting recreation travel on a large scale. The network effect, too, can be heightened by constraints on expanding links within or between systems, such as occur with national boundaries or language barriers.

Despite the regularities noted above, the essentially discretionary nature of recreation movements and the element of unpredictability put some difficulties in the way of developing an efficient and economic system of management for the special characteristics of recreation travel. Particular problems are the incidence of peaking, variability in participation and the heavy reliance placed on the motor vehicle. Patterns of recreation movement display daily, periodic and seasonal peaks and troughs, associated with time of day, weekends, vacations and suitable weather, especially in the summer season in coastal locations, and in winter in alpine areas. Some of these peaks are cyclical, and to that extent predictable. However, the problem remains of providing a transport system which can cope with short periods of saturation set against longer periods of under-utilisation.

The situation is worsened by the pervasive reliance on the automobile as the primary means of recreation travel. The motor vehicle ranks with television as the most powerful influence, positively and negatively, on recreation participation. The reasons are not hard to find. Use of the car allows for the unstructured nature of recreation (and other) trips, and provides for flexibility in timing and duration of the outing, and choice of route and destination. The car is readily available and is a good means of access to most sites, without the necessity for change of travel mode. It combines the function of moving people, food and equipment with shelter, privacy, a degree of comfort, and a relatively inexpensive means of transport.

The expectation of car ownership and its dominant role in outdoor recreation affect more than travel behaviour. The motor car is a fundamental influence on recreation landscapes and on the type and location of recreation facilities. The motor car has significantly affected the morphology and function of places, and has given rise to a completely new series of leisure activities and support industries.

In considering this close attachment of the recreationist and the motor vehicle, it would be wrong to assume that car ownership or access is universal. There will always be a social need to provide for the non-motorist in the community, if recreation opportunities for these, sometimes less mobile members of the community, are not to be severely restricted.

Given that recreational trip-making is largely unstructured and discretionary in nature, it is noteworthy that efforts have been made to isolate common variables influencing decision making, and to use these to explain and predict recreation behaviour and associated patterns of movement.

Studies of trip generation are numerous, using models incorporating a variety of predictive variables to attempt to answer questions concerning: why particular forms of outdoor recreation are selected by different individuals and groups; why certain sites are patronised and others neglected; the expected frequency and duration of recreational trips; and the degree of substitutability between recreation activities and recreation sites.

One of the most popular and frequently applied techniques is some version of the gravity model, which has been used with success in forecasting visitor flows to recreation sites. Essentially, gravity models are based on the premise that some specific and measurable relationship exists between the number of visitors arriving at a given destination from specific origins or markets and a series of independent variables, in particular, population and travel distance.

If these variables can be quantified with reasonable accuracy, predictions can be made as to the likely attendance at selected recreation sites from designated points or areas of origin (e.g. visitation rates to parks from surrounding regions, counties, towns or cities). If the actual, measured levels of attendance match the expected, then the model can be used to predict visits to proposed new parks, to indicate the need for greater efforts in publicity and advertising, or to assess the impact of improved accessibility on the propensity to travel.

The technique can also be applied to delineate the range or impact zone from which a site could be expected to attract visitors. In theory, if this zone was merely a function of the friction of distance, it would consist of a series of concentric zones surrounding the site, with numbers of visitors progressively declining outwards from the centre. However, distortion of the size and shape of the area is to be expected because of the kind of factors noted above. Variations in demographic characteristics, in conditions of accessibility and in the orientation and impact of promotional advertising within the hinterland, as well as competition from peripheral attractions, all help to explain why actual patterns of patronage depart from the theoretical.

Models of outdoor recreation participation can be developed at various levels of sophistication and application (i.e. local to national levels), but all must involve compromise and rest on certain assumptions, because of the complex nature of recreation behaviour. Caution is necessary, then, in the use of models and in the application of the results. In such an unstructured field of choice-making as recreation, where decisions are often more intuitive than rational, and more impulsive than considered, norms are not appropriate.

A further cause for concern in modelling recreation behaviour is the assumption that relationships between the several sets of variables remain constant. Yet, lifestyles and social mores change progressively, as do economic and technological circumstances, so that prediction is difficult and value-laden (e.g. see Lee-Gosselin and Pas 1997). The dynamic nature of many inputs into recreation decision-making can be a source of miscalculation in planning. New trends and fashions, changing values, charismatic leaders and different policies by governments or other institutions, can all act as 'triggers' to release latent participation and bring effective demand more into line with overall demand.

The underlying element of choice in recreation means that individual participants or particular recreation pursuits should not be studied in isolation. Rather, the entire spectrum of leisure activities must be examined as a series of substitutes and complements that are capable of providing a variety of satisfactions, and that act as potential trade-offs for one another (Phillips 1977).

Substitutability and interchangeability are responses to the relationships between the experiences and satisfactions sought in outdoor recreation, and the geographic, social, psychological, economic or physiological barriers which prevent those expectations and satisfactions from being fully realised. The effect of these barriers is to stimulate replication of satisfactions by resort to some other activity. In short, the concept of substitutability implies that recreation preferences and propensities are much more elastic and open to manipulation than is generally accepted, making the recreation choice process that much more complex.

Recreation choice behaviour

'The concept of choice presupposes the availability of alternative courses of action and the ability or freedom to choose among them, and is therefore central to the concepts of leisure and outdoor recreation' (Williams 2003: 51). Predictions regarding recreation behaviour would have greater validity if more was known about attitudes, motivations and perceptions affecting recreation decision-making. This would help explain, for example: (1) why certain activities and sites are favoured; (2) why some recreation businesses are failures, while others provide satisfaction and even draw excess patronage; and (3) how and why alternative recreation opportunities are ranked.

The recreation choice process is influenced by people's perceptions of what recreational opportunities are available (e.g. see Louviere and Timmermans 1990). In every decision-making situation, individuals evaluate selected environmental attributes against some predetermined set of criteria in order to arrive at an overall utility or preference structure (see Aitken 1991). A predisposition or propensity (i.e. demand) for recreation is translated into actual participation through a choice mechanism, heavily dependent upon perception of the recreation opportunity and experience on offer. Perceptions are personal mental constructs, which are a function of the perceiver's past experiences, present values, motivations and needs.

Perception operates over several dimensions and various scales in recreation decision-making, and initial mental constructs may be confirmed or revised as a result of further spatial search and learning. Information levels, as well as the ability to use that information (which may be governed by such factors as personality characteristics and aversion to risk), also help structure evaluative beliefs and mental images concerning the nature and quality of anticipated recreation experiences.

Information sources, and the credibility of the information itself, are key issues in the choice of leisure settings. The validity of some spatial choice models has been questioned because of the assumption of perfect information and the assumed ability of consumers to evaluate completely all alternatives (Roehl 1987) or even construct an accurate image of the real world (e.g. Walmsley and Jenkins 1999). In reality, individuals typically consider only a subset of available alternatives. For example, in any choice situation, an individual's decision will be influenced by his/her awareness set. Larger natural settings (e.g. national parks close to urban populations) with distinctive characteristics, are more likely to be known and considered by potential participants. In an urban context, Roehl demonstrates that smaller neighbourhood parks, with fewer facilities, and designed to serve lower-order needs rather than community- or higher-order needs, are less likely to be in a consumer's awareness set.

Desbarats (1983) notes how the supposedly objective spatial structure of opportunities is narrowed into an 'effective choice set' comprising those (recreation) opportunities that are known to the individual and actively considered. Effective choice sets may represent only a small fraction of objective choice sets, because of the direct and indirect effects of constraints on behaviour stemming from the sociophysical environment. In particular, contraction of the initial choice set may occur because of lack of information about existing options. 'The better the information, the greater the congruence between effective and objective choice sets' (Desbarats 1983: 351). Both the quality and timing of information are important factors in recreation decision-making. Inadequate information and misinformation act as constraints in the process of discriminating between alternatives (Krumpe 1988). The implications for management and policy are obvious and are discussed further below.

Information also helps structure images of the environment to which recreationists respond. 'Cognition is a broad term that covers activities such as thinking and reasoning ... Cognition involves the acquisition, coding, storage, recall, and manipulation of information' (Walmsley 2003: 57). However, the cognitive processes involved in image formation are complex (Beaulieu and Schreyer 1985). As Walmsley notes,

> Two sorts of knowledge thus derive from cognition: figurative knowledge (images resulting from direct contact) and operative knowledge (information that has been structured through a variety of mental operations so that individuals are able to extrapolate from known situations to infer knowledge about situations yet to be experienced.
>
> Walmsley (2003: 57)

The (objective) information flowing from an environment is filtered through the perceiver's set of preferences and values, and cultural interpretations of place meaning. The process is complicated by the personal nature of reactions to external stimuli and by the multifaceted characteristics of the environments being experienced. Studies have, for example, shown how people's estimates of distance to tourist destinations can vary dramatically and have questioned the impacts that such cognition or image-making might have for attracting visitors (e.g. Walmsley and Jenkins 1999). The ways in which human minds develop 'warped and twisted model[s] of the real world' (Downs and Stea 1977: 144, in Walmsley and Jenkins 1999: 299) significantly influence recreational choice.

Dissection is risky. In nature-oriented environments, in particular, it is difficult to reach consensus on what components or attributes (e.g. landform, water, vegetation, distance from home, cost) contribute most to the appeal of the landscape or a resource. In any case, these attributes must be mentally fused to complete the totality of the image, so that the whole is greater than the sum of the parts. 'A landscape is more than the enumeration of the things in the scene. A landscape also entails an organisation of these components. 'Both the contents and the organisational patterns play an important role in people's preferences for natural settings' (Kaplan and Kaplan 1989: 10).

Natural environments as recreation settings

The Kaplans applied this reasoning to what they see as human preference for natural environments. They believe that it is not only the dominance of nature in the scene which is appealing, but that it is also the spatial configuration of landscape elements which is important to people's reactions. Certain natural settings are favoured because of their openness, their very lack of structure and precise definition, their transparency, and the perceived opportunities to enter and move around. Wild environments, impenetrable forests and even built environments, on the other hand, may evoke less positive responses, along with feelings of insecurity.

Research by Driver *et al.* (1987) attests to the importance of the natural setting in achieving the desired outcomes from leisure pursuits. In a wide-ranging study of wilderness users in Colorado, the most important 'experience preference domains' were linked to enjoyment of nature. Many other studies have described the increase in recreational use of natural areas and further explored the importance of natural areas and settings to people's leisure and the factors influencing their experiences (e.g. Kearsley and Higham 1996; Higham 1996; Font and Tribe 2000; Newsome *et al.* 2002). For example, the main motivations for domestic and international visitors to the backcountry of New Zealand in a recent study were scenic beauty/naturalness (92 per cent and 97 per cent respectively), to enjoy the outdoors (96 per cent and 95 per cent) and to encounter wilderness (78 per cent and 82 per cent). Far fewer people wanted to meet new people and make new friends (16 per cent and 18 per cent) or to experience solitude (38 per cent and 39 per cent) (Kearsley 2000). Clearly, the natural environment plays a fundamental

Plate 2.1   Scotland's landscapes are diverse and provide many opportunities for viewing of splendid scenery in remote areas. While people hike for many reasons, the viewing of scenery in these places surely must have some attraction

part in attaining the outcomes and satisfactions sought from participation in certain forms of recreation.

Given the widespread appeal which nature apparently holds for people, its importance in the experience of leisure should come as no surprise. Many of the benefits associated with natural settings are, or should be, fundamental to the realisation of leisure. The opportunity for self-expression and subjective freedom of choice, accepted by many observers as characteristic of leisure, appears to be sought more often in natural, than in created human-dominant landscapes. The intrinsic values derived from experiencing leisure are perceived as being more in keeping with the natural scene and with a minimum of social manipulation. In terms of 'effective functioning' (Kaplan and Kaplan 1989), the natural environment would seem to offer greater scope for personal satisfaction, through integration of mind and body in the leisure activity itself. The links between leisure, recreation and spirituality are being increasingly recognised. Driver *et al.* (1996) suggest managers of parks and natural areas should seek to incorporate spiritual benefits associated with outdoor recreation, as well as other benefits, when planning recreational opportunities.

Undoubtedly, perceived environmental attributes are a powerful influence on recreation behaviour. This is borne out by Schreyer *et al.* (1985), who suggest that the most useful representation of the environment for the explanation of behavioural choice is at the macroscopic or holistic level, rather than at the attribute level. According to Schreyer *et al.* (1985: 16), 'People do not search for specific elements of the environment as much as they search for settings which will allow them to behave in the ways they desire ... which will allow for the attainment of the desired cognitive state'. They go on to stress the importance of the social milieu and the social definition of the physical environment to the totality of the setting in which recreation takes place.

By definition, outdoor recreation is resource-related, and increasing attention is being given to the setting in which action takes place as a prime force in the satisfaction gained from the ensuing recreation experience. The assumption that recreation experiences are closely related to recreation settings is central to the concept of the recreation opportunity spectrum.

Recreation opportunity spectrum

In many ways, the recreation opportunity spectrum (ROS) is an application of behaviour setting analysis from environmental psychology (Barker 1968; Ittelson *et al.* 1976; Levy 1977). This approach suggests that all human behaviour should be interpreted with reference to the environment or behaviour setting in which it occurs. It is further suggested that, given knowledge of the behaviour setting for a specific recreation experience, such as a park visit, it should be possible to identify the human values and expectations associated with that experience. Examination of the human and non-human features of the behaviour setting should then indicate those contributing to or detracting from satisfaction.

As with all human behaviour, response to external stimuli is not always simple or direct. Environmental psychologists see people not as passive products of their environment, but as goal-directed individuals acting upon that environment and being influenced by it (Ittelson *et al.* 1976). All leisure environments affect recreation behaviour in some way; it is the dynamic interaction between the environment and users which is crucial to the outcome.

Within this conceptual approach, a recreation opportunity allows the individual to participate in a preferred activity, in a preferred setting to realise a desired experience (Driver and Brown 1978). The focus is on the setting in which recreation occurs. The ROS describes the range of recreational experiences which could be demanded by a potential user clientele if a full array of recreation opportunity settings was available through time. Clark and Stankey (1979) define a recreation opportunity setting as:

> ... the combination of physical, biological, social and managerial conditions that give value to a place (for recreation purposes). Thus, a recreation opportunity setting includes those qualities provided by nature (vegetation, landscape, topography, scenery), qualities associated with recreational use (levels and types of use) and conditions provided by management (roads, developments, regulations). By combining variations of these qualities and conditions, management can provide a variety of opportunities for recreationists.
>
> (Clark and Stankey 1979: 1)

The basic premise underlying the concept of the ROS is that a range of such settings is required to provide for the many tastes and preferences that motivate people to participate in outdoor recreation. Quality recreation experiences can be best assured by providing a diverse set of recreation opportunities. Failure to provide diversity and flexibility ignores considerations of equity and social welfare, and invites charges of discrimination and elitism (Clark and Stankey 1979). A sufficiently broad ROS should be capable of handling disturbances in the recreation system. These might stem from such factors as social change (e.g. in demographic characteristics) or technological innovations (e.g. all-terrain recreation vehicles) (Stankey 1982).

The ROS offers a framework within which to examine the effect of manipulating environmental and situational attributes or factors to produce different recreation opportunity settings. Clark and Stankey (1979) suggest that the most important of these 'opportunity factors' are:

- access;
- non-recreational resource uses;
- on-site management;
- social interaction;
- acceptability of visitor impacts; and
- acceptable regimentation.

Some of these factors are discussed in greater detail in later chapters. In particular, it should be noted that the weight or importance given to each will vary with individual site and management circumstances.

The range of conditions to which an opportunity factor can be subjected, and the way each can be managed to achieve desired objectives are shown in Figure 2.4. By packaging a recreation opportunity setting in some combination of the six factors described, a variety of recreation opportunities or options can be generated, and the ROS materially enlarged. In their scenario, the authors present only four generic opportunity types, arrayed along a 'modern to primitive opportunity continuum'. However, within each, there is scope for many complex combinations, thus providing even more diversity.

The ROS also allows an examination of opportunity settings with respect to the capability of potential users to avail themselves of the opportunities presented. Limited resources and, perhaps, lack of awareness or imagination mean that, generally speaking, the established recreation system caters for the majority, on the premise, apparently, that everyone is young, healthy, ambulant, educated, equal and possesses the means to participate. The reality, of course, is very different. Reference is made in Chapter 3 to constraints on recreation because of age, lack of income and other factors. Racial and ethnic origins can be a disadvantage, particularly in inner cities and suburbs, where these minorities are often concentrated. Likewise, the spectrum of recreation opportunities for people who are disabled is likely to require special attention if real choice is to be offered.

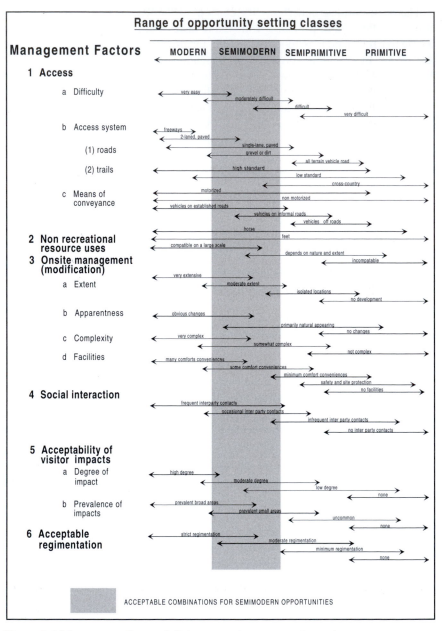

Figure 2.4 Management factors defining recreation opportunity settings

Source: Adapted from Clark and Stanley (1979: 15).

Despite its inherent appeal as a means of facilitating choice in outdoor recreation, specific applications of the ROS approach have attracted some criticism. For some managers of recreation sites, the concept has been treated as a 'blueprint', from which little deviation was possible or desirable. In other situations, there has been a reluctance to amend the range of opportunity settings from that initially created, so as to allow some flexibility, in keeping with the dynamic aspects of the recreation environment and the preferences of users. Indeed, there appears to be relatively little consultation with potential visitors to identify preferred recreation settings. The approach is predominantly 'top-down', reflecting what management feels will be satisfying for visitors, and conducive to managerial convenience. The emphasis, too, has been on manipulation of the biophysical elements of recreation settings, whereas opportunities for social interaction are at least equally important influences on satisfaction and quality recreation experiences (Heywood 1989). Some of these shortcomings are brought out in examples of the application of the recreation opportunity spectrum concept in management situations presented in Chapter 6.

Incompatibility and conflict in outdoor recreation

Further complexity is added to the recreation choice process, when the issue of incompatibility is considered. Most often, this is seen as a problem between outdoor recreation and other forms of resource use. However, conflicts can arise just as readily between groups of recreationists, even when engaged in the same leisure pursuits.

Conflict has been the focus of a good deal of attention in recreation settings, especially over the past 30 years or so. Recreational conflict has been approached from two main directions. The first is at an interpersonal level where goal interference leads to conflict. The second is at a social level where people or agencies have different views about recreational behaviour. So, conflict does not always stem from face-to-face or direct contact, and is not always actual. Sometimes it is perceived. Actual conflict arises when people or groups come into contact. There may not be a collision, but rather, simply 'an invasion of the collision zone in which the person believes that physical contact will occur if evasive action is not taken' (Ravenscroft *et al.* 2003: 68). Perceived conflict, on the other hand, stems from people's thoughts and emotions as they relate to an environment. So people may not experience violence in an urban environment in any way (i.e. see no violence or be attacked themselves), but they may develop a perception (perhaps an emotional response) to media or personal reports. Perceived conflict is a state of mind stemming from many possible factors singularly or in combination: competition for shared resources; escalating annoyance; negative experiences; goal interference; minimisation of expected benefits; mutually exclusive use goals/values/norms; manner and purposes of use; attributions of blame to others or to external factors; perceived control over desired outcome; prior knowledge and experiences; differences in social values (Ravenscroft *et al.* 2003: 68–70). These matters arise in Chapter 7 (e.g. violence in urban parks) and Chapter 8 (e.g. conflicts arising from the goals of different agencies).

The question of compatibility revolves around the degree to which two or more activities can co-exist in the use of a given recreational resource or setting. Goodall and Whittow (1975) point out that the problem is linked with the resource requirements for particular recreational pursuits. Only where recreational activities have similar requirements is there a possibility of shared use of a site, or alternatively, of conflict. Noisy activities, such as those involving off-road recreation vehicles or power boats, conflict with fishing, bird-watching, use of wilderness and other activities requiring peaceful countryside locations. Nor is conflict necessarily confined in space or time; site disturbance can have a lasting effect and can spill over to adjacent areas. Goodall and Whittow stress that the incidence of incompatibility is, in part, a function of the activity, the manner in which it is practised and the characteristics of the setting or the resource involved. Trails, rivers and other constricted linear resources are particularly sensitive to use incompatibility. On the other hand, timbered land may increase compatibility by reducing visual intrusion and noise penetration.

Conflict and compatibility involve a good deal more than simple one-to-one comparisons of selected recreational activities. According to Lindsay (1980), the conflict problem may be summarised as one of recreationists competing for the same physical, social and psychological space during the same time period. Thus, confrontation over use of recreation space should not be interpreted solely as interactivity conflict. The complexities of human behaviour are such that conflict situations can develop between different types of recreationists engaged in the same activity. A commonly identified conflict arises between recreationists engaging in non-mechanised and mechanised activities. Examples include snowmobiling and cross-country skiing; power-boating and fishing or swimming.

Jackson and Wong (1982) examined perceived conflict among skiers and snowmobilers from urban areas. They identified four principles of recreational conflict:

- *Non-mechanised versus mechanised recreation*, where issues such as noise diminished the degree of solitude and tranquillity sought by skiers (non-mechanised recreationists).
- Conflicts were generally considered asymmetrical with the non-mechanised group expressing annoyance at the mechanised group and the mechanised group expressing greater tolerance of the non-mechanised group.
- Conflict stemming from *different recreational motivations* among groups and subsequently low quality experiences as intrusions into others' activities occurs. Hence, competition for resources in and of itself is not an adequate explanation for recreational conflict.
- Conflict also arising from *perceived impacts on the environment and a lack of appreciation of other groups' views* on resource use.

Conflict also often arises between more passive, non-motorised activities (e.g. surfing with board or ski and body surfing or boogie boarding; cyclists and walkers) and between people participating in the same or similar activities. For example, surf rage arises from clashes among surfers. Local surfers of a particular point or break have resorted to violence to discourage non-locals; inexperienced surfers who don't know the unwritten rules of the water 'drop in' on experienced surfers; board riders abuse body boarders and swimmers and ignore repeated requests from lifeguards not to surf between 'the flags' or designated safe swimming areas (e.g. see Young 2000).

Recreation conflict has also been extensively studied in six Boulder County Parks and Open Space Properties (Bauer 2004). In this study, 624 interviews were conducted with visitors. Conflicts were experienced by only 2 per cent of respondents on the day they were interviewed. Approximately two-thirds of respondents had never experienced any conflict. However, approximately 34 per cent of visitors had experienced conflict some time in the past. The most frequent reporters of conflict were hikers, dog-walkers and equestrians. The nature of conflict stemmed mainly from the speed at which mountain bike riders moved, their communication behaviours and their failure to yield; leashing habits and control of dogs by their walkers; presence of horse faeces. On very few occasions were reported conflicts with hikers and other visitors (Bauer 2004: 14). The study provides baseline data which is to be updated every five years as part of a more detailed monitoring and evaluation programme, while standards are developed for acceptable levels of conflict.

Jacob and Schreyer (1980) believe that the key to conflict resolution lies in identifying the 'conflict potential' of recreation resource clientele, rather than in labelling certain activities as conflict-prone. It is not merely a question of skiers not getting along with snow-mobiles, or of 'motor versus muscle'. In this context, four causal factors are identified as conducive to conflict in outdoor recreation:

- *Activity style* – various personal meanings are assigned to an activity. For some, participation may be intense; the activity becomes the focus of life interest, with acquisition of status and achievement of a high quality recreation experience prominent goals of participants. As a participant becomes more specialised or 'involved', the potential for conflict increases with others not so committed or expert (McIntyre 1990).
- *Resource specificity* – some individuals attribute special values and importance to certain physical resources, and develop possessive, protective attitudes to favoured recreation sites – a common trait of skilled fishermen and hunters. Tension can develop with lower status 'intruders', who do not share this appreciation and interpretation of site values, and who disrupt the exclusive, intimate relationship built up with a place. Once again, conflict has little to do with activities themselves, but can occur between divergent classes of resource users.
- *Mode of experience* – the manner in which individuals approach a recreation experience can provide the ingredients for conflict. Jacob and Schreyer (1980)

distinguish between 'focused' and 'unfocused' modes of experience. The latter is concerned with overall spatial relationships and environmental generalities, rather than specific entities within that environment. Thus, the 'focused' wilderness user, intent upon achieving an intimate relationship with specific aspects of the natural environment, has little in common with the 'unfocused', for whom merely being in the countryside is sufficient. The greater the gap between recreationists along this continuum, the greater the potential for conflict.

- *Tolerance for lifestyle diversity* – the suggestion is, that individuals deliberately choose recreation settings and associations which reflect their societal outlook and behaviour, and are unwilling to share resources with other lifestyle groups categorised as deviant, or merely different. Value-laden inferences are made about people indulging in alternative forms of recreation, stereotyped as 'less worthwhile'. Thus, the trail-bike, the power boat and the snow-mobile are seen as symbolic of a society that arrogantly exploits and consumes resources. Ethnic, racial and social class distinctions can also be the basis for lifestyle-based conflicts. Such people are often labelled 'out-of-hand' or 'inferior', so that even when pursuing the same activity and following the same rules, conflict still ensues, especially as the number and variety of people desiring access to recreation resources increase.

Jacob and Schreyer suggest that the degree to which these four factors are present, singly or in combination, represents the extent to which the potential for conflict exists. Conditions for conflict may just as readily occur in the mind and be part of the mental state and attitude of the participant, as in the nature of the recreational activities. The authors conclude with a warning for management:

> Unfortunately, the tendency to define conflict as confrontations between activities has left the sources of recreation conflicts unrecognised. In failing to recognise the basic causes of conflict, inappropriate resolution techniques and management strategies are likely to be adopted.
>
> (Jacob and Schreyer 1980: 378)

The benefits of outdoor recreation

Leisure and outdoor recreation offer many benefits. These include:

- *physiological benefits*, through gains in fitness, health and wellbeing. When people experience nature or become excited because they have climbed a mountain or feel exhilarated through a physical activity, hormones are released (e.g. norepinephrine and endorphins) and stress may be relieved;
- *psychological benefits*, by way of escape, enhanced self-esteem, and opportunities to express themselves or find solitude, privacy, and means of releasing frustrations (Leitner and Leitner 2004a: 379–83);
- *social benefits*, through sharing experiences, promoting group interactions, developing cooperation, and role play.

Table 2.1 The benefits of leisure activities

Physiological	Social/psychological/emotional benefits
Improved circulation	Greater psychological wellbeing
Improved respiration	Higher quality of life
Greater flexibility and balance	Higher morale and life satisfaction
Greater strength and power	Higher self-esteem, self-concept, and self-efficacy
Greater endurance	
More energy	Keener mental abilities
Lower blood pressure	Feelings of achievement and accomplishment
Lower cholesterol	
Improved mobility	Greater optimism
Greater physical independence	Interchange of ideas on leisure
Recovery from illness, disease and injury	Greater levels of social interaction
Greater longevity	Laughter
Improved general health	Educational
Reduced risk of osteoporotic fractures	Lower anxiety and hostility
Reduced risk of falls	Lower incidence of loneliness and depression
Skill improvement	
Better able to cope with diseases and physical strains	Improved perceived health
	Reduced fear of falling
	Sense of control over one's life
	Better citizenship and satisfaction with the local community
	Employment opportunities
	Reduced costs of healthcare and hence more disposable income
	Environmental awareness

Source: Adapted from Leitner and Leitner (2004b: 17); Driver (2003: 32).

Leitner and Leitner (2004b) listed the benefits of a range of leisure activities, including outdoor recreation, for older people (Table 2.1). With few exceptions, the above benefits are relevant to people of almost all ages. In the case of the elderly, however, Leitner and Leitner (2004b) have argued that their participation 'in the more desirable leisure pursuits, such as physical activities, [is] abysmal' (p. 17) (see Chapter 3).

The benefits of leisure and outdoor recreation are being increasingly recognised in the creation of recreational opportunities, but unfortunately, when it comes to recreational provision for those who are unemployed or experience forms of discrimination, governments are too slow to react. In the case of the former, governments are withdrawing from many avenues of recreation supply and are relying on market forces (or the private sector) to supply relevant opportunities. In cases where demand is low or there is an inability to pay a market price, recreational opportunities are often not supplied and governments may be loath to intervene or at the very least they will seek a partnership. In general terms, Osborne and Gaebler (1992: 2) have described the situation well, explaining the ways in which public agencies or institutions are leaner, innovative and decentralised. Market forces, competition, partnerships and choice are some of the familiar buzzwords in the supply of recreational opportunities formerly the purview of government agencies in national parks and in therapeutic forms of recreation where volunteers are critical to sustainable programmes. Many of these issues are taken up in the following chapters.

Summary

Many factors affect recreational motivation and choice, with much debate continuing about the forces affecting recreation decision-making at the individual, group and societal levels. This chapter explored the nature of recreation demand and participation, and the range of influences on recreation choice behaviour. The importance of accessibility and the travel phases in the overall recreation experience was stressed, and an overview presented of the recreation decision process. Reference was made to the factors affecting participation in recreation, in particular, the role of perception of recreation opportunity. The types of decision choices which confront individuals and groups, and how these affect people individually and collectively, are related to the concept of the recreation opportunity spectrum. Further complexity is added to the recreation choice process with consideration of compatibility and conflict between recreation activities and recreationists. If these are carefully managed, the benefits of outdoor recreation are increased greatly.

This review of the relationship between people and the leisure environment reveals some of the dynamics and complexities of the choice process in recreation behaviour. As stated at the outset, it is the unbounded, subjective nature of leisure and its expression in recreation activity which make explanation and prediction difficult. By definition, recreation is discretionary and any suggestion of obligation or compulsion must compromise the experience. Moreover, participants in recreation, as distinct from other forms of human behaviour, can exercise more control over decisions regarding what, where, and with whom, '... in the design of their desired products and thus the experiences they derive from participation' (Williams 1995: 32).

Finally, it is the interaction of such environmentally-related supply factors with demographic, socioeconomic and situational variables, or population-related demand factors, which generates opportunities to participate in recreation. However, recreation decisions depend not on actual objective opportunities, but on individual perceptions of those opportunities. These, in turn, depend greatly on formal and informal social and information networks, and on the personal characteristics and motivations of potential recreationists.

Guide to further reading

- Recreation and travel motivation, choice and behaviour: Mayo and Jarvis (1981); Pearce (1982); Ibrahim (1991); Bammel and Bammel (1992); Garling and Golledge (1993); Ross (1994); Stopher and Lee-Gosselin (1997); Dann (2000).

- Benefits of outdoor recreation: Allen (1996); Manfredo and Driver (2002); Driver (2003).
- Outdoor recreation demand: Hanley *et al.* (2003c).
- Recreational conflict: Driver and Bassett (1975); Jacob and Schreyer (1980); Devall and Harry (1981); Gramann and Burdge (1981); Adelman *et al.* (1986); Schneider and Hammitt (1995a, 1995b); Williams *et al.* (1994); Ramthun (1995).
- Recreational and tourist satisfaction: Ryan (1995, 2003); Leitner and Leitner (2004a, 2004b).
- Recreation Opportunity Spectrum : Trail Bridge Crossings and application of the ROS at http://www.fs.fed.us/na/wit/WITPages/bridgecatalog/ros.htm; recent developments on the ROS at http://www.fs.fed.us/ne/newtown_square/publications/technical_reports/pdfs/2003/gtrne309.pdf.

Review questions

1 Discuss the dimensions of recreation demand.
2 Critically review discussions of the relationship between recreation demand and recreation supply.
3 Present an overview of the factors affecting recreational motivation and choice.
4 Generally speaking, is it enough to derive explanations of recreation motivation and behaviour from only one factor (univariate analysis), or should we consider the relationships between many factors (multivariate analysis)? Explain your answer with reference to appropriate studies.
5 What are the main demographic changes taking place within your country and local region? Identify recent recreation policy and planning responses to such changes at the national and/or regional level.
6 Identify a local recreation site where conflict between recreationists has arisen. Why does/did that conflict exist? Has the conflict been resolved? Why/why not?
7 In small groups, discuss the implications and relevance of the opening quotes for this chapter.
8 List some of the main benefits of leisure and outdoor recreation. Categorise them in your own way and prioritise them. Explain the reasons why some of the benefits are more important to you than others. Or, do you find that the benefits are ranked differently depending upon the recreational activity involved?

3   Special groups and special needs

Recreation need is characteristic of all human beings, but as Veal stated:

> Every individual is unique and so could be said to have unique leisure requirements. In family settings and some organisational settings this uniqueness can be catered for, but human beings are social animals with interests, demands and needs in common … Classifying people into groups and considering their common characteristics and needs is not therefore to deny their individuality; in fact, it has been the failure of providers to consider the common needs of some groups which has, in the past, denied members of such groups their individuality. As a result of campaigns, regulations, research and the spread of ideas such as 'market segmentation' and 'niche marketing', some of these problems are now beginning to be overcome.
>
> (Veal 1994: 189–90)

There are some for whom participation in, and the resultant satisfaction derived from, recreation requires that special services, programmes and/or facilities be provided to ameliorate or remove leisure constraints. These people are commonly regarded as having special needs.

Research on recreation non-participation and constraints to leisure is growing. Such research makes theoretical contributions to our understanding of leisure choice and behaviour, and makes practical contributions by providing information which will generate or affect service delivery by way of policy-making, planning, programming and marketing (Jackson 1990).

From this perspective, this chapter outlines the concepts of leisure constraints and recreation need, and considers factors which may act as barriers to recreational participation and satisfaction. It discusses possible approaches to the assessment of constraints and needs as a basis for future planning and programme development. As society becomes more complex and dynamic, as 'social services' of the state are privatised, and as much recreational need and tourist travel become more discerning, sophisticated and expensive, it becomes increasingly difficult for individuals, acting alone, to satisfy their recreational needs.

Leisure constraints, recreation needs and human rights

'A constraint to leisure is defined as anything that inhibits people's ability to participate in leisure activities, to spend more time doing so, to take advantage of leisure services, or to achieve a desired level of satisfaction' (Jackson 1988, in Jackson and Henderson 1995: 32; also see Henderson *et al.* 1989: 17). The constraints associated with leisure and recreation participation have been studied by several authors in general terms, and in specific terms with reference to special groups, including people with disabilities, youth and adolescents, the elderly, and women with physical disabilities. In the main, two types of constraints have been identified – *intervening constraints* (those that come between a preference and participation and which thereby limit participation) and *antecedent constraints* (those that influence a person's decision and subsequently inhibit preferences).

Research on constraints to leisure behaviour and recreation participation has captured much attention and is growing conceptually, theoretically and in practical application (see Jackson 1988; Stemerding *et al.* 1999). Work on leisure constraints and barriers has a relatively lengthy history, with the Outdoor Recreation Resources Review Commission (ORRRC) studies in the 1960s (e.g. Ferris 1962; Mueller *et al.* 1962) receiving considerable prominence. Since the early 1980s, however, constraints research has proliferated, with perhaps the most notable early models initially put forward by Crawford and Godbey (1987). Their formulation of leisure constraints focused on intrapersonal, interpersonal and structural constraints (see Figure 3.1). In 1991, Crawford, Jackson and Godbey revisited that formulation. In its place, they proposed a hierarchical process model, in which the three types of constraints above (intrapersonal, interpersonal, structural) were integrated. They derived three propositions from that model:

• leisure participation is heavily dependent on a process of negotiating through an alignment of multiple factors, arranged sequentially;
• sequential ordering of constraints represents a hierarchy of importance;
• social class may have a more powerful influence on leisure participation and non-participation than is currently accepted (i.e. the experience of constraints is related to a hierarchy of social privilege).

Crawford *et al.* believe that this more recent model may help to clarify some paradoxical findings that were not fully explained previously. In particular, they noted the more frequent reporting of structural constraints among people of higher socioeconomic status. They go on to point out, however, that this hypothesis is largely speculative, and that research should proceed in three main directions.

First, there is a need for intrapersonal and interpersonal-level data, with investigations encompassing the entire array of constraints – intrapersonal, interpersonal, and structural – simultaneously. This would permit testing of propositions that people negotiate through sequential levels of constraints, and that these levels represent a hierarchy of importance.

Second, the issue of social stratification needs to be examined from a dynamic perspective, particularly given the widespread and increasing disparity between

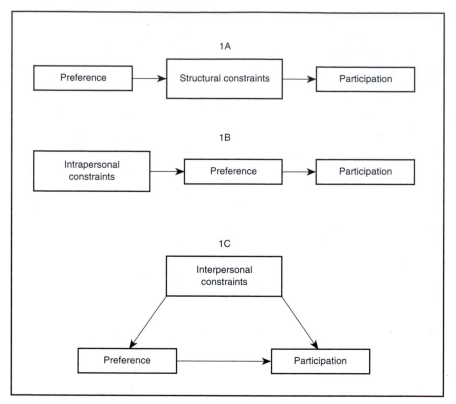

Figure 3.1 Crawford and Godbey's three types of leisure constraints

Source: Crawford *et al.* (1991).

the affluent and the poor. Longitudinal studies would reveal associated changes in recreation patterns and processes.

Third, we should move beyond examining constraints which result in non-participation, to investigate constraints which affect levels of participation. Clearly, then:

> Although for practical reasons it is often useful to conduct research at a high level of detail on separate parts of a system (in this case the system of leisure behaviour), there is a need to integrate leisure constraints research within the mainstream of leisure studies. Leisure researchers cannot afford to investigate the phenomena in which they are interested in isolation from other factors that influence leisure choices.
>
> (Crawford *et al.* 1991: 318)

Jackson *et al.*(1993) went even further and suggested that leisure constraints negotiation (i.e. how a person decides to experience an activity despite the constraints encountered) is the key to understanding constraints. They noted people may be reluctant or resistant with respect to change in their patterns of recreational participation and consumption as a result of their experiences in dealing with or negotiating constraints. Jackson *et al.* (1993) argued that some studies have failed to establish clear relationships between various constraints (e.g. costs) and participation. Some people do respond positively to constraints and will attempt to overcome them. Some respond poorly or not at all, and may not even respond to incentives or the efforts of others to eliminate or reduce constraints to clear the way for participation (e.g. see Kay and Jackson 1991; Coalter 1998). In brief, 'constraints are conceived as phenomena that more likely result in modified participation' (Jackson *et al.* 1993: 9). Ultimately, a balance is struck between constraints and motivations (see Figures 3.2 and 3.3).

In a more recent review of the constraints literature, Ravenscroft and Curry (2004) cited research which now clearly

> suggests that new forms of constraint may be emerging, where the question is not so much whether or not to participate, but when, where and with whom. This reflects Desbarat's (1983) construction of 'effective choice sets', to the extent that people's behaviour may be more associated with social structures than with individual preferences and choice.
>
> (Ravenscroft and Curry 2004: 173)

They go on to discuss that constraints may be little more than an academic construct and that there is a need to critically review the ways in which constraints are studied. They then argue that:

> participation is increasingly contingent, not only on the continued negotiation of constraints themselves, but on the ways in which individuals contextualise – and thus operationalise – those constraints. This leads us to ask important questions about the point at which constraints occur: do they, for example, reduce or eliminate preferences at the outset or do they only become apparent at the point of participation?
>
> (Ravenscroft and Curry 2004: 174)

It seems that some constraints such as time may represent some form of inertia which constantly arises and 'can increasingly become embodied as a barrier to any change in a person's pattern of recreational participation' (Ravenscroft and Curry 2004: 183; also see Jackson *et al.* 1993). Siegenthaler and O'Dell (2000), for instance, explain how long-term relationships between couples bring about a merging of partners' attitudes.

Whatever the case, constraints to leisure (whether they lead to non-participation or to less than optimal participation from the participant's perspective) stem from many factors, including biological, psychological, sociological, political and economic sources. For people to find leisure experiences or to establish desired levels of participation, planning and management to ameliorate or remove

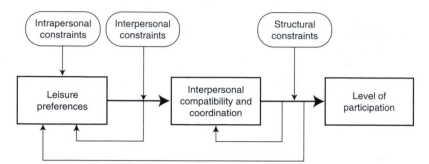

Figure 3.2 Jackson, Crawford and Godbey's conceptualisation of interactions among types of constraints

Source: Jackson *et al.* (1993: 7).

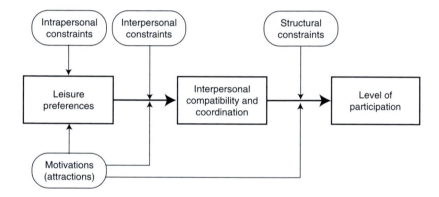

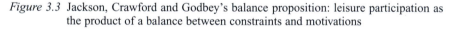

Figure 3.3 Jackson, Crawford and Godbey's balance proposition: leisure participation as the product of a balance between constraints and motivations

Source: Jackson *et al.* (1993: 9).

constraints are required. On the time dimension, free time has come to mean little if 'free time' is simply regarded as time not spent at work, or meeting other basic necessities (also see Chapter 1). Such is perhaps the case for the unemployed, the retired, those in public or private institutions (e.g. hospitals or prisons), or in circumstances where free time is a burden. Leisure activities for these people are not always, perhaps rarely, freely chosen or necessarily enjoyable, and may even involve physical and/or psychological stress. Under these circumstances, individuals may require additional services or information, for instance, by way of education in developing leisure decision-making and participation skills, as well as in identifying opportunities to seek satisfying leisure experiences.

The concern for special needs groups arises from increased societal awareness of and concern for, a more egalitarian society, based on human rights, social equality and accessibility to resources. Put simply:

> ... every person has the innate right to pursue his dreams and must be given the opportunity to fulfil his needs (within societal approval) as he has the capacity to achieve without artificial hindrance or restriction. The only limitations upon individual achievement should be biological potential and social acceptability.
>
> (Shivers 1967: 131)

With respect to recreation, concerns for egalitarian recreational opportunity were given international prominence when the World Leisure and Recreation Association (which was later renamed World Leisure) promulgated the Charter for Leisure (revised in 1981 and 2000), which contains eight articles (see Box 3.1). These articles present an overriding ideal of equality of recreational access, extolling the virtues of leisure, and exhorting governments to make provision for leisure as a social service. Unfortunately, like previous editions of the Charter, they do not, as noted with respect to earlier editions (Veal 1994: 9), declare access to leisure facilities and services as a human right, and this problem is manifested in recent planning and policy. For instance, in the United Kingdom, until 1995, when the government's Disability Discrimination Bill was introduced in July that year and eventually became the Disabled Rights Act 1995, there was no legal framework to protect people with disabilities from discrimination in seeking access to museums and country heritage sites. In the United States, it was not until 1990 that an American Disabilities Act was introduced, requiring all government, commercial and public premises to be readily accessible.

Humans are not created equally, nor do they share equality in life. Moreover, people, and public and private institutions have created (deliberately or otherwise) or contrived, artificial restrictions which may prevent individuals or groups participating in recreational activities which would otherwise be socially acceptable. Such restrictions are based on age, gender, race and ethnicity, religion, socioeconomic status, political affiliation, employment and location (e.g. remoteness). In other words, it is society's definitions, perceptions and attitudes to such factors, and their relationship to recreational need which serve as one basis for inequality.

Recreation need is a multi-dimensional concept (Bradshaw 1972), and there exists a plurality of needs in any community (Hamilton-Smith 1975). Such needs are dynamic, individually and collectively. Taylor (1959) outlined four uses of the term need:

- to indicate something needed to satisfy a rule or law;
- to indicate means to an end (either specified or implied);
- to describe motivations, conscious or unconscious, in the sense of wants, drives, desires, etc.;

Box 3.1   World Leisure: Charter for Leisure

Introduction

Consistent with the Universal Declaration of Human Rights (Article 27), all cultures and societies recognise to some extent the right to rest and leisure. Here, because personal freedom and choice are central elements of leisure, individuals can freely choose their activities and experiences, many of them leading to substantial benefits for person and community.

Articles

1   All people have a basic human right to leisure activities that are in harmony with the norms and social values of their compatriots. All governments are obliged to recognise and protect this right of its citizens.
2   Provisions for leisure for the quality of life are as important as those for health and education. Governments should ensure their citizens a variety of accessible leisure and recreational opportunities of the highest quality.
3   The individual is his/her best leisure and recreational resource. Thus, governments should ensure the means for acquiring those skills and understandings necessary to optimise leisure experiences.
4   Individuals can use leisure opportunities for self-fulfilment, developing personal relationships, improving social integration, developing communities and cultural identity as well as promoting international understanding and cooperation and enhancing quality of life.
5   Governments should ensure the future availability of fulfilling leisure experiences by maintaining the quality of their country's physical, social and cultural environment.
6   Governments should ensure the training of professionals to help individuals acquire personal skills, discover and develop their talents and broaden their range of leisure and recreational opportunities.
7   Citizens must have access to all forms of leisure information about the nature of leisure and its opportunities, using it to enhance their knowledge and inform decisions on local and national policy.
8   Educational institutions must make every effort to teach the nature and importance of leisure and how to integrate this knowledge into personal lifestyle.

Approved by the World Leisure Board of Directors, July 2000. The original version was adopted by the International Recreation Association in 1970, and subsequently revised by its successor, the World Leisure and Recreation Association in 1979.

Source: World Leisure (http://www.worldleisure.org/pdfs/charter.pdf).

- to make recommendations or normative evaluations. These are sometimes difficult to distinguish from the above three uses which are intended as purely descriptive statements.

(Taylor 1959: 107)

There are several frameworks for assessing recreation needs. Mercer (1975) presented a typology of need comprising four categories based on Bradshaw's (1972) work:

- *felt need*: those needs which individuals have and which they want satisfied;
- *expressed need*: those needs which are expressed by people;
- *comparative need*: those needs identified on the basis of comparison of individuals or groups;
- *normative need*: those needs involving external assessments by experts, who identify a gap between what actually exists and what is desirable.

Each of the above dimensions lends itself to different methods of assessment (e.g. see Hamilton-Smith 1975), although Mercer (1975) viewed normative needs assessments with some suspicion. He argued that the 'experts' who make them are largely considered a 'small élite group in our society – the well-educated, well-to-do planners, politicians, engineers and academics'. This leads us to a critical point in identifying and assessing recreational need. The identification and assessment of recreation needs are value-laden activities, open to personal interpretation and subjective judgement, while any single measure of need will be inadequate, and a combination of approaches is needed. Values affect leisure and recreation public policy. As Simmons *et al.* (1974: 457) noted, 'it is value choice, implicit and explicit, which orders the priorities of government and determines the commitment of resources within the public jurisdiction'. These issues go to the heart of the structural constraints and problems identified by Crawford *et al.* (1991) and discussed earlier in this chapter.

As noted above, research concerning the recreational needs of special groups is expanding, in terms of both the types of groups studied and the depth of knowledge with respect to different groups (including variations within groups) (e.g. Aitchison 2000; Veal 2002; *Annals of Leisure Research* 2004). Moreover, the normative aspects of recreational need, namely leisure and recreation policy-making and decision-making processes, which were largely ignored in the 1970s (e.g. see Mercer 1975), began to receive greater attention in the 1980s (e.g. see Henry 1993; Veal 1994, 2002), and deservedly so. Prescriptive-rationalist approaches to public policy, for instance, would see the decisions of government as being part of an inherently rational policy-making process, where goals, values and objectives are identified and ranked, after the collection and systematic evaluation of the necessary data (Wilson 1941). However, this approach fails to recognise the inherently political nature of public policy, and the influences of values, power, institutional arrangements (including interest groups), and other factors (e.g. lack of monitoring and evaluation of policies and programmes) on the policy process.

There are winners and losers with respect to any leisure and recreation policies and programmes. We need to know a lot more about who benefits and who loses out in terms of outdoor recreation, and we need to be able to explain why and how people benefit or lose out.

In fact, people cannot always participate in the recreational activities of their choice. The satisfaction of recreational need requires individuals and groups to successfully overcome 'intervening variables' such as age, income, education and awareness and health status. The differential impacts of barriers to participation mean that some individuals and groups have more difficult barriers to overcome than other groups in society. For those individuals and groups unable to overcome the impediments associated with intervening variables, or unable or unwilling to make use of incentives or encouragement to participate, a case of special need may be identified. In such instances, resources will need to be allocated to services, programmes and facilities, over and above those usually required. In this respect, governments play a crucial role, though increasingly these kinds of responsibilities are being delegated to the private sector or non-government organisations, including charities and other agencies relying heavily on volunteers. Recreation represents people's expression of the need to do things other than work, even though much recreation is institutionalised. That we are able to identify many people with special needs suggests that the institutional arrangements for recreational satisfaction are inadequate. The satisfaction of special groups' recreational needs thus requires institutional action. If this view is not accepted, then we run the risk of further disadvantaging these people.

The question of one's state of mind raises questions, too, about whether activities which may be seen by some as recreational, may be perceived very differently by special needs groups, who require assistance to seek alternative opportunities during their leisure time. It is often the way in which a particular activity is perceived by an individual that will determine whether it is recreational or not (see Chapter 1) and whether the activity is a desirable one. Persistent media attention, for instance, on violence (say in urban parks), dangers (e.g. shark attacks on Australian and South African coastlines), risk (e.g. airline crashes or surf conditions) or congestion (e.g. lengthy traffic delays at particular destinations at particular times of year) will significantly influence at least some people's recreational decision-making and experiences.

Recreational choice and participation are affected by demographic, socio-economic and situational characteristics, external factors, and perceptions of recreational opportunity (see also Chapter 2). People's perceptions of recreational opportunities will change over time, and the ways in which knowledge and experiences accumulate cannot be neglected in our analyses because previous events will influence people's perceptions. The availability of recreation resources in functional terms depends upon such things as quality, degree of development, environmental and social capacity, ownership, distribution and access. These, in turn, reflect economic, behavioural and political factors, which help shape public and private decision-making about recreation provision. In Britain, efforts are being made to increase the quality of life of disadvantaged groups by providing opportunities for enjoyment of the countryside. 'Attending to the (recreation) needs of vulnerable people is not about doing them a favour. It's part of putting into place the vision of an inclusive society' (Ling Wong 2004: 42).

The barriers to access, canvassed earlier in this chapter, apply with even greater force to disadvantaged groups – lack of knowledge of where to go; the cost of transport, activities and entry fees; inadequate provision for the elderly and those with disabilities; fear of racism in isolated places. The Countryside Agency in the UK is allocating resources to meet the diverse needs of a range of socially excluded or marginalised groups in relation to countryside access. A particular initiative of note in Britain is the Black Environment Network which has recently completed two pilot access projects: the 'Mosaic Project' in partnership with the Council for National Parks, and the 'People and Historic Places Project' (Ling Wong 2004). Through such programmes, a network of ethnic community groups have enjoyed a range of experiences and are encouraging others in their communities to venture into the countryside. The responsible organisations, too, have recognised the significance of providing for the cultural needs of ethnic groups with prayer facilities and appropriate food.

The special needs of such individuals and groups should be given due recognition in the context of the more usual recreation provisions of the community. In the past two decades, attention has been increasingly drawn to the problems facing various groups and individuals who might have special needs. We now turn our attention to other specific special groups. 'Recreation for All' (e.g. Smith *et al.* 2001) is a wonderful goal, but contemporary Western societies are well short of achieving it.

People with disabilities

Increasing public attention is being drawn to the problems facing people with disabilities or handicaps, as well as means of enhancing their access to, and experiences associated with, leisure and recreational activities (e.g. Aitchison 2000; Veal 2002; Leitner and Leitner 2004a). Recent developments in legislation, policies and programmes in such areas as health, education, employment, facility design, and leisure and recreation, are evidence of changing attitudes and perceptions in society. The United Nations Year of the Disabled in 1981 was an important precursor to this situation, raising global awareness of people with disabilities.

Two models have substantially influenced, among other things, definitions of disability, the development of policies and programmes and research foci. The World Health Organisation, for instance, initially fostered a medical model based on normative assumptions. In this model, the problem of disability is located 'within the individual as a personal tragedy'. The cause of disability is considered a product of the 'abnormal' body that stems from disease, illness, or trauma (Oliver 1996: 31). In the social model, disability is linked to social attitudes, with its defining element being 'in the transformation of an impaired person into a disabled person as a product of the ways in which society is organised' (Darcy 2003: 115). Thus socially constructed constraints and barriers 'affect an individual's community participation, create disability, and discriminate against people because of their impairment' (Darcy 2003: 116). Aitchison (2000) argued the importance of the intersection or nexus of medical and social models that have dominated discourses in disability studies since about 1980. Aitchison suggests it is at this nexus 'that disabled people's leisure identities and relations are frequently played out' (p. 5). This is an interesting revelation, based on her accounts of definitions and research associated with professionals and academics on the one hand and disabled people and the organisations they control and run on the other hand (for a more detailed and incisive discussion, see Aitchison 2000).

A person with a handicap has been defined as one 'whose physical, mental and/or social well-being is temporarily or permanently impaired …' (Calder 1974: 7.3). It is perhaps proper to distinguish between functional disability as a result of primary impairment, and handicap which is determined by individual and societal reaction to limitations on social roles and relationships. Disability is a defined impairment, which becomes a handicap only when the disability prohibits activity in the pursuit of specific goals (see Dibb 1980).

In 1988, it was revealed that approximately 6.2 million adults (14 per cent of the population) in the United Kingdom had some kind of physical, sensory or intellectual disability. More recent estimates show that: nearly 1 million people are blind or partially sighted; 7.5 million are hearing impaired (of whom 2 million use a hearing aid and 55,000 use British Sign Language); 35 per cent of visually impaired people are also hearing impaired, while 14 per cent are mobility impaired (Blockley 1996). The last two statistics highlight a 'stigma on stigma' phenomenon (discussed below), which demonstrates (1) the diversity within singularly defined groups, and (2) the need for multivariate analysis in examining recreational access, motivations, choices and experiences.

The Australian Bureau of Statistics (ABS) (2003) provides an indication of the number of people with disabilities, as well as the nature of such disabilities. Disability was defined as any limitation, restriction or impairment which has lasted, or is likely to last, for at least six months and restricts everyday activities. Disabilities included such things as arthritis, hearing, mental disorders, respiratory, circulatory and nervous system diseases, disorders of the eye, head injuries, strokes, brain damage and dementia. According to the Survey of Disability, Ageing and Carers (SDAC), approximately 20 per cent of Australians (approximately 4 million people) had a disability. Rates did not vary significantly between men and women, but they did increase with age; in fact, some 81 per cent of people aged 85 years and over had a disability. Almost 6 per cent of people required help with at least one self-care, mobility or communication activity (i.e. had a profound or severe level of core activity limitation), and this applied to about 54 per cent of people aged 85 years and over (ABS 2003). People with disabilities had higher rates of employment and were more likely to be employed part-time.

The above survey distinguished between disability and handicap. The latter results from a disability linked to certain tasks associated with daily living, in relation to such activities as self-care, mobility, verbal communication, schooling or employment. These definitions of disability and handicap were based on the

International Classification of Impairments, Disabilities and Handicaps, published by the World Health Organisation (WHO) (1980).

The therapeutic value of leisure for people with disabilities

Leisure has therapeutic value and provides a means of integrating people with a disability into the wider community (e.g. see Shivers 1967; Patterson and Pegg 1995; Aitchison 2000; Leitner and Leitner 2004a). Therapeutic recreation encompasses 'A process which utilises recreation services for the purposive intervention in some physical, emotional and/or social behaviour to bring about a desired change in that behaviour, and to promote the growth and development of the individual' (Gunn and Peterson 1978: 11). It promotes independent leisure for special groups through remedial, educational and recreational experiences that use various activity and facilitation techniques, and promotes their integration into society more generally (e.g. Veal 2002: 251; see Snead 2003). Nevertheless, integration is a complex concept and practice, which has been undertaken, in some instances, in haste, and in a piecemeal fashion with inadequate resources (see Patterson and Pegg 1995).

Recreational opportunity for people with disabilities

Developing a spectrum of recreational opportunities for people with disabilities should encompass three main principles: (1) strong leadership, which can overcome the stress and rigours of programme formulation, implementation, monitoring and evaluation; (2) appropriate assessment of the needs and skills of participants who are disabled (i.e. their physical, emotive, cognitive and social requirements); and (3) the means of integrating recreation programmes into the wider community.

Chubb and Chubb (1981) present a useful summary of the effects of disabilities on participation in recreation, while tourism and mobility issues have been more explicitly explored recently by Burnett and Bender Baker (2001). The conditions and characteristics listed by Chubb and Chubb range from left-handedness, allergies and aberrations of body size, through impaired manual dexterity and mental retardation, to physical disabilities, including sensory impairment. In the area of outdoor recreation, much emphasis has been given to this last category, especially to those affected by constraints on mobility and access, and by impaired sight and hearing. Recreation assumes great importance in the lives of such people, who often have a greater proportion of leisure time than most others. Yet, opportunities to participate, restricted in the first place by disability, are often worsened by building and design standards, and by regulations and requirements.

Massie (2004) gives examples of how visits to the English countryside can be frustrating for a disabled person. He concedes that it is not possible to make the entire countryside accessible and there will always be some places and some experiences out of the reach of people with a disability. However, simple improvements and enforcement of regulations to improve access for those with disabilities can make a great difference.

The dimensions of the recreation opportunity spectrum for people with handicaps are limited by 'environmental barriers', which are taken to include architectural barriers, transportation problems and societal attitudes (Calder 1974). Recreation participation and spectator opportunities for people who are handicapped are seriously impaired by barriers of one kind or another, built into the design and construction of public and private buildings, national parks and playgrounds, and other recreation sites and facilities. Steps, gravel, escalators and narrow entrances, all effectively deny or restrict access for many classes of people with handicaps. Transportation, likewise, is often inaccessible to people with handicaps, because of unsuitable design, inadequate services or lack of appropriate facilities, especially space.

Technical approaches are only part of the solution. Attitudinal barriers within the community also have a marked influence on the ease with which people who are disabled can participate in recreational activity. Many individuals with disabilities are developing mature leisure attitudes and skills, and are no longer personally handicapped by their disabilities; they have developed adaptive skills that allow them to enjoy meaningful leisure experiences. Possibly the greatest handicaps they confront are the social barriers that prevent them from enjoying their leisure and recreation. Such barriers include inaccessible facilities and services, the absence or lack of specialised policies, plans and programmes, and the attitudes of some sectors of the community, who discriminate against people who are handicapped as a minority group, and who, because of misinformation and misconceptions, stereotype people who are disabled as being incapable, unproductive and in need of protection. The attitude of people who are disabled also has a bearing on their ability to make good use of opportunities. Problems of adaptation, education and retraining, especially where the onset of a person's handicap or disability is sudden (e.g. car accident or stroke), can reinforce the already difficult circumstances which tend to exclude these people from the normal leisure experiences enjoyed by the wider community.

Women

There is considerable debate and a growing field of research on gender and leisure (e.g. see Henderson 1994a, 1994b), and on the differences in leisure participation and constraints between men and women (e.g. see Harrington *et al.* 1992; Henderson 1991; Searle and Jackson 1985; Shaw *et al.* 1991; Jackson and Henderson 1995). More specifically, research has documented the similarities and differences in the leisure patterns and processes concerning men and women in countries such as the US (e.g. Blood and Wolfe 1960; Komarovsky 1967; Schneider and Smith 1973; Stafford 1980; Shaw 1985, 1992; Firestone and Shelton 1994; Hutchison 1994). In particular, the growing participation of women who voice their concerns, has added insight and significant depth to such research, which, in conjunction with feminist thinking, has developed foci concerning comparisons between men and women and the barriers or constraints to women participating in recreational activities.

Generally, women experience unequal access to and participation in leisure, as an inevitable consequence of societal attitudes, perceptions and public policies. Women's leisure is constrained by many factors, including:

- time limitations (e.g. Horna 1991, 1993);
- lack of financial resources (e.g. a socioeconomic system that fails to reward women's labour equitably);
- increased participation in the workforce of industrialised countries by women, exacerbating the inroads into their discretionary time from domestic commitments and reaching the point where there may be little time left for personal pursuits (Levine 2001);
- hegemonic constructions of heterosexual femininity – predominant influences depicting women's heterosexual attractiveness as important;
- traditional family and societal arrangements that give men authority over women;
- a judicial system that trivialises male sexual violence against women;
- structural barriers and lack of broad acceptance of female participation in traditionally male activities (e.g. Shaw 1994; Henderson 1994).

Women's concerns about personal safety and violence in relation to leisure and recreation have received some consideration (e.g. Henderson *et al.* 1996). Fear may arise even if not as a result of personal experience and may act as a barrier or form of social control (Green *et al.* 1987 and Frances 1997 in Bialeschki and Hicks 1998). James and Embrey (2001) reported that approximately 70 per cent of Australian women are fearful of attacks after dark and that this fear constrains their recreational choices at night. According to Bialeschki and Hicks (1998), women do, however, often negotiate constraints to continue their participation. How women might respond to obstacles that limit or prohibit participation in adventure recreation was studied by Little (2002). Little used qualitative research methods (in-depth interviews and diaries) as a means of recording the meanings that forty-two women attached to adventure activities and experiences, and the relationships between those meanings and their adventure recreation behaviour involving activities such as cycling, flying, kayaking and rock climbing. These women identified four constraints:

- socio-cultural constraints such as their knowledge and experiences of outdoor recreation adventure activities;
- family and other commitments such as their responsibilities to partners, children, the home, and their work;
- self-doubt, fear and perceptions of outdoor recreation adventure activities;
- technical skills and equipment specifically, and the overall structure of activities generally.

The ways in which women coped with obstacles or constraints to adventure activities were grouped under four main techniques:

- *prioritising* adventure as an important aspect of their lives and hence quite explicitly making time for adventure activities;
- *compromising* and hence participating in alternative activities;
- *creative adventure* programming or planning whereby a broader range of activities and locale are included in their recreational activities;
- *anticipating* that inability to participate is only a temporary situation.

Women face considerable time and family constraints, and their discretionary time may be severely limited. They may, then, redefine leisure to signify a time when they can combine a leisure activity such as walking in the park or watching TV, with a family or domestic responsibility such as child-care or housework. These patterns are familiar in working-class contexts, where women's access to baby-sitters and household time-saving appliances is more limited, and in social-cultural milieus where traditional views of a gendered division of labour and a 'woman's place' prevail.

Elderly and older age

Given that people are living longer and in better health and are more educated than their predecessors it is not surprising the over 65 age group are becoming active consumers of a diverse array of leisure. Although a negative stereotype still prevails most of this older age group are by no means 'over the hill'. Rather they're taking the hill by storm and actively seeking new experiences and lifestyle opportunities. Ageing is now considered a time for growth and development, not despair and decline.

(Grant 2002: 36, 38)

In several countries of the Western world, some of the most dramatic demographic changes affecting leisure and recreation demand, supply, and planning and programming, have been shifts in age structure. There is an increasingly signifi-cant proportion of elderly and retired people in the population, for whom greater longevity, improved healthcare and better financial provisions generate demands for new leisure opportunities and requirements. The ageing of populations of industrialised societies has caused government and non-government organisations to ensure that physical planning and service delivery of community recreation resources address the needs of the aged. Participation in outdoor recreation specifi-cally, and recreation generally, often enhances the wellbeing, quality of life and physical and psychological health of the elderly, and can lead to reductions in social isolation and in medical/drug dependence.

The extent and nature of participation in leisure and recreation change with a person's age (e.g. Singleton 1985; Hayslip and Panek 1989; Kelly 1990; MacPherson 1991; Dionigi forthcoming 2005). Generally speaking, participation in leisure activities declines with age, although there are variations according to one's 'income level, personality, interest, health condition, ability level, transportation, education level and a number of social characteristics' (Hayslip and Panek 1989: 425).

In rural Britain, the population is expected to grow by up to four million over the next 20 years. Many of these new country-dwellers will be middle-aged or elderly, and expecting to live longer, healthier and more active lives. Unfortunately, a good proportion will also be located in areas remote from services, social networks and leisure facilities. Yet, Durham (2004) believes there is an opportunity here for this influx of older people to inject enthusiasm and vitality into traditional rural communities, so that the countryside gains from their presence and their contribution, including enhanced access to leisure opportunities and recreation.

Much research has focused on the relationships between the leisure/recreation behaviour of the elderly and physical or psychological wellbeing (Iso-Ahola 1988; Coleman and Iso-Ahola 1993; Smale and Dupuis 1993; Kleiber 1999), pleasure and satisfaction in the activity (Kelly *et al.* 1987; Losier *et al.* 1993; Delin and Patrickson 1994; Kleiber 1999), constraints (Mannell and Zuzanek 1991), recognition and self-esteem (Tinsley *et al.* 1987), increased coping skills (Coleman 1993), self-rated health (Delin and Patrickson 1994), continued development (Kleiber 1999) and life satisfaction (Hayslip and Panek 1989; Kelly 1990; Hersch 1991; MacPherson 1991). The latter studies show that life satisfaction for older people who are not engaged in paid employment is very closely related to meaningful leisure and recreation participation.

Most forms of leisure, and indeed recreational participation, involve social interaction, which plays an important role in psychological wellbeing (Smale and Dupuis 1993), and offers many other benefits (see above). Interestingly, research has shown that older single adults, aged over 70, participate to a greater extent in organised social activities than do those who are of similar age and married (e.g. see Thompson 1992). More recent research in Australia indicated that:

> Overall, there appears to be considerable dependence upon relatives and friends as a source of social activities for those aged 60 and over. In addition, there is a trend among those who are single, and who may not have a range of family ties, to be reliant on a wider circle of social contacts.
>
> This is more pronounced for women. Of some question is the social support and network role of the organised groups in facilitating leisure of this age group. Only a small number of respondents (less than 3 per cent) indicated that they relied on organised groups for their participation in recreational activities.
>
> (Simmons and Dempsey 1996: 41)

Leisure behaviour and recreation participation vary between the elderly and the rest of the population. Specific constraints such as lack of transport, poor health, and insecurity (even fear), inhibit participation in community activities, so that home-based activities present a safer, more familiar and comfortable environment. 'The elderly as a category are becoming younger, fitter and more affluent' (Veal 1994: 193), but disability is an important constraint for 45 per cent of the population aged 60–5, increasing by age to 83 per cent of those aged over 85.

Retirement, too, is becoming more common among women. Researchers, planners and policy-makers should be directing attention to gender differences in retirement and retirement recreational activities (Mobily and Bedford 1993), because 'it is abundantly clear that elderly women and men participate in different free-time activities' (e.g. Mobily *et al.* 1986).

Retirement impacts on a person's morale and self-esteem, and meaningful use of time in a person's later life becomes a significant adaptive task (Havighurst 1961: 310). Indeed:

> Staying alive requires effort on the part of the older individual to move beyond mere existence and in so doing he or she must be able to demonstrate a willingness to embrace risk, challenge and adventure. An important dimension of adventure is curiosity – the urge to know self as well as the mysteries of life that encompass our physical and social worlds. Samuel Johnson the author of the first dictionary is thought to have argued that curiosity is one of the most permanent and certain characteristics of a vigorous intellect. Youth do not have sole ownership on risk, curiosity, challenge and adventure. If they do, then it is because older people have relinquished these essential ingredients of a vital existence.
>
> (Seedsman 1995: 33)

So, there are marked variations among the elderly in terms of leisure and recreation. Despite these variations, useful generalisations have been made about the leisure needs of older people. These include the need to:

- render some social useful service;
- be considered a part of the community;
- occupy increased leisure time in satisfying ways;
- enjoy 'normal' companionships;
- be recognised as an individual;
- have regular opportunities for self-expression;
- attain a sense of achievement in leisure and other activities;
- access health protection and care;
- obtain suitable mental stimulation;
- acquire suitable living arrangements and family relationships; and
- achieve spiritual satisfaction

(Hersch 1991)

Clearly, leisure is a realm of human activity for people of all ages, and perhaps 'the most important condition for good adjustment to the role transitions related to aging is the maintenance of meaningful activity' (Parker 1979, in Williamson 1995: 63; Hayslip and Panek 1989).

Youth and adolescents

Adolescence is generally considered a period of transition and change occurring between the early teenage years or puberty and adulthood, thus ranging somewhere between 11 or 12 and around (but certainly no more than) 20 years of age. 'An adolescent is often characterized by dramatic physical, cognitive, emotional, and social changes' (Hurtes 2003: 8). Leisure is a significant component (40–50 per cent) of the life of adolescents (Caldwell *et al.* 1992), and the central role of leisure activities is well-documented (see McMeeking and Purkayastha 1995: 360). According to Willits and Willits (1986: 190), leisure and recreational activities are 'not only ends, providing immediate gratification and enjoyment'. Rather, they are 'part of the learning process whereby the individual seeks to establish his/her personal identity … practises social and cooperative skills, achieves specific intellectual or physical attainments, and explores a variety of peer, family, and community roles'. Adolescence can be a very challenging and confusing period in people's lives. In many Western countries, there is a 'high incidence of negative behaviour among adolescents, such as substance abuse, vandalism, and violent crimes' (Leitner and Leitner 2004a: 329). Risk-taking behaviours and suicide among adolescents are major concerns.

Leisure can provide an avenue for the expression and development of identity, autonomy, intimacy and personal growth. Leisure provides the opportunity for young people to hone or test skills and physical endurance, compete against others or better their own standards, and to broaden their general life experience (Iso-Ahola 1980). 'It is in this life phase that much searching is done as young people attempt to recast the identities which have been moulded for them by their parents, caregivers and other significant people and institutions in their lives' (Lynch and Veal 1996: 332). Participation or involvement in leisure activities in a person's adolescent years, in part, shapes the behaviour and attitudes that lead to more permanent patterns in later adolescence and later life (e.g. see Hultsman and Kaufman 1990), even to the extent that about 50 per cent of adults' ten most important recreation activities were begun in childhood (Kelly 1974).

Of course, some young people may choose not to participate in recreation activities. Nonetheless, access to, and participation in, leisure-based activities are influenced by many factors. Access to outdoor recreation activities varies among young people for many reasons:

- gender-related differences (e.g. Godbey and Parker 1976; Shivers 1967);
- in urban areas, for instance, there is generally better access to art and cultural activities, sporting events, music (including discos and live bands) and eating venues (see Gordon and Caltabiano 1996: 37) than in rural (especially remote) areas;
- family, significant other adults and peers affect leisure choices and behaviour (e.g. see Snyder and Spreitzner 1973; Iso-Ahola 1980; Caldwell *et al.* 1992). For instance, parents are the major providers of advice and guidance during adolescence, to the extent that parental influence is important in early

adolescents' decisions *not to join* an activity (Hultsman 1992; also see Youniss 1980). Parents may even be seen as a 'salient barrier to leisure' (Gordon and Caltabiano 1996: 37) or may influence an adolescent's involvement in delinquent leisure (vandalism, theft) through their failure to play with their children or share in their leisure activities (Robertson 1999);

• access to public and private transport affects mobility in time and space (Hultsman 1993);

• lack of, or decline in, the number of volunteer leaders has led to the collapse of some youth groups, yet such 'significant adults' strongly influence the recreational activities of adolescents (see Stephens 1983);

• employment, among other things, provides money and social contact, and a feeling of worth (Leitner and Leitner 2004a). Unfortunately, for many adolescents who are not studying, finding gainful and rewarding employment is becoming increasingly difficult. For those who do find work, it may very well be only casual or part-time.

Constraints on leisure may lead to leisure boredom and, subsequently, deviant involvement (namely drug use and delinquency) (Iso-Ahola and Crowley 1991; Yin *et al.* 1999; Leitner and Leitner 2004), and smoking and consumption of alcohol (Orcutt 1984). 'Because motivation is needed for active leisure participation, drug use might affect an adolescent's choices, when it comes to what kind of leisure activities he or she likes to do' (Gordon and Caltabiano 1996: 37). Frequent and/ or prolonged drug use could cause physical debilitation, alienation from peers and family, or alter awareness and expectations of life events.

Unemployed youth is a problem which has manifested itself in many industrialised nations. Unemployment may be regarded as a manifestation of enforced free time, with leisure regarded as free time. The free time associated with unemployment is not the equivalent of leisure time. 'Unemployment imposes a number of burdens on individuals and people close to them, however, it also frees up large amounts of time which would otherwise be spent in the workplace or earning an income' (Lynch and Veal 1996: 340).

Unemployed people spend less time on outdoor activities and a great deal more time on home-based activity. If activities are expensive, they are largely curtailed. Activities may also be curtailed because of the social stigma of being unemployed, while diminished income appears to be an important factor in reduced participation in out-of-home entertainment, and in membership in clubs and associations (see Lobo 1995). Furthermore, 'Research on special schemes of public provision showed low participation levels. It is likely that the generally disappointing results of the schemes were due to the consequences of unemployment, namely, psychological, social and financial deprivation' (Lobo 1995: 26).

For McMeeking and Purkayastha (1995), an important consideration, and an issue warranting further research, is the extent to which leisure pursuits for adolescents are mediated by their experience of place. If we extend the accessibility of leisure and recreation opportunities to a person's opportunities to travel, then experiences of place may well be wider for those of higher socioeconomic status (individually, or through their family's wealth), those who are better educated, and those with greater social networks and access to marketing and travel information. As recent research has shown, 'the more leisure opportunities available to individuals, the more they want to participate' (Gordon and Caltabiano 1996: 41). Clearly, then, as adolescents experiment, change, adapt and seek 'a stable sense of self', recreation programming for adolescents ought 'to facilitate or provide a "safe" atmosphere for the resolution of developmental issues by allowing for the positive use of free time, the development of healthy relationships, and social and life skill development' (Hurtes 2003: 8). In the United States, examples of such programmes include organised camps (see Box 3.2) and teen centres, the Redcliff Ascent – Wilderness Treatment Programme for troubled teens aged 13–17 years (http://www.redcliffascent.com/), and the Spring Ridge Lodge Retreat in Western Montana (http://www.familyfirstaid.org/spring_creek.htm). 'The intent and purpose of the RedCliff Ascent Outdoor Therapy Program is to provide an impacting, therapeutic experience' (http://www.redcliffascent.com/).

Stigma upon stigma

Just as the context of, and constraints to, leisure seem to differ somewhat between males and females, between people of different ages, and between people of different socioeconomic status, so differences occur within such groupings. Put

---

***Box 3.2* Organised camps**

Organised camping is 'a sustained experience which provides a creative, recreational, and educational opportunity in group living in the outdoors' (Ball and Ball 1996: 3). The structured and programmed nature of organised camps differentiate them from the more informal forms of camping that take place in outdoor recreation settings.

Organised camping is a term sometimes used synonymously with holiday camps, youth camps, and school camps. However, these terms identify particular characteristics of differing types of organised camps. Organised camps are extremely diverse in terms of the timing and length of programs offered, the focus of the program, the social and demographic characteristics of camper groups, and the sophistication and location of the camp facilities. Organised camps may be owned and managed by non-profit organisations, private businesses, or local, state or federal governments.

It has been suggested that those who participate in organised camps are likely to experience physical, psychological, social and spiritual growth (Ball and Ball 1996). Supporters of organised camps argue that participants are encouraged to embrace a range of socially positive values that help them become responsible citizens of the world (Slater 1984).

Source: Lyons 2003.

simply, any understanding or explanation of leisure constraints must incorporate many diverse variables. What of fathers separated from their children as a result of divorce (see Box 3.3)? What of the growing number of men, who, either through economic circumstances or choice, decide to assume the role of primary care-giver to children, while the female partner, in a heterosexual relationship, pursues

---

Box 3.3 Non-resident fathers and leisure with their children

Leisure and recreation are important aspects of most people's lives and are fundamental to quality of life concerns. Veal and Lynch (2001: 400) contend that it is men who have 'traditionally enjoyed the bulk of privileges in a variety of social contexts, including leisure'. However, 'Analyses of men's leisure, and scholars (primarily male) who might examine these areas, are missing' (Henderson 2003: 304). Research on fathers and leisure is particularly lacking.

Research concerning leisure and families more generally has been directed to such matters as marital leisure patterns, with involvement in leisure and recreation activities linked to factors such as joint leisure experiences, family bonding and strength (e.g. Hawks 1991). Shaw's (2001, in Henderson 2003) work on constraints suggests that families sometimes see family recreation as a form of purposive leisure. While the traditional two parent family has been the focus of a good deal of attention, wider recognition is gradually being given to leisure in non-traditional families (e.g. sole parents), and with good reason. According to the Australian Bureau of Statistics (ABS 1997), about 1.1 million children under 18 years of age are living with one of their natural parents, usually as a result of marriage breakdown. Of these children, approximately 88 per cent are living with their mothers. About 20 per cent of Australian fathers do not live with their children. Approximately 18 per cent of children live in one-parent families, and 8 per cent are in step families (i.e. if the lone parent re-partners) or blended families (i.e. when the lone parent re-partners and a new child is born to the new couple). Put simply, the notion of family is a complex one. So, too, are the concepts of 'father' and 'fatherhood'.

Fathers play important roles in their and other children's development and are important figures in children's lives (Amato 1997: 32). Yet, there have been no systematic and detailed studies of the extent and nature of the leisure activities undertaken by non-resident fathers with their children. We do not know what periods of time fathers who do not live with their children (i.e. non-resident fathers) actually spend with their children, the extent and nature of the leisure activities they engage in together, how patterns of leisure involving non-resident fathers and their children change over time, the benefits these fathers derive from leisure with their children, and the constraints they face in initiating such leisure. However, care must be taken so as not to limit research to the constraints to leisure arising from fathers not living at home with their children. Research should be balanced, ensuring that investigations seek to discover the ways in which separation and divorce may provide some fathers with opportunities to expand their leisure options.

An Australian study is presently analysing non-resident fathers' leisure with their children, investigating: how, why, when and where these fathers initiate and experience leisure with their children; the benefits and constraints relating to this leisure; and the significance of leisure to family relationships and quality of life. The study is based on a small sample of separated or divorced fathers who live in the Hunter region (approximately two hours drive north of Sydney) but who do not reside with their children.

In the present study, particular attention will be given to the frequency and nature of fathers' leisure activities with their children, fathers' aspirations, attitudes and experiences, the leisure benefits they derive, and the constraints they face. More specifically, the study will seek to address the following key research questions: What do non-resident fathers consider leisure to be? What forms of leisure do non-resident fathers presently engage in without their children and why? What are the aspirations of these fathers for leisure with their children? Are their aspirations met? Do non-resident fathers feel that they spend enough leisure time with their children? Do they have feelings of guilt or contentment when they are at leisure with their children and when do such feelings generally arise and why? What constraints do non-resident fathers face in engaging in leisure with their children? How are their leisure aspirations influenced by their personal and other circumstances? What benefits do fathers derive from leisure with their children?

an income and career in the paid workforce? What of men who are labelled househusbands, and who soon find themselves occupying a status which has been the traditional preserve of women? How do these men manage to negotiate the values and practices of conventional masculinity (e.g. see Morrison 1994; Lynch and Veal 1996)? What of young children with disabilities, where research has shown 'the lack of social interaction with friends and peers was of great concern for young disabled people and their parents' (Aitchison 2000: 17)? In brief, we need to challenge many existing ideas about freedom of choice in leisure and what freedom means. According to Aitchison,

> There seemed little doubt that the lack of accessible leisure, together with non-inclusionary attitudes towards disabled people, meant that many young disabled people endured a high degree of isolation. The contribution that appropriately devised and managed leisure provision can make to enhancing social inclusion, however, was equally evident.
>
> Aitchison (2000: 17)

What of women with disabilities, where research (Henderson *et al.* 1993, 1995) indicates there is a magnification of leisure constraints for such women? Another study (Davidson 1996) demonstrated that women with young children do not have uniform holiday experiences or perceptions of those experiences, while mobility is also particularly limited for these women if they do not have access to a car (Woodward and Green 1988).

Different characteristics may result in different leisure experiences among men and women: race, socioeconomic status, marital status, sexual orientation and physical ability (Henderson *et al.* 1995). The issue of sexual orientation also raises an important issue. While gay and lesbian studies appear to have gained increasing research legitimacy in some countries such as Australia and New Zealand, little attention has been afforded (1) the place of leisure in the lives of gay men and lesbians, or (2) the meanings attached to leisure by these groups (Markwell 1996: 42). According to Woodward (1993, in Markwell 1996: 43), 'sexual behaviour in general, and sexual pleasure in particular, has received insufficient attention in the leisure studies literature'.

Clearly, the opportunities for investigations concerning special needs groups are enormous. Specific data on manifestations of disability, gender, race and age (among other dimensions of special needs) are growing, but will never provide answers to all the questions which can arise in modern industrialised societies.

There is a clear need for continued questioning of the values that underpin recreational services and facilities, and, no doubt, recreation providers will perceive, and rightly so, many interests in any planning and development processes. If recreation is a fundamental human right, educators, planners and policy-makers must continue to probe the depths of accessibility in all its dimensions, and promote an egalitarian recreation ethic which fully accepts the recreational needs of people whatever their age, race, sex or sexual preference. However, this will only be possible if there is sufficient depth of understanding of constraints to leisure, accessibility to leisure opportunities, and the resources which the public sector and communities (e.g. associations and volunteers) are willing to provide.

Summary

The discussion of the recreation needs and opportunities of special groups illustrates the broad potential for application of the recreation opportunity spectrum concept as a technique in recreation resource planning and management. However, it needs to be noted that interaction of people with resources is two-way. Understanding recreation behaviour and participation patterns certainly calls for changes in personal and institutional dispositions involving attitudes and values, if we are to witness a more qualitative dimension to the human condition, and to the leisure component of human existence.

Guide to further reading

- For broad overviews and/or conceptual frameworks concerning constraints to leisure and means of addressing constraints, see: Wade (1985); Crawford and Godbey (1987); Jackson (1988, 1990, 1991, 1994); Jackson and Burton (1989); Crawford *et al.* (1991); Jackson *et al.* (1993); Veal (1994); Lynch and Veal (1996); Stemerding *et al.* (1999); Patterson and Taylor (2001).
- For an excellent overview concerning women, gender, and leisure, see *Journal of Leisure Research* (1994, vol. 26, no. 1).
- The following suggestions for further reading serve to provide information on a wide range of conceptual and applied issues with respect to groups with special needs. The quite specific nature of many constraints- and special needs-related research means that the title of each work (see Bibliography) is often quite specific about the special group under investigation: Lopata (1972); Rapoport and Rapoport (1975); Deem (1982); Hendry (1983); Roberts (1983); Poole (1986); Ferrario (1988); Iso-Ahola and Weissinger (1990); Stokowski (1990); Atkinson (1991); Driver *et al.* (1991); Sullivan (1993); Spinew *et al.* (1996); Smith *et al.* (2001); Sylvester *et al.* (2001); Stumbo (2002); Stumbo and Peterson (2004); *Annals of Leisure Research* (2004) vol. 2.

Review questions

1 Distinguish between the concepts of 'disability' and 'handicap'.
2 What is meaningful leisure? Does the concept of meaningful leisure take on different meanings for different groups, or is it a generic concept dictated by an individual's circumstances?
3 Apart from the special groups discussed in detail in this chapter, what other special groups can you identify? What makes those groups 'special'? What outdoor recreation planning and policy questions and issues do those groups raise?
4 Identify any policies which have been designed for a special group in your local area. Critically examine the extent to which the impacts and outcomes of those policies have met the needs of their intended audience.

4   Outdoor recreation resources

In a perfect world, demand for outdoor recreation activities would be matched by an ample supply of attractive and accessible recreation resources. Barriers to participation would be absent or easily negotiated, satisfactions sought would be realised, and quality recreation experiences would be the norm. A broad spectrum of recreation opportunities would be presented to potential participants, so that selection of desired opportunity settings was readily achievable, and real choice in the recreation experience was assured.

In reality, interaction between demand and supply factors is qualified by spatial, social/institutional/political, psychological, economic and personal impediments. These impediments prevent or inhibit satisfaction, and detract from the quality of the recreation experience. Thus, the *supply* of recreation resources in quantity and quality, and in space and time, is a fundamental element in creating and structuring fulfilling recreation opportunities. However, understanding of the factors which impinge on or enhance the adequacy of supply of recreation resources, calls, first, for consideration of some basic concepts underlying resource phenomena.

Resources – a functional concept

For many people, the concept of resources is commonly taken to refer only to tangible objects in nature which are of economic use (e.g. material substances, including mineral deposits, waterbodies such as dams, forests and agricultural land). An alternative view is to see resources not so much as material substances, but as *functions* which such substances are capable of performing. In this sense, resource functions are created by human society through selection and manipulation of certain attributes of the environment. The physical existence of coal, iron ore or fertile soils does not constitute a resource; such elements *become* resources as a result of society's subjective evaluation of their potential to satisfy human wants relative to human capabilities. This functional approach to resource phenomena was set out formally many years ago by Zimmerman (1951) and restated by O'Riordan (1971), who defined a resource as:

An attribute of the environment appraised by man to be of value over time within constraints imposed by his social, political, economic and institutional framework.

(O'Riordan 1971: 4)

In these terms, resource materials of themselves are inert, passive and permissive, rather than mandatory, prescriptive and deterministic. Creative use of resource potential requires the existence of a cultural and socioeconomic frame of reference, in which elements of the environment acquire a function as a means of production, or for the attainment of certain socially valued goals.

The existence of a body of water does not necessarily represent a resource in functionally useful terms. Indeed, in some circumstances water might be regarded as a hazard, and even dysfunctional to the utilisation of other more vital resource materials. Any number of attributes or constraints (e.g. size, depth, quality or accessibility) may inhibit the resource functions which water is capable of fulfilling. Creative use of resource potential requires that certain prerequisite conditions be met, among them:

- recognition of the functional possibilities of resources;
- the will to exploit them; and
- the ability (and technological know-how) to put them to use.

The global environment offers many examples of materials with functional promise, but which must await the appropriate circumstances before being harnessed for human use. Consideration of resource phenomena in functional terms, also helps explain their changing roles and fluctuating values over time and space. To a marked degree, resource functions of environmental attributes are dynamic, reacting to changes in economic, social, political and technological conditions. Presently valued resources can lose their function as circumstances alter, and previously neglected resource potential may be put to use to meet new and complex demands. The salt resources of biblical times, and charcoal in the era of the industrial revolution, are just two substances which have only limited functional resource value in today's world. Exploitation of Australia's extensive deposits of uranium, coal and other minerals are subject to the political whim of successive governments and pressure from the 'green' lobby and other interests, so that mines are not operating from a great number of potential resource deposits. Indeed, approximately

18 per cent, 25 per cent and 27 per cent (compared with 19 per cent, 26 per cent and 30 per cent in 2000) of Australia's EDR of ilmenite, rutile and zircon respctively, are unavailable for mining. Areas quarantined from mining and now largely incorporated into national parks include: Moreton, Bribie and Fraser Islands; Cooloola sand mass north of Noosa; Byfield sand mass and Shoalwater Bay area in Queensland; and Yuraygir, Bundjalung, Hat Head and Myall Lakes National Parks in New South Wales.

(Geoscience Australia 2002: 35)

These dynamic characteristics of resource phenomena can be readily demonstrated by further reference to water resources, and to the range of functions identified with particular streams or waterbodies through time. For example, a river, perhaps initially valued merely as a convenient water supply for drinking, may subsequently acquire a function as a means of transport, a source of power or irrigation, or even as a waste disposal site. The Connecticut River in the US is a good case in point (see Box 4.1). The emerging roles seen for water in outdoor recreation, and as a focus of environmental interest, are further evidence of the way in which changing perceptions of this resource are reflected in pressures to adjust its function.

Box 4.1 **The Connecticut River**

The Connecticut River stretches about 650 kilometres from north-eastern Vermont's Canadian border, through New Hampshire and Massachusetts and to the Connecticut Coast in the US.

People have lived along this River since it was formed at the end of the last ice age, about 11,000 years ago. 'The first people to inhabit the Connecticut River Valley were Native Americans (Paleo-Indians) who hunted caribou, woolly mammoth, and other cold-adapted animals. Native American populations flourished in the valley as the climate became progressively milder and more temperate' (Libby Klekowski undated). Presently, there are more than 1,000 dams along its length, and the River serves as a critical element in agricultural and other industries, is widely used for outdoor recreational activities (see http://www.crjc.org/recreation.htm), and hosts many species of flora and fauna. Multiple use conflict is not uncommon.

> Excess is the enemy of a place like this. Too many people or too much exploitation can destroy the equilibrium which exists between the present and the past, between people and the sustaining environment of clean air and water, productive farms and forests ... A 1951 government report called the upper Connecticut River 'damaged' and described its load of untreated domestic sewage from thousands of homes, and of untreated industrial wastes from pulp and paper mills, milk processing plants, other industries and similar loads from 24 tributaries. The poor quality of the river blighted the valley and was even considered a limitation for further industrial use. The river had earned its reputation as the 'best landscaped sewer in New England'.
>
> (http://www.crjc.org/corridor-plan/plan-riverwide1.html)

As a result of this misuse and degradation, environmentalists and others argued for legislation and laws to rehabilitate and protect the river. Among other things, the Vermont Connecticut River Watershed Advisory Commission and the New Hampshire Connecticut River Valley Resource Commission met in 1989 and joined together to form the Connecticut River Joint Commission. At the 1989 Conference, an 'Agenda for the Year 2000' was set and has led to many beneficial actions including the development and implementation of the Connecticut River Corridor Management Plan. The goals of that Plan are:

Goals for the Connecticut River and its environment

- That continued progress is made toward restoring and maintaining a fishable, swimmable river and healthy ecosystem with no degradation as a consequence of human activities;
- That plants, migratory birds, anadromous fish, and other native birds, fish, and wildlife continue to find the Connecticut River corridor and watershed hospitable to their unique needs for clean water and connected, protected open lands and forests;
- That river shores and floodplains remain undeveloped and that a wise public gives the river room to be a river;
- That prime agricultural lands are permanently secure from development and are farmed to meet the food needs of their New England neighbours;
- That the rural character, scenic quality, and historic heritage of the valley are appreciated and maintained;
- That valley residents and visitors can continue to enjoy the refreshment of outdoor recreation without spoiling the resources they enjoy.

Goals for the people of the valley

- That local planning boards and commissions, historical groups, conservation commissions, land trusts, and corporations live up to their potential and exercise responsibility in acting to safeguard resources for the future;
- That voters and property owners understand their responsibility to the river and its watershed, and practise conservation out of enlightened self-interest;
- That farmers and other property owners apply best management practices to their activities on the land, and receive the assistance they need from supportive state and federal agencies;
- That visitors to the valley enjoy its heritage while respecting local property rights and land ethics;
- That agriculture prospers from expanded markets, and forests are a strong element in the economy because of significant value added to forest products;
- That people who come here to live, or to vacation, accept this river valley on its own terms and not try to homogenise it into suburbia or resorts that exist elsewhere;

continued...

Box 4.1 continued

- That a sustainable economy is developed in a manner that does no harm to our river;
- That local leadership is supported by partnerships of federal and state agencies, private organizations and philanthropists.

Source: Connecticut River Corridor Management Plan (http://www.crjc.org/corridor-plan/plan-riverwide1.html).

Equally fascinating is the existence in space of contemporaneous, though contrasting, interpretations placed on an essentially homogeneous resource base. Again, water resources are a good illustration. The same physical attributes of a river valley, for example, can take on different dimensions in the minds of inhabitants aware and capable enough, and prepared to take advantage of the opportunities offered. Different groups of people, occupying that same environment, may have literally different resources. For some, the valley and its waters represent, perhaps, a tranquil setting in which to carry on traditional farming pursuits; for irrigators and other primary producers, the river is seen as providing the means of introducing intensive agriculture. Others may value the waterbody for active or passive recreation pursuits. Contrasting perceptions of what are taken to be appropriate resource functions help explain conflicts which arise over allocation, distribution and use. This theme will recur frequently in the issues considered in later chapters that deal with recreation resource use.

Recreation resources

As with the examples noted above, identification and valuation of elements of the environment as recreation resources will depend upon a number of factors (e.g. economics, social attitudes and perceptions, political perspectives and technology). Problems can arise in the identification process because, given the appropriate circumstances, most environments are, in some sense, recreational. Thus, resources for outdoor recreation can embrace a wide spectrum of areas and settings, ranging over:

- space itself (airspace, as well as subterranean and submarine space);
- topographical features, including tracts of land, waterbodies, vegetation and distinctive ecological, cultural or historical sites;
- the often neglected climatic characteristics of an area.

Hart (1966) used the term 'recreation resource base' to describe the total natural values of countryside or a particular landscape. He included in his definition such attractions as the view of a quiet agricultural scene, along with more tangible phenomena such as sites for picnicking, camping and boating. Recreation resources, then, embrace areas of land, bodies of water, forests, wetlands and other features of the natural or built environment in use for recreation. Current use identifies actual recreation resources, while the probability of use indicates potential recreation resources, rather than the characteristics of an area or site.

The process of creation, use and depletion of resources for outdoor recreation differs little from that in other areas of human activity, such as agriculture, forestry or mining. As Clawson and Knetsch (1966) put it:

> There is nothing in the physical landscape or features of any particular piece of land or body of water that makes it a recreation resource; it is the combination of the natural qualities and the ability and desire of man to use them that makes a resource out of what might otherwise be a more or less meaningless combination of rocks, soil and trees.
>
> (Clawson and Knetsch 1966: 7)

Recreation resources include natural attributes of the environment, as well as facilities and attractions such as sporting complexes and theme parks. This continuum is implicit in Kreutzwiser's (1989: 22) definition of a recreation resource '... as an element of the natural or man-modified environment which provides an opportunity to satisfy recreational wants'. As with resources in general, the supply of recreation resources depends, initially, on human recognition or perception of the environment as capable of satisfying those wants. However, society must also wish to use the environment for that purpose and have the ability, appropriate technology, organisational and administrative arrangements and will, to create an attractive, accessible and functional environmental setting for recreation.

Again, in common with the functional approach, recreation resources are not static or constant, but take on a dynamic character varying in time and space. Resources can become redundant; just as changing economic, social and technological conditions can reveal new recreation potential in previously neglected areas. Natural resources are cultural appraisals, and what is recognised as a recreation resource by one group of people at one period of time may be of no conceivable use or value to them or others in different circumstances.

The renowned surfing beaches of the Australian coastline, for instance, have really only achieved prominence for outdoor recreation in the past half century, with the relaxation of attitudes to public bathing. To the Aboriginal inhabitants of the continent, they were an important source of food, whereas the early European colonists found the surf a formidable hazard in coping with the isolation of coastal settlements. Moreover, the gleaming sand itself, which to most Australians is an integral and attractive component of the recreation resource base, represents a very different kind of resource function for the rutile miner or the building contractor. Put simply, 'The coast is not one resource, but many' (Patmore 1983: 209). This reasoning applies just as much to mountain areas and other regions or landscapes.

Contrasting perceptions of environment help explain conflicts concerning recreation resource use. Forest and wild land recreation, for example, is largely a product of the conservation movement of the twentieth century, and claims on countryside and water resources can conflict with more traditional uses of rural land (see Chapter 8). Further attention will also be given, in later chapters, to the potential for conflict resulting from differing perceptions of the resource functions of water resources and the coastal zone. In the same way, scenic roads, walking tracks and trails of various kinds represent important resources for popular forms of outdoor recreation, but not all of these uses sit comfortably or compatibly with other demands made on such linear resources or adjacent areas. Even the extensive network of public footpaths, so much (and for so long) in demand by ramblers or hikers through the English countryside, sometimes brings recreational users into conflict with neighbouring landholders.

It is important to recall from Chapter 2 that conflict in outdoor recreation need be neither resource-based nor activity-based. Conflict can just as readily arise from the attitudes and mindsets of recreationists, as from competing claims on a common resource base, or incompatible recreation activities. At the same time, many forms of outdoor recreation do not require exclusive use of land or water, but lend themselves to multiple use, in harmony with other resource functions.

Outdoor recreation and multiple use of resources

Outdoor recreation often imposes relatively non-aggressive and benign claims on the resource base, so that it is possible to envisage and actually plan for situations of multiple use as demonstrated above. Forest lands and waterbodies are perhaps the most common examples of outdoor recreation existing as a compatible partner with the primary role for the resource, while recreation in protected areas is also progressing sympathetically in many countries. However, given the right circumstances, recreation activities can also coexist with agriculture and grazing land (Swinnerton 1982). Although more common with publicly owned resources, opportunities for multiple use can also be found in areas in private ownership.

From a social perspective, multiple use makes a lot of sense, especially where resources for outdoor recreation are limited, or where prevailing conditions limit their recreation resource potential. In economic terms, multiple use is justified if the combined benefits arising therefrom are greater than those from a single use, and are sufficient to cover any additional costs. This is generally accepted, in the sphere of forest management, to include outdoor recreation. In Australia, the New South Wales National Parks and Wildlife Act, 1974 (No. 80) (at http://www.austlii. edu.au/au/legis/nsw/consol per cent5fact/npawa1974247/) provides for multiple use of parks in that State. In the US, for example, the Multiple Use Sustained Yield Act of Congress, provides for recreation as one of the main objectives of national forests. This Act declares that the purposes of the national forest include outdoor recreation, range, timber, watershed and fish and wildlife. In this Act, multiple-use refers to

> The management of all the various renewable surface resources of the national forests so that they are utilized in the combination that will best meet the

Plate 4.1 Multiple use – cycle and pedestrian trails are a major recreational attraction in the Australian Capital Territory, Australia. They have been carefully planned and are one of the most extensive urban network of trails in any Western nation. Around Lake Burley Griffin, these trails experience high levels of passive and competitive use

needs of the American people; making the most judicious use of the land for some or all of these resources or related services over areas large enough to provide sufficient latitude for periodic adjustments in use to conform to changing needs and conditions; that some land will be used for less than all of the resources; and harmonious and coordinated management of the various resources, each with the other, without impairment of the productivity of the land, with consideration being given to the relative values of the various resources, and not necessarily the combination of uses that will give the greatest dollar return or the greatest unit output.

(http://www.fs.fed.us/emc/nfma/includes/musya60.pdf)

By contrast, in Australia, managers of public forest lands have been slower to endorse recreation use alongside timber production. It is only in recent years that outright opposition to outdoor recreation in state forests has changed to guarded tolerance, and, now, to commitment to recreational use of forests and specific inclusion of recreation opportunities in management plans. Particularly enterprising developments can be seen in the early and continuing efforts in Tasmanian forests by the State Department of Parks, Wildlife and Heritage who engaged in a 'Minimal Impact Bushwalking' (MIB) campaign to combat the environmental impacts of bushwalking as visitor numbers grew. This was remarkable foresight given the recent and rapid recognition and development of Tasmania's nature-based tourism resources and their growing popularity. In this campaign, ecotourism principles were being applied well before the term was coined in the late 1980s. Aside from the usual hardening and management of walking trails, the Department developed a range of materials to educate walkers. These materials included pamphlets illustrating minimal impact walking (Walking the Wilderness), posters, videos (Walk Softly), activity sheets (Minimal Impact Bushwalking Teacher's Kit), and comic posters (Minimal Impact Bushwalking). The Department presented articles in newspapers and bushwalking magazines and supplied copy to electronic media, and deployed rangers to field sites. These actions were evaluated by way of questionnaires and focus groups (see O'Loughlin 1993).

Unfortunately, not all Australian States share this enlightened approach. Reported moves to consider the sale of publicly owned pine forests in New South Wales have met resistance (*Sydney Morning Herald*, Editorial, 18 October 2004). These pine forests, along with native forests managed by Forests NSW are seen as valuable recreation resources for mountain bikers, walkers, horse-riders, volunteers and picnic groups. Moreover, their value is increasing as large areas of former State forest are being transferred to the national park estate with more restrictive recreation policies. Recreation groups are urging that any sale of the pine forests must guarantee continued access for outdoor recreation.

With water resources, the situation can be more complex. There are many different ways in which streams and waterbodies can function to satisfy recreation wants, and there are different forms (sometimes overlapping) of ownership and management of water resources, where responsibilities and functions often overlap. Expanding resource potential through multiple use is a challenge given the diverse interests and requirements of recreational fishing, swimming, boating and passive shore-based recreation (see Table 4.1). Management approaches, based on multiple use of water resources, are explored further in later sections, along with the issue of operation of water storages to provide for outdoor recreation opportunities.

Recreation resources in the built environment

Up to this point, the emphasis has been on examining recreation resources as attributes of the natural environment, many of which are publicly owned and managed. However, a significant part of the recreation resource base comprises components of the built environment, which provide for incidental and perhaps opportunistic forms of outdoor recreation. Some of these come under what Ibrahim and Cordes (1993) call 'private recreation resources', which include private residences, second homes, clubs and organisations of various kinds, shopping centres, and industrial sites. In addition, plazas, malls, school grounds and parking lots, can all offer recreational opportunities in urban settings.

To these should be added purpose-built facilities and attractions which play an important role as recreation resources. Whereas a good proportion of these (e.g. urban parks, sporting facilities and community recreation centres) is the responsibility of government at various levels, many are commercial operations

Table 4.1 Examples of water-dependent and water-related recreation activities

Water-dependent activities

Aesthetic appreciation of water	Powerboat racing
Beachcombing	Rafting
Canoeing	Sailing
Crew racing	Shell collecting
Driftwood gathering	Shellfish gathering
Fishing	Small boat cruising
Houseboating	Snorkel or scuba diving
Ice fishing	Surfing
Ice hockey	Swimming
Ice skating	Voyages in cruise ships
Ice yachting	Wading
Model boat sailing	Waterfowl hunting
Playing in water	Waterskiing

Activities that are frequently water-related

Beach games	Pleasure driving
Birdwatching	Relaxing
Camping	Rock or fossil collecting
Hiking	Seasonal homes
Nature study	Sightseeing
Painting and sketching	Snowmobiling
Photography	Sunbathing
Picnicking	Walking

Source: Chubb and Chubb (1981: 314).

offering diverse attractions and services such as food-and-drink outlets, sports venues, accommodation and theme parks. As pressure on 'natural' recreation resources grows, these created or 'artificial' additions to the resource base will help take the pressure off the natural environment. The success of theme parks, for example, supports the notion that substitution of the distinctive (physical) attributes of a recreation setting might be possible without impairing satisfaction, so long as *functional* similarity is maintained (Peterson *et al.* 1985). Given that the desired attributes of nature can be identified and replicated, or simulated in a less pristine setting, pressure on authentic, nature-oriented environments may be relieved. Moreover, less demanding types of recreation might well make do with more tenuous links with nature; a bush barbecue need not be sited in a national park.

Ditwiler (1979) takes the consideration of substitutability further by questioning whether particular resources or environments are necessarily a prerequisite for the leisure experience desired. He argues that the experiences people seek from a natural setting, for example, could well be obtained from an artificial environment designed to include those characteristics of the natural environment required for the purpose. If Ditwiler is correct, and many supposed wilderness recreationists are more interested in diversion, excitement or challenge than in nature *per se*, it should be possible to substitute the utility inherent in specifically nature-oriented settings by creating artificial environments. Examples of such substitutions are already numerous, for example, in simulated settings such as Disney's Epcot Center, Florida. Despite scepticism and, perhaps, resistance from 'purists', there could be a useful role for technological ingenuity in helping to alleviate pressure on the natural resource base.

One of the most innovative examples of a created recreation environment is Europe's largest leisure resort, Tropical Islands, which opened in December 2004. Located in a converted airship hangar, just an hour's drive south of Berlin, Germany, the resort offers winter-weary people the chance to bask in temperatures of 21 degrees Celsius amid palm trees and sandy beaches with a golden sunrise projected on a 140 metre long screen. The site created for the project is a hall of 5 million cubic metres, the world's largest free-standing building more than 100 metres high. It could fit six football fields on two levels, with space for 850 deck chairs on two beaches, and stays open 24 hours a day. A further positive note is that the Tropical Islands resort has already created some 800 jobs in an economically depressed area of the former East Germany.

Space, location and accessibility

Identification, assessment, allocation and use of elements of the natural environment as recreation resources will depend upon a number of technological, socioeconomic, political and perceptual factors. Physical characteristics are, of course, fundamental; water must exist for water-related tourism. However, such variables as space, location and accessibility have a direct bearing on functional effectiveness. Whereas some of these aspects may be offset or foreseen in the creation of 'artificial' inputs to the recreation resource base, they remain important in the ongoing function of all recreation resources to provide satisfying quality recreation experiences.

Outdoor recreation necessarily has its focus in space-consuming activities. Certainly, it is in the spatial distribution and frequent locational imbalance of leisure opportunities where much of the resource manager's interest and emphasis are focused. Space, then, is a critical resource for outdoor recreation, and certain kinds of activities require space with specific attributes, dimensions and qualities.

As recreationists increase in number and mobility, there is greater pressure on recreational activity space, on service space for ancillary facilities, and on access space (e.g. parking areas and routeways such as roads, walking and cycling trails). These latter considerations can have a marked bearing on the effectiveness of recreation resource space. Pressure on capacity, too, may stimulate multiple use of space, over time, for varied activities, day and evening, week and weekend and year-round rather than seasonal.

Conditions of location and access are basic to the definition of certain types of recreation space. Indeed, these aspects were a major part of the rationale supporting the early classification of recreation areas as user-orientated or resource-based (Clawson and Knetsch 1966). For some outdoor recreation areas, isolation, to various degrees, is vital to maintain the individual experience sought. In this respect, an inherent characteristic of resource-based recreation areas is their locational immobility; they are site-specific to a particular environmental setting. Wilderness areas, for example, are largely delineated by their remoteness and difficulty of access, so that their primitive natural qualities will not be impaired by over-use.

On the other hand, valuation of the spatial element for user-orientated recreation resources depends very much on the location relative to population concentrations. In heavily populated areas, an adequate supply of readily available recreation space is especially valuable. Yet, urban centres often have to 'make do' with fragmented pockets of relict land and water, otherwise useless for the purpose of economic return. In many cases, these are legacies from past subdivisions, where the statutory proportion of an estate set aside for open space has been selected on the basis of its inherent poor quality, either locationally or for residential purposes. The result is recreational space which is not appropriately sited for present needs, nor flexible or adaptable enough in quality to cater for the changing character of dynamic urban populations.

Awareness of deficiencies in current allocations of land for outdoor recreation has beneficial connotations in terms of planning for the more effective selection and location of new recreation areas in both intra- and extra-urban environments. Some constructive efforts in this regard have been made towards the provision of an improved environment for human recreation needs in new towns in Australia, Britain and the US. In such decentralised communities, the emphasis is on co-ordinated planning for the new city, together with its surrounding region. The maintenance of environmental values as a basis for recreational amenity, and provision of a wide range of accessible sites and settings, are seen as essential strategies.

At a finer scale, the relative disposition of spatial components into access space and viewable space, may contribute to a fuller realisation of the recreation experience. The complementary nature of these fundamental spatial elements can be illustrated with reference to Banff, Alberta, Canada and resorts and villages in mountainous regions. The mountains and lakes (viewable space) in Banff rate low as occupiable space, but greatly enhance the appeal of the valley landscape, within which the resort and its access routeways are situated.

A similar distinction is made by Gunn (1988) in his study of the planning of tourist regions. Gunn suggests that a tripartite approach should be adopted in the design of tourist attractions. Stress is placed on the spatial relationship between the prime element, or *central attracting force* (e.g. a waterfall) and its essential setting, or what Gunn calls the inviolate belt. The function of this *setting*, or *entering space*, is to condition the visitor's anticipation in an appropriate fashion, so as to enhance the subsequent recreation experience. The third conceptual element is the *zone of closure*, or outer area of influence, containing service centres, circulation corridors and transport linkages. This functional component completes the tripartite concept and contributes to the wholeness of an attraction.

The example of Canada's Niagara Falls illustrates the importance of a complementary environmental setting in adding to, or detracting from, a prime tourist attraction. Niagara Falls is categorised as one of the 'wonders of the world', with the immediate vicinity being enhanced by an attractive reserve. However, only the most insensitive visitor could remain indifferent to the garish vulgarity of the

Plate 4.2 Garish street developments near Niagara Falls, Ontario, Canada

outer approach to the Falls, marked as it is by inappropriate and poorly designed services and facilities. It is almost as if human beings had set out to mask the natural splendour of this magnificent feature with a veneer comprising the worst aspects of landscape design and commercial display.

A comprehensive approach to the planning and design of recreation space, therefore, should give attention to the regional background, the approach and means of access, and the immediate setting, all of which complement, or detract from, the satisfaction visitors and users derive.

Resource-based recreation areas are, by definition, located at the site of the prime element or attraction. Yet, even here, scope exists for manipulating the attributes of resource elements to enhance or inhibit their recreational function. In so doing, the 'location' of recreation resources and hence recreation opportunities is essentially being arranged to meet management objectives. A forest or a water-body, for example, may have several areas with recreational potential. However, it is the selective provision of access and facilities that will determine the location of those sites to function as recreation resources. In the same way, and on a larger scale, an agency may attempt to correct imbalances in the location and spatial distribution of visitors to national parks by strategically allocating ancillary facilities to selected sites (see Chapter 10).

Accessibility is a fundamental concomitant of locational aspects of recreation environments and of the functional concept of recreation resources. The presence or absence of roads, trails, parking space, boat ramps, airports or helicopter pads, can all impinge on the functional effectiveness of the recreation resource base. However, it is difficult to generalise when conditions of access for special groups, such as children, the elderly and people with disabilities, are considered. Moreover, it is not so much a question of physical access as of legal, institutional, and perhaps socioeconomic constraints on movement into and through recreation space. Such situations raise complex questions regarding public rights and access to common property resources; some of these are examined in Chapter 8.

The question of what constitutes a recreation resource and what factors add to, or detract from, the quality of the leisure environment can best be answered by a systematic assessment of resource potential. The task begins with identification and classification of elements of the recreation resource base.

Classification of recreation resources

An important initial stage in the resource creation process is an inventory and assessment of the quantity and quality of resource materials; those presently valued as resources, and those which may function as recreation resources in different socioeconomic and technological circumstances. Such stocktaking is necessary before the significance of stocks can be evaluated. However, inventories themselves are of doubtful value; what is required is more than a simple listing of resource materials. The resource elements must be described and classified according to some recognised and agreed system in order to determine categories of resource deficiency and surplus as an input to recreation planning.

Classification of recreation resources can be approached from several angles. Economists, for instance, might be inclined to 'sub-divide the resource base into depletable resources and renewable resources' (Morgan 2003: 431). Ecologists will be more interested in the composition, structures, functions and processes associated with the ecosystem. So, resources can be classified in many ways. What are they? Are they renewable (in a physical sense) and if so to what extent? Do they have cultural or spiritual meaning to particular people or communities? Who uses them or might use them? What are the benefits and costs associated with their use? Cutter and Renwick (1999) classify resources as:

- *Perpetual* resources which, regardless of their levels of exploitation will provide a constant supply. The sun as a source of solar energy and warmth is a good example.
- *Renewable* resources can be used and their stocks depleted, but in the medium to long term they will supposedly recover or replace themselves. Of course, while this may happen, other aspects of the environment may have been dramatically affected. For instance, removal of old trees from an environment will dramatically change wildlife habitats, while intensive recreational fishing in an area may deplete some fish species and allow others to proliferate beyond sustainable levels.
- *Non-renewable* or *stock* resources are limited in supply (i.e. their supply is finite). Wilderness is a finite resource (as are fossil fuels and minerals).
- *Potential* resources are things that may become a resource in the future. Improved technology for purifying water has paved the way for use of a broader

range of water reservoirs for recreation and for reuse of 'dirty' water for irrigation and other purposes.

One of the earliest systems for classification of resources based on recreational aspects was devised by Clawson *et al.* (1960), who distinguished between recreation areas and opportunity on the basis of location and other characteristics, such as size, major use and degree of artificial development. Under this system, recreation areas were arranged on a continuum of recreational opportunities from user-orientated through intermediate to resource-based.

The Clawson system of classification has been widely applied, although the terminology can be confusing. All recreation areas must be user-orientated to some extent if they are to satisfy the functional concept of resources. Exclusion from the resource-based category of urban and near-urban recreation sites reflects a narrow interpretation of the term 'resources'. Obviously, there is scope for considerable overlap, and city parklands can be just as much 'resource-based' as remote wilderness. Moreover, large national parks, such as those which ring the city of Sydney, Australia (see Chapter 9), and which are close enough for casual day visits, actually qualify as user-orientated recreation areas. Obviously, too, there can be interchangeability over time, as resource-based areas, for example, come within the recreation opportunity spectrum of an increasingly mobile and affluent user population. Alternatively, as the population demographics of a local area change, some resources may be used on only rare occasions, become run down as most people have little interest in their retention and maintenance, and may ultimately be removed or replaced.

Various adaptations of the Clawson system of classification of recreation resources have been devised, using combinations of location, physical attributes, facilities, and type of recreational experience and use. An example presented in Table 4.2 encompasses both resource-based and built recreation facilities. Chubb and Chubb (1981) distinguish between the following classes of recreation resources:

- *Undeveloped Recreation Resources*, including land, water, vegetation and fauna;
- *Private Recreation Resources*, taking in 'personal resources' such as residences and second homes, private organisation resources (e.g. clubs), resources of quasi-public organisations (e.g. conservation groups), and farm and industrial resources;
- *Commercial Private Recreation Resources*, including shopping facilities, food-and-drink outlets and sports facilities, amusement parks, museums and gardens, tours, stadiums, camps, and resorts of various types;
- *Publicly Owned Recreation Resources*, covering local and regional parklands and sports facilities, state and national parks and forests, trails, tourist facilities and institutions;
- *Cultural Resources* in the public and private sector, including libraries and facilities for the Arts;

Table 4.2 A taxonomy of leisure facilities

Facilities not existing primarily for leisure	Resource-based facilities adapted for leisure	Built facilities adapted for leisure	Built facilities designed for passive leisure	Built facilities designed for active leisure
Agricultural land	Woodland parks	Historic houses	Museums	Marinas
Commercial woodland	Urban/rural parks	Ancient monuments	Galleries	Leisure/sports centres
Watercourses	Golf courses	Redundant churches	Libraries	Dance halls
Water masses	Beaches	Warehouses/industrial buildings	Cinemas	Arts/community centres
Private dwellings	Cruising waterways		Restaurants	Squash/tenniscentres
Workplaces	Public footpaths		Hotels	Gymnasia
Streets	Canal towpaths		Shopping malls	Swimming pools
Moorland/mountains	Watercourses			Holiday camps
Reservoirs	Water masses			Snooker halls
	Zoos			Sports stadia
	Theme parks			Playgrounds
	Open air museums			All weather sports pitches
	Holiday camps			Sports clubhouses
				Theme parks
				Shopping malls

Source: Adapted from Ravenscroft (1992).

- *Professional Resources*, which can be divided into two broad areas: administration (the organisation of recreation systems, policy-making and provision of financial support) and management (research and planning, design, construction and maintenance, resource protection and programming).

Clearly, such a listing of recreation 'resources' is all-embracing, consisting of a broader recognition of resource phenomena than those typically associated with outdoor recreation. Whereas all of the 'resources' included in the classification can undoubtedly function to satisfy recreation demands, the intention, here, is to focus on those elements of the natural and built environment (whether in the public or private sector), which are in use, or which have potential for use as *outdoor* recreation settings.

At a finer scale, classification systems have been developed for specific categories of recreation. Gold (1980) classified recreation space on the basis of location, function, capacity and service area, ranging from 'home-oriented space', through 'neighbourhood' and 'community space' to 'regional space'. Such classifications have much in common with Mitchell's (1969) hierarchy of urban recreation units, and have merit in exposing and correcting deficiencies in providing urban recreation opportunities.

Recreation resource capability and suitability

Increasingly, static classifications, specific to a particular time period, are being replaced by resource capability assessments of the potential of an area for a specified use. The assessment may be for one purpose or a combination of purposes. Perhaps the most ambitious and exhaustive scheme for classification of recreation potential has been carried out in Canada as part of the Canada Land Inventory (CLI), a comprehensive project to assess land capability for five major purposes – agriculture, forestry, ungulates, waterfowl and recreation. The inventory has been applied to settled parts of rural Canada (urbanised areas are excluded), and is designed for computerised data storage and retrieval as a basis for resource and land-use planning at local, provincial and national levels.

In marking the first 25 years of the Canada Land Inventory in 1988, Environment Canada highlighted the various features of this unique planning tool (see Box 4.2).

Box 4.2 **Canada Land Inventory highlights: 1988**

- One of the largest land inventories ever undertaken in the world, the CLI covers about 260 million ha of southern Canada for five resource sectors.
- Prime agricultural lands in British Columbia comprise only 5 per cent of the province's total area, and are under continuing pressure from competing uses. In the 1970s, the CLI was instrumental in helping this province and others move quickly to designate agricultural lands for protection.

- Acid rain is a critical threat to Canada's environment. Analysis using the CLI has demonstrated that more than 70 per cent of Eastern Canada's prime resource lands receive acid rain in levels threatening sustainable productivity.
- Waterfowl are an important economic, recreational and ecological resource. In support of the North American Waterfowl Management Plan, the CLI has helped screen out areas where habitat maintenance might conflict with agricultural production.
- The CLI led to the development of the Canada Geographic Information System, an innovative technical accomplishment in handling resource information.
- Not simply a valuable planning tool of the past the CLI, at 25, can make a vital contribution to efforts to understand and resolve urgent environmental problems.

Source: Environment Canada (1988: 1).

Despite its long life, the Canada Land Inventory continues to be relevant to a variety of land use and resource management problems across the country. Key questions in regard to outdoor recreation are: How much good quality land is available for this purpose?, and Where is that land located in relation to potential users? The example 'New Parks with Water Access?' demonstrates how CLI information can assist in the planning of new recreation opportunities.

Example: **'New Parks With Water Access?'**

Suppose that park planners wanted to determine the potential for establishing, within an hour-and-a-half drive of our major cities, more parkland on the shorelines of lakes, rivers and oceans. While the eventual decisions will be based on such factors as land ownership, price, accessibility and current uses, the first step, obviously, is to find out 'how much' there is and 'where it is'.

The CLI outdoor recreation inventory provides exactly this information. Table 4.3 summarises data on capability Classes 1–3 inventoried shoreline within a 121 km (75 mile) radius of selected cities. CLI maps will show the location of these sites, as well. The figures indicate that the potential for new shoreline parks varies greatly from city to city. Sudbury and Ottawa top the lists, while Regina and Calgary have limited potential.

The CLI has not identified the precise location of new park sites. But it has enabled planners to focus their search very quickly, allowing them to target detailed planning on the most promising shoreline sites.

continued...

Table 4.3 CLI recreational capability classes 1–3: inventoried shoreline within 121 km of the centre of selected metropolitan area

Metropolitan area	Class 1	Class 2	Class 3	Total
Halifax	–	29	433	462
Montreal	12	253	1,785	2,050
Ottawa	80	471	3,076	3,627
Sudbury	24	221	4,181	4,426
Regina	5	19	285	309
Calgary	4	59	132	195
Vancouver	69	456	1,588	2,113

Source: Environment Canada (1988: 4).

Source: Environment Canada (1998) Canada Land Inventory Report No. 14 (1978), Land Capability for Recreation: Summary Report.

Again, the Canada Land Inventory can help planners choose among options where potential conflict exists. Targeting of areas of conflict can be indicated by a simple comparison, using overlays of single sector maps of the same areas. These techniques were the forerunner of the much more complex computer analyses and Geographic Information Systems (GIS) (see Chapter 12) in use today. The example 'Ducks and wheat?' highlights sites of potential conflict between competing resource functions for the same area of land either as prime agricultural land or a prime habitat for waterfowl and the basis of extensive and valuable recreation activity.

Example: **'Ducks and wheat?'**

Suppose wildlife managers want to protect prime waterfowl habitat in a particular rural municipality. They know that such habitat is frequently located throughout good farmland. Their goal is to locate those prime waterfowl areas that do not conflict with agricultural production.

As a first step, they obtain the CLI agricultural capability and waterfowl capability maps covering their rural municipality. Then they identify all the Classes 1–3 agricultural land and all the Classes 1–3 waterfowl area. The comparison identifies the general location and extent of areas where there are overlaps, and areas where there are no conflicts with agriculture. This latter group will provide candidates for protection (Figure 4.1).

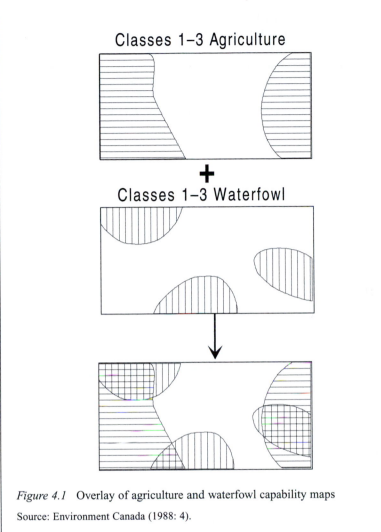

Figure 4.1 Overlay of agriculture and waterfowl capability maps

Source: Environment Canada (1988: 4).

In this way, the CLI system permits inter- and intra-sectoral comparisons to delineate suitable locations for recreational development, and to define priority areas between competing uses, as well as opportunities for compatible multiple resource use. However, the methodology does have some deficiencies. Although the classification is designed to accommodate a wide range of (then) popular outdoor recreation activities, inevitably it excludes certain pursuits (e.g. hang-gliding) which have emerged since, as a result of technological advances and increasing levels of specialisation in recreation activities. Nor does the scheme have much to say about the *quality* of a recreation experience. Recreation resource capability is equated with the *quantity* (or level of use), which does not always coincide with satisfaction.

In Australia, classification schemes focusing on outdoor recreation capability have been applied to Crown (public) lands. The classification can apply to a range of land categories, from 'remote natural' to 'urban', reflecting the area's physical, social and management characteristics relative to its capability to provide opportunities for land-based or water-based recreation activities (Table 4.4).

Resource *capability* is a measure of the feasibility of allowing a range of specified resource uses on an area of land, reflecting both the likely productivity and resilience of the site. Whereas resource *capability* is based on mainly natural physical attributes, resource *suitability* is a socioeconomic and political evaluation of the acceptability or desirability of a particular resource use.

In terms of the *suitability* of a resource for recreation use, relevant issues include community demands and expectations, government policy, and conflicting needs of different user groups. The concept of resource suitability has four basic components:

- *Economic efficiency* – the allocation of resources to the use which yields the greatest financial return (e.g. if assessment shows that a parcel of land possesses a high capability for cultivation, preference should be given to cultivation, not an alternative use such as grazing);
- *Social equity* – ensuring the equitable distribution of social benefits and social costs by adopting a particular form of land use. For example, one issue at the centre of past rainforest logging disputes in New South Wales, Australia, was the conflict between the tangible timber production and job security associated with the logging industry, and the more intangible loss to the National Estate of the rainforest resource itself;
- *Community acceptability* – including such aspects as: changing social attitudes (e.g. concerning environmentally incompatible land uses and external effects, such as adverse noise and visual impacts); regional and community needs for land resources; public participation and political influence;
- *Administrative practicability* – the adequacy of existing infrastructure and services (e.g. accessibility to roads, education facilities, community services, power supply and sewerage) and the economic efficiency of providing services to undeveloped sites.

(New South Wales Department of Lands 1986: 50–1)

Recognition and classification of resource phenomena for outdoor recreation use call, first, for identification of the capability of the resource base to provide for a range of recreation experiences. However, this is only an indication of what recreation activities the area may support. From these possibilities, it is essential that the most desirable option/s or preferred resource uses be selected. Resource suitability takes account of physical capability, but focuses on choosing the recreational use which best satisfies community demand and government priorities.

Environmental cognition is the mental process of making sense out of the environment that surrounds us. Cutter and Renwick (1999) identify five broad factors which affect resource-use cognition and which may be applied to individuals or society generally:

- *Cultural backgrounds* – decisions and actions concerning land arise from social and political processes and these can vary greatly among cultures. These processes involve 'the *values* of individuals, groups, and organisations' in struggles for *power* (Henning 1974: 15). However, values change. So, 'Any conception of human values, if it is to be fruitful, must be able to account for the enduring character of values as well as for their changing character' (Rokeach 1973: 6).
- *View of nature* – there are wide-ranging world views, though one classification pits ecocentricism at one extreme and anthropocentrism and technocentrism at the other. Hence, our value systems relate to matters such as environmental and resource ethics, the ways in which we are concerned about the welfare of future generations and whether humans are merely part of the ecosystem or are somehow superior to other forms of life.
- *Social conditions* including our education, gender, age, income, socioeconomic status, and friendships, which influence environmental management. Gender roles influence attitudes to the environment. Higher levels of education open opportunities for greater awareness of environmental issues. A person's level of income or employment status might influence their attitudes to recreational development depending on the nature and extent of employment opportunities as opposed to environmental impacts.
- *Scarcity* – the economic value of a resource increases the more scarce it becomes. Conversely, for some resources, scarcity may lead to conservation efforts as we see in efforts to save various flora and fauna from extinction;
- *Technological and economic factors* are important in that technological developments can lead to access to new resources, whereas the ability to exploit a resource is very much related to the economics of doing so.

(Cutter and Renwick 1999: 1–3)

Palmer (2004) highlighted the tensions arising in Kakadu National Park, Australia, between non-Aboriginal Northern Territory residents and Aboriginal (*Bininj*) traditional owners. Both groups fish in the park, but Palmer contrasts their different attitudes to land and water and the environment generally. She concludes that there are some similarities in the way fishing is conceptualised as part of the way of life for Aborigines and non-Aborigines. However,

it is in the discourse of resource use and management that *Bininj* and non-Aboriginal knowledges appear the least compatible. While Western science conceives of fish as a resource to be managed sustainably, *Bininj* view fish and their harvest not only as a significant source of food, but also as part of a wider system of interconnected socio-physical relationships and identity.

(Palmer 2004: 74)

Table 4.4 Outdoor recreation capability classes

Classification	activities		Physical setting	
Remote natural	**Land-based** Viewing scenery Bushwalking and hiking Camping Nature study, interpretive services Abseiling and mountain climbing		**Size** **Access**	1,000 ha or more (may be smaller if contiguous with natural areas; namely National Parks and State Forests etc.). Area at least 1 km from all roads and tracks, which are usually open to motorised use.
	Water-based Canoeing – rowing Swimming Fishing Surfing (body)		**Man-made modification**	Essentially an unmodified natural environment which may contain a limited number of tracks. Buildings are very rare and isolated.
Natural	**Land-based** Viewing scenery Picnicking Bushwalking, hiking, horseriding Camping Nature study, interpretive services Off-road vehicles Fossicking		**Size** **Access**	No set criteria (generally over 20 ha) but may be isolated pockets in urban settings. Coast lands variable but contiguous with natural lands. Area normally within 1 km from primitive roads and tracks, which are usually open to motorised use.
	Water-based Canoeing – rowing Sailing Swimming Diving Fishing Surfing (all)		**Man-made modification**	A natural area with subtle-to-dominant modification. Serviced with primitive to sealed roads. Buildings are rare and generally isolated.
Rural	**Land-based** Viewing scenery, arts and crafts Individual and team sports Picnicking Walking, bicycling, horseriding Sightseeing drives Club and kiosk services Showgrounds and cemeteries Motorised transport (automobiles, 4-wheel drives, motorcycles) Camping and caravans; cabins Racing; games and playgrounds; fossicking		**Size** **Access** **Man-made modification**	No set criteria (generally over 5 ha) Coast land variable but contiguous with rural lands. Area serviced by a formed road. A culturally modified environment reflecting past or current practices in agriculture, clear fell forestry, extractive industries, utility corridors etc. Area is serviced by formed roads to highways. Buildings are common and may range from scattered to small clusters, e.g. power lines, towers, resorts, marinas, pit heads and farm buildings.
	Water-based Canoeing – rowing; sailing; power boating; water skiing; fishing; diving; swimming; surfing (all); marinas; water sport.			
Urban	**Land-based** Viewing scenery, arts and crafts, bands Individual and team sports; picnicking Walking, bicycling and horseriding Sightseeing drives; club and kiosk service Motorised transport (automobiles, 4-wheel drives, trains, buses, motorcycles) Camping and caravans; games and playgrounds		**Size** **Access** **Man-made modification**	No set criteria (generally ¼–20 ha) Coast land variable but contiguous with urban lands. Area normally accessed by foot, bicycle or, if necessary, vehicle. A structure dominated environment serviced with sealed roads to highways. Natural or 'natural appearing' elements may be present but buildings and building complexes are dominant, namely resorts, towns, industrial sites, residential areas.
	Water-based Canoeing – rowing; sailing; power boating; water-skiing; fishing; diving; swimming; surfing (all) Wharfs, jetties; marinas; water sports			

Source: New South Wales Department of Lands (1986: 36).

Management setting (controls may be physical, e.g. barriers, or regulatory, e.g. laws)	Social setting (applies to a typical day's recreation. Peak days may exceed these limits)
Little or no on-site management control.	Frequency of contact with other users low on tracks and very low at campsites.
Management control is subtle-to-noticeable but in harmony with the natural environment.	Frequency of contact low-to-moderate on roads, tracks and developed sites. Low elsewhere.
Management controls obvious and numerous. Largely in harmony with the man-made environments.	Frequency of contact is moderate-to-high on roads, tracks, and in developed sites. Moderate away from developed sites.
Numerous and obvious on-site management controls.	Large number of users on-site and in nearby areas.

Evaluation of recreation environments

Measurement of the capability and suitability of the resource base to support various forms of outdoor recreation is a more difficult undertaking than if the task is confined to classification or assessment for a single purpose. Yet, an area seldom provides for only one kind of recreation, and it is more realistic to consider several activities, with due regard for the complex relationships between outdoor recreation and other resource uses. The complexity of the task is typified by an early investigation undertaken in central Scotland to identify and evaluate recreation environments at a regional scale, on the basis of functional connections between different recreational activities, resources and users (Duffield and Owen 1970; Coppock *et al.* 1974).

The approach adopted was to make four separate, independent assessments of the components of resource capability for outdoor recreation, and then to combine these into one single assessment. The components used were: suitability for land-based recreation; suitability for water-based recreation; scenic quality; and ecological significance. The basic spatial unit was a 2 kilometre (approx. 1.25 miles) square grid overlay, covering the study area. Data on the four components were obtained from existing maps, aerial photographs and other published sources, and each grid square was evaluated separately for each category. The combined grid scores were then graded into six classes and mapped to indicate composite recreation environments (Figure 4.2).

Experimental techniques, such as used in the Scotland example, can always be subject to scrutiny, and adaptations to suit specific areas under study are advisable. The authors conceded that the assessment might be altered if other features and forms of outdoor recreation which attract visitors to the Lanarkshire region had been considered. The somewhat arbitrary choice of the four components, and the decision to give them equal weight, can also be challenged.

Projects and their resources can also be evaluated by balancing their costs and benefits, with some indication as to the externalities associated with projects. Cost–benefit analysis and contingency valuation are widely used in public sector management decision-making, especially with regard to large-scale projects (see Box 4.3).

No means of assessment or evaluation of recreation environments is compre-hensive or able to predict the future with great certainty, so in addition to matters already discussed, factors such as the changes in access to resources, or changes in the attitudes of management authorities and the local community to a resource, need to be recognised and addressed at a subsequent stage of resource assessment and development. As noted earlier, the ways in which we value and use resources may change over time and space, so resource evaluation is an ongoing process (also see Chapter 12).

In a recent study of historic pocket parks or internal reserves in Melbourne, Australia, Nichols and Freestone (2003) found that local communities surveyed in 1979 and 2002 did not value these spaces highly. Local residents were ambiguous about their existence and held negative perceptions of them. Yet these internal

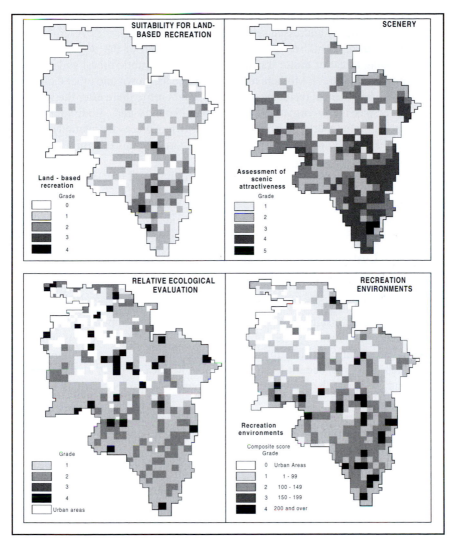

Figure 4.2 An assessment of the recreation resources of Lanarkshire

Source: Duffield and Owen (1970: Appendix 1).

reserves, which were planned open space at the rear of suburban houses or residential allotments and without street frontages, were particularly popular in urban planning in Australia between 1910 and the early 1930s. These reserves grew out of the garden suburb idea and have much in common with the mews in older suburbs of inner London and other large cities in England:

> In Australia, the internal reserve seemed well placed to address the need in new suburban estates for both children's playgrounds and adult recreation

space. When conceived as one unified 'backyard', shared by residents, it neatly and economically addressed several perceived planning issues. A larger space at the rear kept children away from roads and out of the way of both the increasing threat of motor traffic, if not the more subtle moral and physical dangers of the street. Mothers might supervise their own and others' children over the back fence; residents of all ages might retire there after the working day for passive or organised recreation. Planners and subdividers often proposed tennis courts, club buildings, and bowling greens for internal reserves. This was classic social engineering via environmental determinism and a mechanism for building community.

(Nichols and Freestone 2003: 116)

Box 4.3 Cost–benefit analysis and contingency valuation

Cost–benefit analysis

Cost–benefit analysis is a formal technique by which the benefits of a project are weighed against its costs. It is of particular use in the appraisal of investment decisions made by the public sector and/or for large projects where there are wider considerations to be taken into account other than just profitability. Cost–benefit analysis extends the idea of costs and benefits beyond those which affect individuals or businesses to those which affect society as a whole.

Investment appraisal for a private sector project such as a new hotel is relatively straightforward. If the project yields the required return on capital employed then the investment will go ahead. The different nature of public sector and large private sector leisure projects makes investment appraisal more complex. The former includes projects such as national parks while the latter includes projects such as sports stadia.

Public sector leisure investments are often made for reasons of wider social benefits and large private sector projects may give rise to externalities that cause public nuisances or environmental impacts such as noise, or congestion. These factors make private sector methods of appraisal inappropriate since they only measure private costs and revenues. An example of the appraisal of a canal restoration scheme illustrates this.

Private sector investment appraisal of such a scheme would only include the private costs and benefits of the project. The private costs would include the construction costs of the project, for example materials, labour costs and professional fees. The private benefits would include revenue from the project, for example craft licences and charges, fishing licences and rentals from renovated buildings. Since the private costs would almost certainly exceed the private benefits, the investment would not proceed.

However, cost–benefit analysis would extend the decision frame to include the wider social costs and benefits or externalities. Social costs such as noise and congestion associated with the construction phase might be identified. Social benefits of the scheme would include lives saved through improved canal safety, greater public wellbeing caused by improved aesthetics from the project, and the effects on the local economy of new industries and employment attracted to the area because of the project. It might well be the case that total social and private benefits would exceed total social and private costs. Thus there may well be an argument for public sector investment in the project. It is generally the case that social costs and benefits are more difficult to measure because they often do not have an obvious market price.

Source: Tribe (2003: 83–4).

Contingency valuation

Contingency valuation, or 'willingness-to-pay' (or willingness to accept compensation), is a method used in cost–benefit analysis to assess the value which individuals place on things which are generally provided free or at a subsidised price. It is an alternative to the Clawson, or travel-cost, method and the hedonic pricing method. Contingency valuation has been developed largely in the context of natural resource planning, and hence is particularly suited to use in relation to natural recreational open space areas, such as national parks and coastal areas, but is equally suitable for use in relation to historic heritage and the arts. The method relies on surveys of actual or potential users of a facility or resource, or individuals, who, even if not direct users, nevertheless have a concern for or interest in the resource. Survey respondents are asked to indicate how much money they would be willing to pay to use the resource or to preserve it as a natural resource. Grossing up the results for the population as a whole provides an estimate of the overall economic value of the resource to the community. For example, if 50 per cent of the population is willing to pay, on average, $5 a year to preserve the resource, then, in a population of 60 million, this would produce an overall valuation of $150 million a year. Since the survey question is purely hypothetical, there is a possibility that respondents may exaggerate their willingness-to-pay, perhaps if they do not take the question seriously, or might under-estimate it, perhaps if they think there is a possibility of a user charge actually being introduced. However, all methods of arriving at an economic valuation of non-market resources are subject to criticism; the advantage of the willingness-to-pay method is that it is relatively easy to implement and to understand.

Source: Veal (2003: 81–2).

Implicit in the discussion of recreation resource classification and evaluation has been the existence and even acceptance of a strong element of subjectivity in the assessment process. One of the most difficult areas to contend with from the point of view of subjectivity is that of landscape evaluation.

Landscape as a recreational resource

Until relatively recently, landscape has been largely ignored as part of the recreation environment. However, growing concern for environmental quality has led to recognition of the scenic quality of landscape as a major recreational resource in its own right, rather than merely as the visual backdrop for other recreation pursuits. This, in turn, generated interest in systematic attempts to evaluate scenic beauty, and to examine the features of landscapes which contribute to their attractiveness and to their resource value in outdoor recreation (Robinson *et al.* 1976).

That said, difficulties remain with assessment procedures because of the intangible and multi-faceted nature of landscape, which does not permit precise measurement. The resource function can take on several dimensions, depending upon which senses are being satisfied and the characteristics of the population involved. These difficulties are compounded by the assessment of recreational values. Whereas most landscapes probably have some recreation potential, this fact is not easy to establish with any agreement, because of the personal nature of recreation and the subjective manner in which it is experienced. Generalisation and interpersonal comparisons are of doubtful validity, and the multiple characteristics of landscape make dissection and evaluation a risky undertaking.

Despite the essentially subjective nature of the variables involved, efforts are being made towards identification and measurement of scenic landscape values in response to competing resource users. In England and Scotland, a major step forward has been taken with the development by the Countryside Agency of a system for Landscape Character Assessment (LCA) (www.countryside.gov.uk/cci/guidance; www.snh.org.uk/strategy/LCA). The LCA is a means of identifying the features that give a locality its 'sense of place' and pinpointing what makes it different from neighbouring areas. A map of landscape character has been produced along with indicators of landscape change (Countryside Agency 2002). The Agency has launched a online data base to widen the use of LCA by permitting easy access to local reports and other key information (Wharton 2005). The national database is accessible through the Countryside Character Network (www.ccnetwork.or.uk).

Landscape Character Assessment is essentially a framework for decision-making which can be tailored to address a range of issues relating to the use and management of rural and urban landscapes. Understanding the character of an area is seen as the essential first step in making landscape-related decisions, either at the scale of a village street or an entire region (Campaign to Protect Rural England (CPRE) 2003). The first stage of the assessment process is characterisation to document and enhance what makes a place distinctive. The second stage focuses on making judgements about future land management activity and landscape guidelines for new development, including designation of landscapes of particular significance, importance or quality.

A variety of methods and procedures has been suggested to manipulate and rank landscape attributes in order to establish preferences. However, criticism has been directed at particular approaches which claim to present an *objective* measurement of landscape quality. Such refinement is impossible with such an inherently subjective process, no matter what sophisticated analytical methods are used. Landscape evaluation can never be divorced entirely from subjective interpretations, and the best that can be achieved is some appropriate balance between operational utility and scientific elegance.

Apart from problems of subjectivity and the appropriate mix of landscape attributes, assessment procedures are unable to cope, as yet, with the internal visible arrangement and spatial composition of landscape (as opposed to its resource content). With all the difficulties and reservations regarding methodology, why bother evaluating landscape?

The answer can be found in the four broad objectives designated for landscape management by Penning-Rowsell (1975):

- *Landscape preservation* – identification of areas of landscape worthy of preservation and deserving of priority for conservation;
- *Landscape protection* – as a basis for development control decisions, to guide the direction of development, monitor environmental impact and provide for planned landscape change;
- *Recreation policy* – aimed at enjoyment of the landscape and realisation of its long-term potential for appropriate forms of outdoor recreation;
- *Landscape improvement* – identification of visual features which detract from landscape quality, so that such 'eyesores' can be removed or modified.

More recently, the CPRE (2003) identified several possible outputs of the Landscape Assessment Process:

- Information awareness raising and appreciation of landscape;
- Informing development plans and development potential, including location and design of projects such as wind farms;
- Input to environmental impact statements; and
- Providing the basis for landscape conservation and management and enhancement strategies.

To these, could be added the requirement to satisfy a growing body of environmental law in many countries. In the US, for example, all government agencies must ensure that 'environmental amenities' are given appropriate consideration in planning decisions. In Australia, too, there is a growing commitment to landscape conservation, in both the natural and built environment. Over the years, several urban renewal projects have been delayed or abandoned because of 'green bans' imposed by trade unions, or as a result of the actions of conservationists anxious to preserve parts of the national heritage and landscape. Some resort developments have been successfully resisted by local communities. More generally, though, environmental planning legislation in the States and Territories contains guidelines concerning environmental impact assessment and public participation in urban and rural development. However, environmental legislation can be thwarted by governments and developers.

> Secret dealings between State government officials and private developers have always been a feature of coastal resort development and land speculation in Australia, though there have been marked variations between the different States at different times in terms of the extent of the practice.
>
> (Mercer 1995: 173)

In brief, development fast-tracking, statutory amendments, special deals and political favours are not uncommon in Australian tourism and related recreational development.

In New Zealand, the national government is a key resource manager through the Department of Conservation, while the operational role of local authorities has been redefined through the introduction of the 1991 Resource Management Act (RMA). This Act represents an important attempt at regulating for sustainable development, which encompassed a redefinition of the structure and roles of local government. The RMA replaced many pieces of legislation with a more comprehensive framework for the allocation and management of resources (NZTB 1994, in Kearsley 1997). It outlines the responsibilities of central and local government, and is implemented by way of a hierarchy of policies and plans. However, perhaps the most significant change in planning methods is in the explicit departure from prescriptive criteria for resource allocation (or 'zoning' approach), to a system focusing on the effects of activities rather than their intrinsic nature. 'The suitability of a particular use of land is determined by what its environmental outcomes might be rather than by what it is' (Kearsley 1997: 57).

Evaluation of recreation sites

At a finer scale, attention can be directed towards evaluation of the potential of the resource base to support a specific recreation activity or experience at a specific site. Evaluation at this level calls for a different approach from that used in broad regional assessment, especially where questions of land tenure, access and management can often be disregarded.

Site evaluation assumes knowledge and understanding of the detailed resource requirements for each type of recreation involved. The following is a list of the kinds of questions which might need to be answered:

• What kind of topography is most suitable for bushwalking, horse-riding or trail bike riding?

- What river conditions are ideal for white water canoeing, trout fishing or bathing?
- What types of vegetation are preferred for orienteering or children's adventure play?
- What snow conditions are best for cross-country skiing?
- What characteristics make a rock face good for climbing, or a pool suitable for fly-casting? (Hogg 1977: 102)

For certain activities, conditions are necessarily more specific and closely defined than for others, which are more flexible. Competitive activities, for example, are generally more demanding than less formal recreational uses of countryside. Physical and natural circumstances will be most important for some forms of recreation, whereas for others, social factors may need to be taken into account, and created facilities and infrastructure may be mandatory for effective functioning of the recreation resource base.

Recognition of recreation site potential involves synthesis of an 'identikit' specification (incorporating all the relevant site factors), which conforms most closely to the ideal. However, in the evaluation process, it is important to distinguish between *minimum* and *optimum* site requirements. The former represent a threshold or entry zone concept, in that they describe the set of obligatory conditions within a narrow range of acceptability that is essential if the activity is to take place at all. Unless such bare minimum standards are satisfied, the type of recreation envisaged cannot be accommodated. Optimum requirements imply a preferred situation such as might be demanded or experienced at an Olympic site of national or international repute. Such standards are obviously much more demanding and precise, and are applicable only to exceptional situations.

Once the necessary minimum site conditions have been established, a method of ranking or rating is needed which reflects the relative importance of each requirement, and which indicates whether it is considered an asset or a constraint. In an Australian study, Hogg (1977) lists the natural and cultural factors he considers important for overnight bushwalking or hiking (Table 4.5). Unless *all* site (route) requirements are held to be *equally* fundamental, the points allotted and rating scales used, need to be adjusted to reflect their greater or lesser importance. Without some weighting of this kind, serious deficiencies in more critical factors can be largely overcome by high ratings for more trivial aspects. Moreover, a zero rating given to indicate *total* unsuitability on the basis of a single vital factor can be swamped in the additive process by high ratings for more mundane requirements.

Thus, in Hogg's example, presence of quality campsites is assigned the highest maximum positive points value (25), whereas refuge huts are apparently considered of passing importance to bushwalkers and are lowly rated (5). On the other hand, the presence of hazards is judged to be a most serious constraint (–50), and some factors such as restrictions on access, are considered so critical as to be allotted the most negative value of –00. In Hogg's evaluation system, a recreational unit in which any *single* factor is rated –00 will also have a *total* rating of –00, and,

Table 4.5 Factors affecting suitability for overnight bushwalking

Factor	Maximum value	Minimum value
Natural factors		
1 Topography (steepness and variability of terrain, length of uphill climbs)	15	−00
2 Rockiness of terrain for walking	0	−5
3 Weather characteristics during walking season	15	−15
4 Ease of negotiation of vegetation (denseness of scrub, fallen timber, blackberries, nettles, etc.)	0	−30
5 Presence and quality of campsites (ground suitable for pitching tents, firewood and drinking water, general environment)	25	−00
6 Extent of area	15	−00
7 Proximity to users	15	−5
8 Scenic quality (general attractiveness, variety, special features)	10	−5
9 Availability of drinking water between campsites	5	−8
10 Miscellaneous attractions (wildlife, swimming holes, historical features, etc.)	10	0
11 Undesirable features (snakes, leeches, mosquitoes, bushflies, etc.)	0	−00
Cultural factors		
12 Access to suitable starting points	20	−10
13 Tracks suitable for walking (related to 5)	15	0
14 Unnecessary or undesirable tracks and roads	0	−00
15 Refuge huts	5	0
16 Unnecessary or undesirable buildings	0	−00
17 Adequate track marking and signposting (including snowpoles)	8	0
18 Escape routes for emergency use	8	0
19 Hazards (e.g. mineshafts)	0	−50
20 Presence of conflicting recreational activities	0	−00
21 Presence of other conflicting land uses (e.g. logging, mining, grazing)	0	−00
22 Restrictions on access or certain activities (e.g. camping, fires)	0	−00

Source: Hogg (1977: 106).

therefore, is judged as totally unsuitable on the basis of that one factor, no matter how favourable other factors may be.

Actual evaluation of a specific site or recreational unit involves scoring the resource endowment according to the degree to which it satisfies each of the user requirements identified, and how it matches up to the conditions stipulated. This gives an indication of the potential of a site for a specified form of recreation in terms of the presence or absence, and quality, of certain features. The evaluation, therefore, provides a kind of inventory and appraisal of the site's latent potential to supply particular recreation resource functions, although this does not mean that development of this potential will necessarily occur.

Evaluation is not always a straightforward field-checking procedure, and suitability scores should not be accepted without qualification. Some conditions need to be sustained over time, and others may only be ephemeral or present intermittently (e.g. wave conditions for surf-board riding). It is important for the assessor to be able to recognise in a low-scoring site, latent potential which can be realised if certain shortcomings are remedied by provision of additional features and sound management. Conversely, such insight is just as vital in the detection of inherent disadvantages (e.g. ground cover with low tolerance to trampling, or an unreliable water supply), which may become obvious with use and create a problem for subsequent management.

Evaluation of streams and routeways

Linear recreation resources such as streams and scenic routeways frequently call for specialised application of evaluation techniques. These methods seek to identify and measure or rank those physical, cultural and aesthetic attributes of a river and its environment which are considered significant when assessing its recreational value. Typically, the schemes divide the river into manageable segments, for analysis from maps, air photographs and on-site inspections and research. An element of subjectivity, again, is inevitable in judgements concerning the features to be assessed, the recreation activities envisaged, and the scoring and weighting procedures adopted. Most of the methods focus on relatively remote river resources, although some attempts have been made to develop and apply criteria for evaluating urban settings for recreational use, close to the centre of the river recreation opportunity spectrum.

Efforts should also be made to incorporate concepts such as carrying capacity and visitor management models and frameworks (Chapter 6) into evaluation and classification schemes, *and* to provide for user perceptions and public participation in the process. It is clear that there is still some way to go before development of an effective technique that will allow for the dynamic nature of the river resource and its potential users, and that is capable of being replicated in many different river situations.

One of the first systematic attempts to identify and assess scenic routeways was made by Priddle (1975) in Southern Ontario. Priddle's approach was also to break each selected routeway into segments, based on intersections or major changes in landscape. Each segment was then traversed and evaluated in terms of the distance that could be seen, the alignment of the road, and the scenic features and variety present.

The method was refined and developed further by Prior and Clark (1984) in an Australian study in the Hunter Valley region. In this study, the authors identified the essential components of a scenic road system and their relative importance, through a survey of likely users. The responses were used to develop a weighting system to evaluate standard and scenic quality in the valley. The method represents a rapid quantitative assessment of the major components of a rural scenic/ recreational route network to accommodate pleasure driving. Whereas some of the parameters used might be clarified, and additional aspects considered, evaluation of linear recreation resources in this way can provide valuable input to decision-making. In Canada, rails to greenways or rails to trails projects have been prominent. These involve the conversion of disused railway lines and associated corridors and bridges to develop walking and cycling trails (see Marsh 1994). Thousands of kilometres of railway lines and associated corridors have been abandoned in Canada. Such former railway lines are also gaining prominence in countries such as Australia, but their conversion to recreational opportunities or other functions has been slow.

Summary

Classification and evaluation of resource potential are critical elements in recreation planning and management, but make up only one phase in the formulation of a rational strategy for recreational development. Comparative evaluation of resources provides valuable input to the process of informed, effective choice and decision-making. However, strict adherence to evaluation procedures and an over-rigid application of the findings, may obscure opportunities for substitution between sites of recreation activities with more flexible user requirements, and, hence, higher spatial elasticity. Scope should always be allowed for interpretation and sound judgement by management in the incorporation of assessment data into recreation resource development programmes.

Guide to further reading

- The study of outdoor recreation and resources clearly has many dimensions. Liddle's (1997) work provides a very comprehensive coverage of the ecological impacts of outdoor recreation and tourism, giving detailed insights into resource appraisal, capability and evaluation. Readers should also consult Wall and Wright (1977); Cloke and Park (1985); Hammitt and Cole (1987, 1998); Broadhurst (2001); Newsome *et al.* (2002).
- The need for public participation in recreation and tourism planning, including the development of visitor management strategies, is well recognised, e.g. see Murphy (1985); Haywood (1989); Dredge and Moore (1992); Ryan and

Montgomery (1994); Simmons (1994); Hall and McArthur (1996); Newsome *et al.* (2002).

* Shackley (1998) examines visitor management issues at World Heritage Sites.
* Sources concerning recreation resource assessment include: Leatherbury (1979); Kane (1981); Cloke and Park (1985); Mather (1986); Hammitt and Cole (1987, 1991); Countryside Commission (1988); Thomson *et al.* (1995); Countryside Commission (1995); Fraser and Spencer (1998); Harmon and Putney (2003).

Review questions

1   How might an outdoor recreation resource be defined? Discuss the importance of considering resources generally, and outdoor recreation resources specifically, in functional terms.
2   Select a local outdoor recreation site. List the major uses and users of that site. Attempt to identify the current and potential conflicts which relate to that site. To what extent do you think multiple use of the site has been achieved and has been successful?
3   With reference to case studies, discuss the concept of accessibility in terms of recreation resources.
4   Resource assessment and evaluation require considerable subjective judgement in their application. What are the implications for decision-making? How, if at all, can subjectivity be controlled or offset?

5   Outdoor recreation and the environment

Identification of recreation resource potential and classification and evaluation of resources for outdoor recreation are necessary, but only initial steps in the process of creating recreation opportunities. The real challenges for environmental management arise from human use of the recreation resource base. In the final analysis, concern is for the quality of the recreation experience and the degree to which that experience contributes to the physical, psychological and spiritual wellbeing of participants. The quality of the recreation experience is largely a function of the environment in which it takes place, but there is nothing deterministic or inevitable about the relationship.

The primary concern of recreation resource managers is undesirable change in environmental conditions (Hammitt and Cole 1991). However, the mere presence of human beings in a recreation setting need not be the trigger for degradation, any more than 'a bull in a china shop' necessarily spells catastrophe. If the bull was led, docile, well-trained, and did not bellow, get upset, or otherwise disgrace itself, and if the shop fixtures were well-spaced and the china secure, then perhaps the occasion would be without incident.

So it is with outdoor recreation; the use-impact relationship is not straight-forward. A certain amount of recreation use in a particular environmental setting will lead to a certain level of impact, depending on a combination of factors, including the weather, the resistance and resilience of soils and vegetation to trampling, soil drainage, the extent and nature of recreational use, and management strategies (e.g. Stankey *et al.* 1984; Liddle 1997; Hammitt and Cole 1998).

Recreation and environmental relationships, impacts and assessment

It is unwise to rush to conclusions about the impact of outdoor recreation on the environment, or to accept, without qualification, predictions of undesirable or irreversible consequences of human use. As noted above, the outcome is a function of the attributes of the environment, the extent and nature (i.e. volume, intensity, behaviour of participants) of the recreation taking place, and resource management strategies. The ensuing discussion focuses on the relationships between outdoor recreation and the natural, biophysical environment. Hence, there is a good deal of reference to research and applications from the field of recreation ecology. Recreation ecology is a 'field of study that examines, assesses and monitors visitor impacts, typically to protected natural areas, and their relationships to influential factors' (Marion 1998, cited in Leung and Marion 2000: 23; Hammitt and Cole 1987; Liddle 1997).

This relatively recent emphasis in recreation research seeks, first, to identify the type and extent of recreation resource impacts and then to evaluate the relationships between recreation use and environmental and managerial factors (Marion 1998). Yet, an ecological understanding of the environmental impacts of outdoor recreation and tourism are seen to be deficient (Buckley 2005). Sharp differences occur in the data and understanding of recreation impacts between activities, ecosystems and regions. Buckley stresses the need for more recreation ecology research to enhance the effectiveness of management of recreation impacts. Considerable effort is being directed towards expanding research in recreation ecology.

Forms of recreation impact in wilderness (and indeed other nature-based recreation areas) have been described by Leung and Marion (2000) (see Table 5.1). These impacts have been identified through four types of studies: descriptive surveys of recreation sites; comparisons of used and unused sites; before-and-after natural experiments; before-and-after simulated experiments (Cole 1987). Adapting Cole's work, Leung and Marion summarised the research methods applied in studies of trampling, trail impacts, campsite impacts, and indicators and indices, which comprise a large proportion of recreation ecology research (see Table 5.2)

Table 5.1 Common forms of impacts in wilderness

Effects	*Ecological components*
Direct effects	*Soil* – soil compaction; loss of organic litter; loss of mineral soil
	Vegetation – reduced height and vigour; loss of fragile species; loss of flora; vegetation damage; introduction of exotic species/weeds
	Wildlife – loss or alteration of habitat; introduction of exotic or feral species; behaviour modification and harassment of wildlife; displacement
	Water – introduction of exotic species and pathogenic bacteria; increased turbidity; increased nutrient inputs; altered water quality
Indirect/derivative effects	*Soil* – loss of soil moisture and pore space; accelerated erosion; altered microbial activities
	Vegetation – accelerated soil erosion; composition change; altered microclimate
	Wildlife – reduced health, fitness and reproductive rates; increased mortality; composition change
	Water – reduced health of aquatic systems; composition change; excessive algal growth

Source: Adapted from Leung and Marion (2000: 24).

Table 5.2 Four common study designs employed in recreation ecology research

Study design	Description
Descriptive surveys	Estimates or measures used to assess current resource conditions at recreation sites
Comparison of used and unused sites	Comparisons of measurements on recreation sites and nearby undisturbed sites in order to compare extent (amount) and nature of impact
Before-and-after experiments	In order to infer extent and nature of impacts due to change, measurements are taken after (1) commencing or ceasing use of sites, or (2) applying management actions
Before-and-after simulated experiments	In order to infer extent and nature of impact due to treatment of a site, measurements are taken before and after treatments are applied

Source: Adapted from Leung and Marion (2000: 28).

(also see Buckley 2004a; Cole 2004). Other recreation impact research includes effectiveness of management actions, impact indicators for management frameworks such as Limits of Acceptable Change (LAC), grazing, abseiling and rock climbing, human waste, hut construction and use, birdlife, terrestrial wildlife and flights over natural and scenic areas (e.g. Leung and Marion 2000: 35; Knight and Gutzwiller 1995; Buckley 2004b, 2004c; Cole 2004). Interestingly, the impacts and the ways in which they have been studied are also applicable to urban parks and urban bushland environs, though the complexity of impacts (i.e. the factors causing those impacts; species' resistance and resilience) may be greater because of the nearby location of industries, urban development and large scale infrastructure. It should be noted, too, that visitor satisfaction is affected by impacts such as loss of wildlife, erosion and lack of ground cover at campsites (e.g. see Hollenhorst and Gardner 1994).

The social setting, too, is important in its effect on the recreation experience, even in wilderness settings, and, in turn, can be transformed by the recreation activity taking place or minor modifications to an environment. This relationship is explored further under social carrying capacity.

Environmental assessment generally refers to the process of describing

> a natural environment and existing human modifications in a given area, as a basis to evaluate its suitability for various recreational uses. Such information may be used, for example, in zoning an area for different uses; in planning infrastructure or monitoring programs; or in establishing visitor capacities.
>
> (Haas 2002, in Buckley 2003a: 149)

Commonly, this approach may include some form of environmental compatibility analysis. Buckley argues that biological research programmes are 'notoriously under-funded worldwide' and that data collection is frequently limited to either large mammals or endangered species (p. 150). Unfortunately, we have only rudimentary knowledge of the recreational impacts on flora and fauna, except where detailed studies have been undertaken at specific sites, and very rarely have studies focused on lizards, birds, frogs, terrestrial and aquatic invertebrates and terrestrial and aquatic plants (Buckley 2003a). Exceptions include wetlands, which are areas of 'captive' interest for researchers and recreationists.

Attributes of the biophysical environment differ from place to place. Geological and edaphic conditions vary, as do terrain, hydrology, fauna and flora. The biophysical characteristics of the natural environment can also be materially altered by ephemeral or transitory aberrations in weather and seasonal conditions. A simple illustration of this is the difference in effect on a recreation setting of the same type and volume of recreational activity in summer and winter or in wetter or drier periods. The ability of a site to withstand use and to recover over time varies with the season and the weather.

Comprehensive overviews and documentation of biophysical changes to the environment from recreation use are presented by Wall and Wright (1977), Hammitt and Cole (1987, 1998), Kruss *et al.* (1990), Liddle (1997) and Leung and Marion (2000). Buckley (2005) compared the extent of research among several countries and found US recreation ecology research far outweighed that in Australia. Depending on the circumstances, outdoor recreation may affect the type and diversity of vegetation species, soil properties, wildlife populations, habitat, air and water quality, and even the geology of the recreation setting. Sensitive environments, such as parts of the coastal zone, are particularly prone to disturbance.

Environments differ in their ability to withstand use and to recover after use. Hammitt and Cole (1987: 23) make the distinction between *resistance* and *resilience*: 'Resistance is the ability to absorb use without being disturbed (impacted); resilience is the ability to return to an undisturbed state after being disturbed. Resistant sites may or may not be resilient and *vice versa*'.

Environments that are sensitive to disturbance may quickly reflect the effects of human incursion, but just as quickly recover after use. A rock surface is highly resistant, but once scarred by graffiti or other undesirable forms of 'recreation' (i.e. vandalism) or by overuse, the damage may be permanent. Research on the resilience of plants in America's eastern states showed that areas of high fertility and moisture (e.g. riparian lands) recover more quickly than plants in alpine and subalpine environments (e.g. Hartley's 1999 study of subalpine meadows in Glacier National Park, Montana). Nonetheless, the physiological and other characteristics of the plants themselves also influence their resilience and resistance to trampling effects. Attributes of resistance and resilience can also vary according to seasonal and climatic conditions. The best sites to use for outdoor recreation are those that are *both* resistant and resilient in the short to long term.

When comparisons are made between sites, it becomes clear that some ecosystems are more tolerant of recreation activity than others. Some areas are virtually indestructible, while others are so fragile as to permit only minimal use. Goldsmith and Manton (1974) suggest that the ecosystems or habitats most vulnerable to recreation impact are:

*   coastal systems, such as sand dunes and salt marshes characterised by instability;
*   mountain habitats, where growth and self-recovery are inhibited by climate;
*   ecosystems with shallow, wet or nutrient-deficient soils.

Several fragile environments could be added to this list, including marine environments such as coral reefs, rivers, and Arctic and Antarctic regions. Fragility in an environment may also refer to the site or region's cultural dimensions.

A comprehensive survey of the ecological impact of outdoor recreation was presented by Wall and Wright (1977, in Mathieson and Wall 1982), who itemised the consequences of recreational activities for specific attributes of the environment (see Figure 5.1). Most emphasis was directed towards interrelated components of the environment, namely soil, vegetation, water, and wildlife. The authors were able to show that even geology can be affected by certain forms of recreation, and that complex interrelationships exist between types of recreational impact.

A second group of factors which has a bearing on the recreation–environment relationship is related to the nature of the recreation activity and the characteristics of users. A recreation setting may well be able to withstand use by any number of sedate picnic groups, but would quickly deteriorate if subjected to an informal rugby match. Not only is the latter activity inherently harder on the ground surface, its concentrated nature has greater potential for impact than the typically more passive and dispersed pattern of picnicking. Moreover, rugby and similar forms of outdoor recreation can take place under rain-affected conditions, for example, where a natural playing surface is even more susceptible to damage.

Some recreation activities, too, rely on certain types of specialised equipment, which add greatly to their potential for environmental disturbance and can allow users to penetrate deep into sensitive areas, not otherwise accessible. Off-road transport such as all-terrain vehicles (ATVs), snowmobiles, dune buggies, and the increasingly popular four-wheel drive, have become a significant feature of the recreation scene. With this trend has come greater potential for degradation of the recreational environment and conflict with other users and uses. Much of the problem rests with the use of these vehicles in sensitive environments such as coastal sand dunes, arid zones, steep slopes, alpine areas and wetlands.

Off-road recreation vehicles (ORVs) can also be responsible for the spread of noxious weeds and the invasion of despoiled areas by exotic vegetation, normally unable to compete with indigenous vegetation. This is a particular problem in parts of Australia, where seeds from weed-infested roadside reserves are easily spread by tyres and mud on vehicles. Damage has also been experienced at sites of archaeological and scientific significance in coastal areas of Australia, where Aboriginal relics and middens have been destroyed or disarranged. The impacts of ORVs on wildlife have rarely been studied, yet even very responsible use in

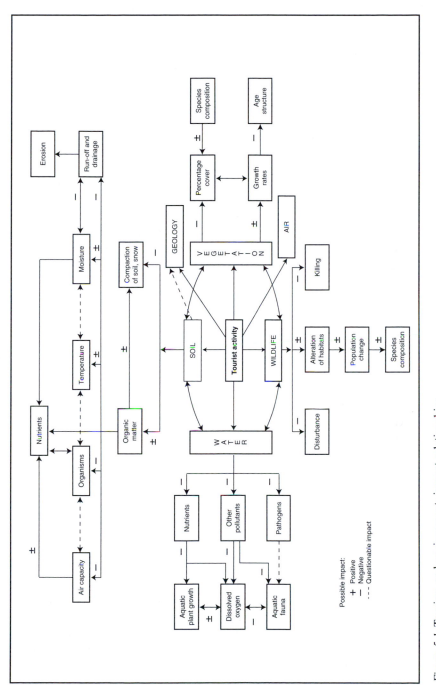

Figure 5.1 Tourism and environment: impact relationships

Source: Wall and Wright (1977, in Mathieson and Wall 1982: 131).

protected areas and on public lands can lead to environmental impacts and destruction. Irresponsible use, however, is seen as a growing problem.

In Nova Scotia, Canada, the Public Lands Coalition (2005) identifies the unrestricted use of ATVs as one of the greatest threats to established and future wilderness areas. Even allowing for discrepancies in registration statistics, the Coalition estimates that there are approximately 40,000 ATVs in use in Nova Scotia, one of the smaller and less developed provinces of Canada. The Coalition believes that:

> Simply put, ATVs and other types of ORVs are incompatible with wilderness conservation, both on and off public land … The explosive growth in sale of ATVs means that the problems experienced at present can only get worse in the future.
>
> (http://www.publicland.ca/issues/atvsandorvs.html)

In the much more densely populated and developed areas of southern Ontario, it is estimated that there are about 250,000 ATVs and many thousands more are sold in that province each year. On 31 July 2003, increased road access for ATVs was created by Ontario Regulation 316/03, specified under the Highway Traffic Act. These new regulations in Ontario allow ATV users to access a range of facilities (fuel, shops, accommodation, attractions) and move more easily between areas, increasing the prospects of an expansion of trails and subsequent impacts. Weak restrictions and lack of action by authorities have seen problems in regulation of use emerge not only in Ontario, Canada, more generally, but also in countries such as the US, Australia and New Zealand. In a study of the effects of off-road vehicles on the biota of the Algodones dunes, in Imperial County, California, Luckenbach and Bury (1983) found that their study site had approximately 5 per cent as many herbaceous plants compared to their controlled/unused sites. In Australia, the New South Wales National Parks Association (NSWNPA) advocates limiting access of ordinary cars, four wheel drives, dune buggies, trail bikes, snow vehicles, amphibious craft and hovercraft used on natural lands to designated roads and parking areas. The Association argues:

> Motor vehicles can have a heavy impact on unsealed bush roads, and off roads in bushland, particularly in or bordering wetlands and on vegetated sandmasses. Associated biological problems can be introduced: e.g. plant pathogens such as *Phytophthora cinnamomi* (Cinnamon Root Rot) and various weed seeds. There are also increased risks of fire and littering.
>
> On beach backshores, severe damage to sand-binding vegetation can occur, resulting in dunes becoming prone to excessive wind erosion and migration. Even on the intertidal zone, it is likely that some sand-dwelling fauna are affected by compaction of the sand in heavily trafficked areas, where seabirds are forced to rise and resettle constantly. The value of these natural lands as habitat and places of aesthetic appeal and human escape is reduced.
>
> (NSWNPA, 1999)

Quite apart from the physical effects of off-road recreation vehicles, a most persistent criticism is the noise associated with their use. Trail bikes and power boats, in particular, can be heard over great distances. Snowmobiles are criticised for being excessively noisy, and can add to site disturbance and adversely affect the chances of restoration. Other ancillary impacts attributed to off-road vehicles are the spread of litter and the risk of fire in otherwise inaccessible areas. Finally, there are considerable hazards and risks with use of these vehicles; deaths and injuries are not uncommon.

Off-road recreation vehicles are one of the more obvious sources of nuisance associated with outdoor recreation activities. However, if time and space patterns of potential effects can be identified, preventive or remedial measures can be undertaken. Perhaps the most constructive response is to set aside special areas for all-terrain vehicles, motor-cross enthusiasts and dune buggies, where their use can be controlled and environmental repercussions minimised. The provision of skateboard tracks in urban areas is an example of a similar trend to manage the emerging problem of this potentially intrusive outdoor recreation pursuit.

Horse-riding causes extensive damage via trampling. A horse and rider with recreational equipment can weigh more than several hundred kilograms. Horses' hooves exert enormous pressure, far greater than hikers' feet, thereby compacting soil, cutting into and damaging tracks and deepening them, enhancing erosion, and damaging tree roots (Buckley 2003; Hammitt and Cole 1998). When horses are tied to trees, their straps cut into tree stems, while compaction and damage to tree roots increase with intensity of use and length of stay. Horses will graze on many native grasses, and they spread weed seeds, consumed perhaps hundreds of kilometres from the site, through their faeces.

Characteristics of participants likewise influence the interaction between recreation and the environment. The attitude and behaviour of visitors can be as important as the pressure of numbers. Some recreationists act responsibly and leave a site in the same condition in which they found it; others are not so conservation-conscious, and make unreasonable demands on the resource base. The problem is heightened by non-uniform patterns of recreation use and the manner in which participants distribute themselves within a site. Visitor pressure tends to be concentrated in space and time (Glyptis 1981). Gittins (1973) documented the differential intensity of recreational activity in Snowdonia National Park in Wales, where patterns of use vary with the time of year, seasonal and daily conditions and popularity of certain features of the park (Figure 5.2). Clearly, at the time of the survey, the park was by no means crowded in the overall sense, but the intensity of recreational use and the potential for environmental disturbance were concentrated in a series of linear routeways and nodal points. This is not an uncommon feature of nature-based recreation resources.

A comparable study was carried out by Ovington *et al.* (1972), on the impact of tourism at Ayers Rock – Mt Olga (Uluru) National Park in central Australia. The study established that, although contact areas for tourists within the park were restricted, each showed evidence of environmental change in terms of topography, soil, drainage patterns, flora, fauna, odour, noise and waste material

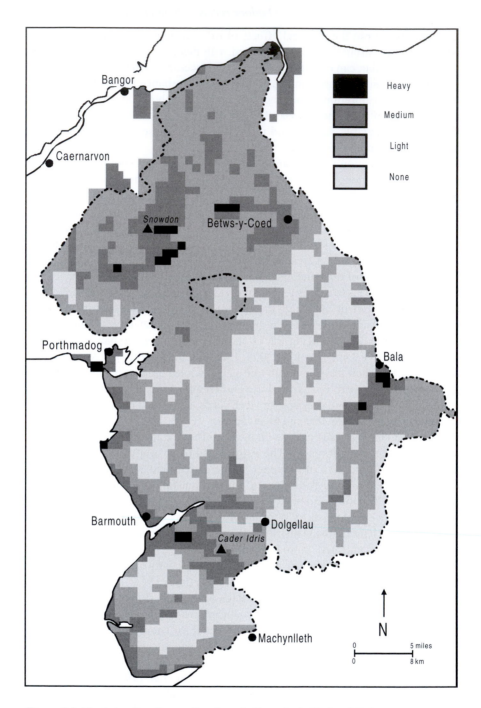

Figure 5.2 The intensity of recreational use in Snowdonia National Park
Source: Gittens (1973, in Patmore 1973: 238).

accumulation. Ecological impacts included soil compaction and erosion, and destruction of vegetation and wildlife habitat. Even the massive monolith of Ayers Rock itself did not escape environmental damage from climbers, including the well-intentioned, but intrusive, installation of chain-railings and lines painted on the rock surface to assist visitors to reach the summit. Most recently, climbing of Uluru has been discouraged (though not denied) because of its spiritual significance to Anangu (traditional owners) (http://www.deh.gov.au/parks/uluru/).

The pattern and extent of wear-and-tear by recreationists on campgrounds, picnic sites and sand dune vegetation have been demonstrated in increasingly sophisticated research on such sites (LaPage 1967; Settergren and Cole 1970; Boden 1977; Slatter 1978; DeLuca *et al*. 1998; Manning 1979; Marion and Cole 1996; Stohlgren and Parsons 1979; Taylor 1997; Zabinski *et al*. 2000). Various methods have been used to record changes in ground cover and species composition which have been subsequently correlated with levels of visitor use. LaPage used sequential photographs on a systematic grid system to reveal a progressive reduction in vegetative cover and number of species, closely associated with concentrations of use around fixed site facilities such as picnic tables and barbecues. With continued use, LaPage found a gradual rearrangement of plant species composition, leading to a relatively recreation-tolerant soil cover.

The ecological environment responds in different ways to visitor pressure, and the possibility of such beneficial changes should not be dismissed. Some observers consider that soil compaction around the roots of trees, for example, has a useful effect in terms of forest viability. Low intensities of trampling can stimulate plant growth, and the opening-up of forests with nature trails allows more light through the canopy and can contribute to an altered, but enhanced, recreation landscape.

Some trampling studies (e.g. Cole 1995) indicate that there is a linear relationship between certain types of vegetation covers and the logarithm of amount of use. Cover loss increases rapidly with initial increases in use. Beyond some threshold the rate of loss slows dramatically (Cole 1995). The suggestion, then, is that unless visitor use is very limited in numbers and behaviour is carefully managed or closely regulated, impacts cannot be reduced substantially. In order to minimise impacts in areas of extensive wilderness or national parks or other extensive recreation resources, it would seem to make sense to concentrate recreational activity to designated sites and trails in natural areas, and in doing so, to carefully select sites which are resistant or very resilient (Hammitt and Cole 1998).

Hendee *et al*. (1978) identified various groups that could be ranked according to their levels of environmental impacts in wilderness areas (from highest to lowest):

1   Large parties of horse users.
2   Small parties of horse users.
3   Large parties of overnight campers.
4   Small parties of overnight campers using wood fires.
5   Large parties of day hikers.
6   Small parties of overnight campers using camp stoves and not building wood fires.
7   Small parties of day hikers.

Plate 5.1   Trampling and erosion, the Giant's Causeway Coastal Walk, Ireland

Generally, however, trampling and use of vehicles damage the environment, producing direct and indirect effects on soils, vegetation, wildlife and water quantity and quality. Hammitt and Cole (1998: 65) produced a conceptual model of trampling effects. This model was partly based on the works of Liddle (1975) and Manning (1979) and demonstrates the complexity of recreation ecology and recreation resource management. Types of impact and possible indicators for a broader park environment are set out in Table 5.3. Box 5.1 provides an overview of trampling research.

Participation in outdoor recreation is increasing, for all the reasons noted in earlier chapters, and with it, the inevitability of environmental change and possible degradation. Such change is seldom sharp or catastrophic, but more usually, incremental and cumulative; the result of many individual actions. Impacts on the ecology of a recreation site often receive the most attention, and these, in turn, can detract from the quality of the recreation experience.

Despite the probability that recreation opportunities could be impaired by such changes to a site's ecology, visitors to a site appear to be more concerned about impacts that decrease its functionality or desirability (Hammitt and Cole 1991). Moreover, the *same* change can be seen as a problem or an advantage, depending upon the environment in question and its use for recreation. Hammitt and Cole (1991) offer the example of conversion of natural vegetation to introduced species of turf. In a pristine wilderness this change is considered undesirable; in an urban park the change to the playing surface may be beneficial.

The complexities of the relationship between recreation and the environment are matched by difficulties in detecting and identifying cause and effect. Not only

Table 5.3 Selected environmental impacts and potential indicators

Selected impacts	Potential indicators
Sewage discharge	Total phosphorus
	Faecal coliforms
	Streptococci
Solid waste disposal	BOD in leachate
	Air quality
	Wind blown litter
Accelerated erosion	Gullying
	Turbidity
Compaction	Bare soil
	Exposed tree roots
Vegetation disturbance	Area disturbed
	Special change
Wildlife disturbance	Habitat change
	Changes in animal sightings
Noise	Decibels
Traffic congestion	Delay times
Litter	Visual assessment
Introduced plants and animals	Species change
	Feral animal populations
Perceived crowding	Number of contacts
	Number of campsites
	Visitor satisfaction

Source: Adapted from Turner (1994: 135).

Box 5.1 Trampling

Trampling is perhaps the most heavily-studied environmental impact of outdoor recreation, particularly in relatively undisturbed natural environments such as national parks (Liddle 1998; Buckley 2001; Cole 2003). The term is often interpreted rather broadly to include the physical and biological impacts of wheeled and mechanised vehicles and of horses' hooves, as well as hikers' feet. Such impacts may commonly include erosion or compaction of soil, and death and damage to plants and to soil- and litter-inhabiting animals. Sufficient research has been done to establish a number of general patterns, as follows. Off-road vehicles (e.g. four-wheel drives), mountain bikes and horses cause far greater trampling damage than hikers, and typically 15 times as much. Damage is commonly greater on slopes and while accelerating, braking and/or cornering; up to 10 times as much. Hikers with heavy boots cause more trampling damage than those wearing light footwear or none. Different soil and vegetation types differ considerably in their susceptibility to trampling damage.

In practice, trampling is rarely a single isolated event, but a series of successive events as areas along tracks or at campsites are trampled

continued...

Box 5.1 **continued**

repeatedly. The degree of damage may depend on weather conditions and time of year, and on whether trampling is concentrated during a short period or spread out over the seasons.

Given these complexities in ecological responses to trampling, comparisons between different environments are not straightforward. One crude but commonly-used indicator for such comparisons is the number of passes, by a medium-weight person wearing hiking boots, required to reduce plant cover to 50 per cent of its pre-trampling value. This measure has now been examined experimentally for a number of different ecosystems, following a protocol established by Cole (1995). Such information can be highly relevant for visitor management in national parks and wilderness areas. It can determine, for example: where to camp and hike with least impact, in lightly- and heavily-used areas respectively; whether a large group should walk single-file or fan out; and whether an eroded section of track will recover if closed to hikers.

In areas open to users with horses, mountain bikes or motorised vehicles, the range of trampling damage and the complexity of management increases accordingly. Horses' hooves exert ten times the pressure of a hiker's boot and a typical four-wheel drive tyre five times. Trails open to horse riding, in consequence, suffer far more trampling damage and soil erosion for a given number of users, than those open only to hikers. Similarly, use by off-road vehicles can rapidly cause orders of magnitude more trampling damage than use by hikers alone. For example, they cause 10–20 times as much soil erosion, and 5–30 times as much vegetation damage. Management of such users, commonly by confining them to designated formed tracks, is hence particularly critical to protect natural areas from trampling damage. A vehicle's weight and tyre pressure, footprint and tread, as well as the way it is driven, can make a considerable difference to the degree of damage it causes, but it will always be much more than a person on foot.

Direct physical damage to soils and vegetation is only one of the environmental impacts of hikers, horses or off-road vehicles. Boots, hooves and tyres can also spread weed seeds and fungal spores, sometimes with conservation consequences far outreaching the localised physical trampling damage. Weeds and pathogens can also be spread in horse dung and horse feed. Mechanised vehicles, in particular, can cause significant noise disturbance to a wide range of animal species, as well as other users.

Source: Buckley (2003: 510–11).

does the effect on sites vary with the type and intensity of recreational use, visitors to a site vary in their reaction to change and to the presence of others. These differences are discussed further in the consideration of social carrying capacity.

The relationship between recreation and the environment can be direct or indirect; immediate or delayed. Although biophysical impacts of recreation may be easier to detect than disturbance to the social setting, precise measurement can be just as elusive. Wall and Wright (1977) point out that it is almost impossible to reconstruct the environment minus the effects induced by recreation, or to establish a base level against which to measure change – the environment is dynamic, with or without direct human intervention. The problem then arises of disentangling the role of recreation from the role of nature. Spatial and temporal discontinuities between cause and effect can further obscure the environmental impact of outdoor recreation. Erosion in one location may result in deposition elsewhere, and considerable time may elapse before the full implications are apparent.

Moreover, the recreation–environment relationship is reciprocal. Visitors have an effect on the environment which, in turn, affects users. For satisfaction to be maintained, environmental values must not be used up faster than they are produced. The capability of the resource base and the recreation setting to continue to provide for recreational use, raises the concept of carrying capacity.

Recreation carrying capacity

Like many concepts in outdoor recreation management, the term 'carrying capacity' is bedevilled by varying and sometimes conflicting interpretations. The concept of 'recreation carrying capacity' derives from the practice, in livestock and wildlife management, of referring to the estimated number of animals an area of rangeland or a given habitat can support. In its initial application in outdoor recreation, the concept was seen as a technique to limit use to the maximum number of visitors a recreation resource or site could tolerate, without damage to the biophysical or social conditions.

Most definitions of recreation carrying capacity attempt to combine this notion of protection of the resource base from overuse with, simultaneously, the assurance of enjoyment and satisfaction for participants. Thus, in broad terms, recreation carrying capacity involves both the biophysical attributes of the environment as well as the attitudes and behaviour of users. An early definition of recreation carrying capacity by the Countryside Commission (1970: 2) reflected this duality: 'The level of recreation use an area can sustain without an unacceptable degree of deterioration of the character and quality of the resource or of the recreation experience'. Leaving aside the vagueness of its definition, the Commission went on to identify four separate types of recreation carrying capacity – physical capacity; economic capacity; ecological capacity; social carrying capacity.

Physical carrying capacity is concerned with the maximum number of people or equipment (e.g. boats or cars), which can be accommodated or handled comfortably and safely by a site. In many ways, it is a design concept, as when referring to the capacity of a car park, a spectator stand or a restaurant. In other circumstances, it could relate to safety limits (e.g. for ski slopes or specific numbers for participation in sports). As will be seen later, restriction of the physical capacity of ancillary facilities can be a useful management tool for applying indirect control over visitor numbers. It is easier to limit boating activity on a lake, for instance, by deliberately reducing the physical capacity of on-shore facilities such as access points, boat ramps and trailer parks, than by regulating boats on the water surface.

Economic carrying capacity relates to situations of multiple use of resources, where outdoor recreation is combined with some other enterprise. Economic compatibility might be a better description, because the term is concerned with getting the right mix of resource uses, so that benefits and costs of recreation do not reach a point at which interference with non-recreational activity becomes economically unacceptable from the management viewpoint. This could happen, for example, at a domestic water supply reservoir, where recreation is permitted, but where the consequent costs of supervision, or of water treatment, cannot be justified. Similarly, with a farm or a forest, the demands and depredations of recreationists may push the costs of efficient production too high for economic management.

The final two components of carrying capacity – ecological and social – are of greatest relevance to outdoor recreation management and receive the most emphasis in the ensuing discussion.

Ecological carrying capacity

Ecological carrying capacity (sometimes confusingly referred to, also, as physical, biophysical or environmental capacity) is concerned with the maximum level of recreational use, in terms of numbers and activities, that can be accommodated by an area or an ecosystem before an unacceptable or irreversible decline in ecological values occurs. This concept has been the subject of much controversy (e.g. see Butler 1996; Newsome *et al*. 2002; McCool and Lime 2001; McCool 2003), especially in subjective judgements of what is 'unacceptable', or 'irreversible decline'. *Any* use of an ecosystem will result in some change, but over-restrictive management could negate the recreation resource function altogether.

It could be argued that an area's ecological capacity is reached when further recreational use will impact the site beyond its ability to restore itself by *natural* means. However, such a viewpoint ignores the essential flexibility of the carrying capacity concept and the scope for, even the presumption of, sound management practices to stretch carrying capacity beyond so-called natural limits (Godin and Leonard 1977). Technological and financial considerations are obviously also relevant to the question of irreversibility.

According to Brotherton (1973: 6–7) any estimate of ecological carrying capacity must take account of:

• the nature of the plant and animal communities upon which the recreation activity impinges;
• the nature of the recreation activity and its distribution in space and time.

Plate 5.2   Hanauma Bay, Oahu, Hawaii. How much recreational use can a resource withstand? Recreational use is monitored via access at the Visitor Information Centre, where interpretive information is also given. However, even under these management conditions, the coral is threatened and the environment is overridden by people

Plate 5.3   Ecological impacts at Half Dome, Yosemite National Park, in the United States, are now reaching extreme levels

Plate 5.4   Even the most resistant surfaces can succumb to ecological and irreparable impacts. Climbing of Half Dome by an increasing number of recreationists is leading to noticeable erosion of the rock face

Several writers have warned against the misconception that capacity levels are somehow inherent or site-specific (Brotherton 1973; Ohmann 1974; Bury 1976; Manning *et al.* 1995). Bury is especially critical of the notion of a fixed, uniquely correct, recreation carrying capacity for a site, and suggests that the concept may be hypothetical in terms of managerial usefulness. He demonstrates the various components of biological and ecological carrying capacity, and the inter-relationships between them which inhibit generalisation. Bury gave the example of Big Bend National Park in southern Texas, which had about reached its *hydrologic* carrying capacity under 'existing standards of water use'. Lower standards, or elimination of certain forms of water use, could increase the hydro-logic carrying capacity of the park. Similarly, capacity for sewage and waste disposal is, to some degree, a function of whatever mandatory regulation or standard is adopted.

With all the contrasting physical characteristics possible within any particular site, development of strict measures of ecological carrying capacity, capable of general application, appears pointless and even counter-productive. As with the livestock grazing analogy, precise setting of the carrying capacity of a site at conservative levels, could be uneconomic and wasteful of legitimate recreational opportunities (Hammitt and Cole 1991). Over-restrictive limits could also reflect unrealistic management objectives in terms of maintaining the pristine integrity of a site. On the other hand, adopting levels that are too liberal for carrying capacity, may provide a short-term revenue windfall, but could lead to longer-term environ-mental degradation and ultimate closure of the facility.

Lindberg and McCool (1998) conclude that the carrying capacity framework may work well in situations where there is widespread consensus concerning management objectives and extensive data regarding use-impact relationships. The inference is that such situations are rare in outdoor recreation and tourism and therefore the carrying capacity concept is inadequate to deal with the complexity of tourism development. In any case, the setting of carrying capacities is only one component of an overall recreation management programme, and must be accompanied by systematic monitoring of environmental conditions and the flexibility to respond quickly to indications of stress. Moreover, generalisation is not feasible. Each recreation site has a range of carrying capacities, depending upon the nature of the recreation activity, characteristics of participants, background environmental conditions and the management objectives adopted. Lindberg and McCool (1998) recommend that various management-by-objectives be preferred, such as the Limits of Acceptable Change or Visitor Impact Management framework (see Chapter 6).

In particular, concern for ecological carrying capacity *alone* is inappropriate for outdoor recreation management. Recreation carrying capacity is evolving from its original emphasis on ecologically based use limits to an understanding of the complex relationships between environmental disturbance and participant satisfaction. The final test of whether a site measures up, rests with the minds of visitors, and their perceptions of, and reactions to, both the biophysical and social conditions (and perhaps the cultural context) of the recreation environment. Perception thus plays a key role in setting and managing the social carrying capacity of a recreation site.

Social carrying capacity

Outdoor recreation involves people, and the social environment in which recreation takes place has a good deal to do with the level of satisfaction experienced. Social carrying capacity (also referred to as perceptual, psychological or even behavioural capacity) relates primarily to visitors' perceptions of the presence (or absence) of others at the same time, and the effect of crowding (or in some cases, solitude) on their enjoyment and appreciation of the site. Social carrying capacity may be defined as the maximum level of recreational use, in terms of numbers and activities, above which there is a decline in the quality of the recreation experience, from the point of view of the recreation participant (Countryside Commission 1970).

The concept has much to do with tolerance levels and sensitivity to others, and, as such, is a personal, subjective notion linked to human psychological and behavioural characteristics. Put simply, social carrying capacity represents 'the number of people (a site) can absorb before the latest arrivals perceive the area to be "full" and seek satisfaction elsewhere' (Patmore 1973: 241). It is the least tangible aspect of recreation carrying capacity and the most difficult to measure. Not only does it vary between individuals, but also for the same person at different times and in different situations.

Bury (1976) suggests that visitor satisfaction is linked to the notion of 'territory' and 'living space', so that social carrying capacity is derived from the number and types of encounters with other humans in the recreation area. Manning *et al.* (1995) support the contention that trail and camp encounters, for example, are key variables in determining the quality of a wilderness experience. Bury (1976) makes the interesting point that it is not merely the *actual* number of times an individual meets other recreationists, but the *potential* number and type of such encounters which are important.

> ... recreation satisfactions may be impaired even before any encounters occur if the number and density of people *seem* higher than the visitor would prefer, or if the potential encounters seem likely to be more intense, or closer, than the visitor wishes.
>
> The condition may also be reversed – as when teenagers go to a beach to see, be seen, and interact with others. In this case, the desire of the visitor is for high densities of human use.
>
> (Bury 1976: 24)

The link between social carrying capacity and the type of recreational experience is illustrated graphically in Figure 5.3. The satisfaction derived from a wilderness experience is reduced, even at very low levels of use and social interaction – 'two's company and three's a crowd', indeed. The canoeists in Lucas's (1964) study of Boundary Waters Wilderness on the US–Canadian border, had no wish to see fellow humans. On the other hand, being the *sole* visitor to, say, Disneyland, would hardly be an enjoyable experience. In fact, the satisfaction gained from such essentially gregarious occasions increases with the level of use, at least until the

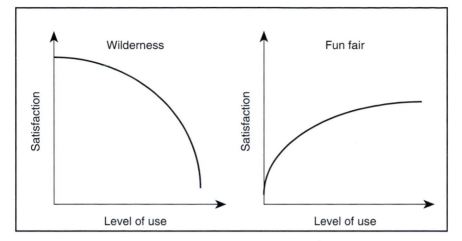

Figure 5.3 The effect of crowding on recreational satisfaction

Source: Adapted from Brotherton (1973: 9).

point where crowding and congestion begin to irritate. It could well be the waiting and the queuing which then become exasperating, rather than the numbers of people in attendance.

Certainly, numbers of people alone do not cause visitor dissatisfaction. Reaction to crowding is variable and, to some extent, self-regulating. This makes any measurement of social carrying capacity just that much more difficult, because the non-gregarious individuals may be absent or may have redistributed themselves in space and time so as to avoid peaks in recreation use.

How a person reacts to the presence of others is influenced by underlying psychological factors such as personal values, goals, attitudes, expectations and motivations. The level of satisfaction is also affected by other events or conditions incidental to the recreation experience, for example, vehicle troubles or traffic problems on the trip, illness, or even the weather.

Social circumstances, too, help shape people's perception of a particular situation, and the way they receive and interpret information about a recreation environment. Human perception is, in part, a function of the psychological factors noted above, but is also a result of demographic characteristics and the socioeconomic backgrounds of participants. Once again, it is not so much the size of the crowd, but similarities or contrasts in social status, behaviour or composition of the group which become a source of frustration and conflict (see the discussion on incompatibility and conflict in Chapter 2).

Perception of the quality of a recreation experience also reflects the charac-teristics of the physical environment or situation in which the activity takes place. Site features such as location, size, configuration, terrain, vegetation, proximity to compatible activities and the type of support facilities, can all influence satisfaction levels. In particular, they may affect the capacity of the landscape to 'absorb' users. It is the actual awareness of others which is crucial to social carrying capacity, so that any objective measure of the density of use may not be a true reflection of crowding. Out of sight *is* out of mind, and if others present are not visible because of certain site characteristics, social carrying capacity may be considerably enlarged. Bury (1976) points out that carrying capacity generally increases with increasing density of vegetative cover at or between sites (e.g. between camping areas). If visitors cannot see or hear one another, the area *seems* less crowded. Wilderness above the timberline, for example in heathland, has a smaller social carrying capacity than wilderness at lower altitudes, where in forest valleys participants are screened by both topography and vegetation. This associated notion of 'landscape absorption' has obvious implications for management of existing recreation sites and design of proposed sites and facilities.

In this discussion of recreation carrying capacity, most emphasis has been placed on ecological and social aspects. Consideration of these two components, separately, and sequentially, does not imply any order of importance, nor should it obscure the complex relationships between them. Both resources *and* people must be taken into account when considering carrying capacity. It is important for decision-makers to be aware of the dynamic, multidimensional nature of the capacity concept, in order to adopt a balanced approach to managerial responsibilities.

This comprehensive viewpoint was emphasised in a comment by Lindsay (1980: 216), who conceptualised outdoor recreation carrying capacity as '… a function of quantity of the recreation resource, tolerance of the site to use, number of users, user type, design and management of the site and the attitude and behaviour of the users and managers'.

Thus, a competent recreation management programme would incorporate both environmental considerations *and* human needs and desires. It remains a matter of judgement as to when degradation of the resource base or deterioration in the quality of the recreation experience reach the point where action is called for. The difficult task of determining carrying capacities is matched by that of deciding when and where they are, or might be, exceeded, and, ultimately, the choice of remedial and pro-active management procedures.

Since the first rigorous application of carrying capacity to the management of parks and recreation areas in the 1960s, the concept has expanded from its original focus on resource impacts to include emphasis on the social setting and the quality of the recreation experience. More attention, too, is being given to the carrying capacity of the developed portions of parks, and to private lands, rather than exclusively to back country and wilderness settings (Manning *et al.* 1995). Despite its shortcomings, the concept of recreation carrying capacity continues to evolve as an important component of a more comprehensive and holistic approach to environmental management. When viewed in its proper perspective, recreation carrying capacity remains useful as 'an organizational framework for making rational judgements about appropriate conditions and public use of parks and recreation areas' (Manning *et al.* 1995: 337).

Recreation carrying capacity in review

Widespread application of the concept of recreation carrying capacity led to growing scrutiny of its effectiveness as a management technique. As noted above, the concept derived from use in the pastoral industries and wildlife management. Whereas the fixing of carrying capacities for recreation sites may appear intuitively simple, managing recreation use for people differs considerably from determining the forage requirements of cattle and food supplies for wildlife. Even there, carrying capacities are not 'set in stone' but vary with climatic and vegetative conditions and other considerations. The same is true of recreation. The relationship between use and impact, typically, is not direct, and is affected by the type of recreation activity; its timing and distribution on the one hand, and the attributes of the environment where use occurs, on the other.

Although national park agencies worldwide continue to fix and observe carrying capacity levels for protected areas in their care, questions have been raised increasingly concerning the appropriateness of applying a supposedly simple model of range land management to the recreational needs and desires of people. Implementing a recreation management strategy that identifies acceptable and desirable sociocultural and biophysical conditions, and standards of quality which reflect and sustain these, will be more effective than relying on specific numerical carrying capacities (McCool and Lime 2001). Recent years have seen the adoption of a number of alternative approaches to monitor and manage the impacts of outdoor recreation on the environment. The concern is determining appropriate levels of change and development of policies, plans, programmes and indicators to manage recreation environments (see Chapter 6). McCool's (2003) view was:

> The change in the character of the question driving concern about impacts from recreational use reflects a different paradigm defining recreational impact issues. It focuses not on one of the principal input variables – use level – but rather on the output or results of management. This conceptualization of use-impact relationships renders the theory of carrying capacity, as applied to recreation, out-of-date and of little practical utility to managers or theoretical value to scientists ... In summary, the carrying capacity theory has probably outlived its usefulness.
>
> (McCool 2003: 44)

McCool's statement is perhaps rather harsh and runs the risk of dismissing a historically useful concept for thinking about both environmental capacity and relationships. Brown *et al.*'s (1998: 293) response to a comment on their work applying the concept of ecological carrying capacity in the Maldives and Nepal by Lindberg and McCool (1998) offers a more cautious view:

> carrying capacity may be conceptually controversial, is problematic in its application and is of limited use in devising policy guidelines ... Our conclusions remain that whilst carrying capacity and open access were useful as concepts in analysing the environmental and social impacts of tourism,

defining a carrying capacity is difficult. However, we are circumspect on the use and application of the carrying capacity concept, and we argue that whilst ecological and tourism capacity may be of use, that use is limited, and indeed social and cultural factors may be more important in determining the extent and impacts of tourism development.

If applied in a dynamic way, carrying capacity invites us to think about how different physical and human conditions can affect the ability of an environment to sustain different uses. Indeed, the concept of carrying capacity has more than one dimension (e.g. ecological, social, cultural), and these dimensions are now more recognisably interrelated, particularly in the application of the concept in recent research. Brown *et al.*'s comment acknowledges this situation.

The emphasis in understanding the relationships between recreation and the environment has changed from visitor levels as such, to the inevitability of change accompanying recreational use; from there to managing the amount and type of change resulting from human activity, on top of that occurring in nature. This approach reflects the view that variation in human behaviour and the nature of resource management is probably as influential as the actual number of visitors, in bringing about change to that environment. Moreover, it does not follow that change equates with degradation, or that impact is the same as damage. Some forms of impact can be tolerated by users and managers alike. Specific judgements are required on how much change or impact can be accepted before it becomes 'damage' and requires intervention by management (Turner 1987).

Summary

The relationship between outdoor recreation and the environment in which it takes place is complex and as we will see, there are many ways of assessing, monitoring, evaluating and managing sites, even a specific site. Clearly, the quality of the recreation experience will be affected by the recreation setting, whether natural or created, and the environment, in turn, will reflect the presence of recreationists and their activities. Whereas carrying capacity in its various forms remains an important aspect of managing the recreation–environment relationship, it is now realised that the concept must be applied with care. Generalisation is not possible, and the adoption of arbitrary limits to use ignores both the biophysical attributes of a site and contrasts in the nature and scale of recreation activity. A more positive approach concedes that some change is inevitable with recreation use; the challenge is to monitor conditions and keep change within acceptable limits.

In addition, reference to carrying capacities is only one component of an overall recreation management programme. Systematic monitoring and feedback on environmental conditions, and a rapid and flexible response to indications of stress are essential elements in a more comprehensive and holistic approach to managing the reciprocal relationship between outdoor recreation and the environment. Hence, we now turn to recreation and resource management planning processes and frameworks adopted in countries such as Canada, New Zealand, Australia, the UK and the US.

Guide to further reading

- Recreation–environment relationships and impacts of outdoor recreation receive thorough treatment in Wall and Wright (1977); Hammitt and Cole (1987, 1998); Liddle (1997); Leung and Marion (2000); Buckley (2005).
- Recreation carrying capacity, and recent qualifications and trends, are canvassed in Hammitt and Cole (1987, 1998); Kruss *et al.* (1990); McCool (1990a, 2003); Manning *et al.* (1995); Butler (1996); Liddle (1997); Lindberg and McCool (1998); Brown *et al.* (1998); McCool and Lime (2001).

Review questions

1   How might seasonal conditions affect carrying capacities? Explain the link between human 'action space' and social carrying capacity. What basic variables influence ecological carrying capacity? Why is social carrying capacity so difficult to measure? What resource attributes have a bearing on recreational impact? What is the significance of visitor distribution on the impact of outdoor recreation?
2   How might cultural, ethnic or racial attributes affect recreation carrying capacity?
3   How can recreation management objectives affect the setting of carrying capacities?
4   What are some of the risks and advantages of bringing user preferences into the recreation management process?

6 Recreation resource management
Approaches, frameworks, models

Resource management has been defined by O'Riordan (1971: 19) as 'a process of decision-making whereby resources are allocated over space and time according to the needs, aspirations, and desires of man within the framework of his technological inventiveness, his political and social institutions, and his legal and administrative arrangements'. O'Riordan argues that the emphasis in resource management should be 'upon flexibility and the minimisation of long-term environmental catastrophes, while maximising net social welfare over time', and that resource management is 'becoming increasingly concerned with the protection and enhancement of environmental quality and the establishment of new guidelines for the public use of such common property resources as air, water, and the landscape'. Thus, O'Riordan's definition and related perspectives of resource management are closely aligned with the concept and application of carrying capacity and the Recreation Opportunity Spectrum, as well as the models and frameworks presented in this chapter.

Of all the resource management conflicts in the countryside, recreation offers perhaps the greatest opportunities for multi-functional resource use. Many recreational activities are compatible with other resource uses, while some are not and may have to be specially catered for. Resource use conflicts are inevitable, but vary in their extent and nature. Thus, innovative recreation planning approaches may be called for.

Recreational resource management is a process which requires strategic planning for visitor management, generally, and for site selection, design, use, monitoring and evaluation, specifically. This chapter gives an overview of important recreation management issues and activities. It begins with an introduction to the recreation resource management process and an early model of that process, examines several approaches to managing resources for recreational use and concludes with a critique of some of the most recently applied models employed by agencies in the US, UK and Australia.

The recreation resource management process

The primary aim of outdoor recreation management, presumably, is to bring together supply and demand to attempt to equate resource adequacy with human recreational needs and desires. In so doing, the manager must obviously have regard for the character and quality of the resource base, ensuring that capacity is not exceeded, and that environmental degradation is minimised. At the same time, the managerial role extends to consideration of visitor enjoyment and satisfaction. Action must be taken to reduce conflict and to maximise the quality of the recreational experience. These dual responsibilities hold, whether for the economic success of commercial enterprises, or for the protection of public investment in parks and recreation areas.

A first step in the management process is the establishment of management objectives. From these will flow the selection of specific management benchmarks, approaches and procedures. Modification of the system may well follow implementation of the management approaches decided upon. An important element in this phase is evaluation of the system, based on monitoring of its operation by managers and feedback from users.

An earlier model of the recreation management process is presented in Figure 6.1 and described in detail below. A set of objectives is articulated, first, with reference to the capabilities of the resource base. Information on resources should indicate which activities are physically possible, as well as some of the resource constraints on recreation opportunities. Institutional, economic and other constraints, as well as personal and other circumstances, also have obvious implications for management, and set limits on the range of recreation opportunities possible.

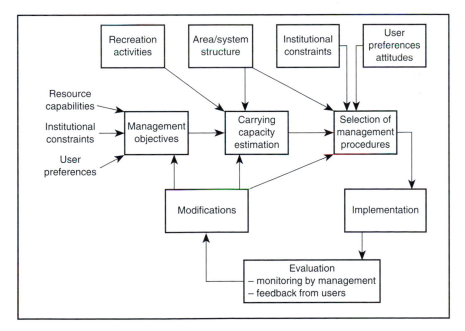

Figure 6.1 The recreation management process

Source: Adapted from Brown (1977: 194).

Legal restrictions and standards, administrative policies and guidelines, and budgetary and personnel considerations, can all influence the selection of realistic management objectives. In some countries and regions, problems for management stem from overlapping political and/or agency jurisdictions, each of which may have different interpretations of what is appropriate recreational use.

Ideally, management objectives should reflect user preferences if they are to receive support at the implementation stage. A good example is the 'battle' fought some years ago in Yosemite National Park between hundreds of young campers and a combined force of park rangers and state police. The dispute centred on what were seen as restrictive zoning regulations and excessive fees as demonstrators felt they were discriminated against by the park administration. A significant outcome of the confrontation was the admission that park planning authorities were out of touch with public opinion on many matters relating to national parks. As a result, provision was made for a much greater degree of public involvement in the park planning process for Yosemite and other sites (Mercer 1980a).

More and more people are demanding the right to participate in decision-making, and public involvement is increasingly seen as a necessary and desirable input to management, although it is sometimes costly (due to time delays and consultation, and even court costs stemming from public challenges). This does not mean that the process must be totally democratic, in the sense that the user population 'calls the tune'. Identifying the population to be consulted is always a problem and, in any case, it would be foolish to disregard entirely the expertise of management in reaching a decision. As is often the case, compromise in the form of 'guided democracy' is probably the best approach. Whatever the case, public participation is a means of making decision-making processes more meaningful, successful and inclusive, and a means of enhancing the acceptability and longevity of decisions. Agencies and citizens develop more trusting and workable relationships. Consultation can be undertaken at any time (though there is usually a prescribed schedule in government decision-making), should involve affected individuals and agencies, and may take many forms (see Table 6.1) (for more detailed discussions

Table 6.1 Public participation formats

1	One-on-one meetings/interviews
2	Media notices
3	Competitions
4	Letters inviting comment – to residents and/or organisations
5	Public exhibitions (static or mobile) with opportunity for comment
6	Public meetings
7	Production/distribution of printed and/or video material
8	Focus groups
9	Attendance at group meetings (e.g. clubs)
10	Establishment of working parties with outside membership
11	Postal questionnaire survey of organisations
12	Community surveys
13	Nominal group technique

Source: Veal 2002: 109.

see Lawrence and Daniels 1996; Selin and Chavez 1995; Veal 2002: 107–9; Schindler *et al*. 1999).

User preferences regarding resource attributes, social characteristics of the recreation environment and preferred management approaches should certainly be canvassed, but interpreted in the light of managerial experience of what is desirable and possible. However, it is important that the process of public involvement be seen as more than just good public relations. Whereas user preferences are only one of many inputs to the formulation of management objectives, there should be clear evidence that they have been considered and integrated into the final decision. That said, it must be conceded that many factors inhibit and distort the clear expression and articulation of user preferences. Thus, a range of approaches may be needed to encourage participation, and combat apathy and indifference (Jubenville 1978; also see Veal 2002).

Even when success has apparently been achieved in provoking a constructive response from communities affected by management proposals, there are risks involved in the process of public participation. Care must be taken to ensure that it is not only vocal pressure groups and politically active professional lobbyists who receive attention. The 'squeaky wheel syndrome' may not truly reflect majority preference. Conversely, the imposition of elitist managerial attitudes on users, could lead to management objectives which are unrealistic and unacceptable. A good measure of public participation is essential, but a balance should be struck between uninformed reactions, perhaps merely reflecting trends and fashions, and a more objective appraisal by supposedly detached management experts.

The second stage in the outdoor recreation management process (i.e. setting appropriate carrying capacities or limits of acceptable change in a resource, consistent with management objectives) should be related to the structure of the management area or system. As was noted in the previous chapter, limits on ecological and social carrying capacity are, in part, a function of the natural features of the site and the built facilities and amenities, and the recreation activities to be accommodated. It is also worth reiterating that carrying capacities, once set, are not inflexible but remain open to manipulation by management; hence, the feedback loop in the chart from management procedures.

Several approaches to managing recreation sites are discussed below. In Brown's (1977) view of the management process, some of the factors which contribute to the formulation of management objectives also influence the choice of specific tactics or tools to achieve those objectives. It is not always a case of what managers may see as necessary, or desire, so much as what they are able to do.

Once again, institutional directives can act as constraints in the selection of management procedures. Obviously, also, the characteristics of the recreation area or site set limits on which combination of approaches is likely to be successful. Linear sites (e.g. walking tracks) call for a different approach from those with more regular dimensions, and remote forested areas with rugged terrain may not need such strict regulation as would more open, accessible sites with fewer natural deterrents to, or restrictions on, recreational use. Brown (1977) also stresses the value of knowing user preferences and attitudes with regard to the choice and effectiveness of specific management tools. Some people may respond kindly to appeal and inducements; others may react more favourably to direct regulation.

A desirable feature of effective management is flexibility, so that, if in the implementation of the management procedures, selected deficiencies are detected, an adaptive management process should allow for modifications. The need for adjustments may be discerned by management or become apparent in feedback from users. Subsequently, modifications can be made at various points in the system, and the management process becomes self-regulating.

Once objectives have been formulated and estimates made of carrying capacities, the primary task of outdoor recreation management emerges – that of selection, implementation and modification of on-site management procedures. However, effective management begins at an even earlier stage, with proper site selection, planning and design. If these preliminary considerations receive adequate attention, the most appropriate sites and the more resilient components of the environment (in terms of low vulnerability and high tolerance to visitor use) will already have been set aside for recreation, and developed so as to minimise management problems.

Recreation site selection

This is a most important step to which much of the ease or difficulty in subsequent operations can be attributed. Fundamental considerations are ownership, boundary definitions, user access and the suitability of the resource base for the recreational activities envisaged. This should already have been established by application of the site evaluation techniques described in earlier chapters. However, the fact that a site apparently meets the basic suitability criteria laid down, may conceal shortcomings in specific resource attributes which will prove costly to management in later use. Where more than one site meets basic requirements, a detailed examination of site characteristics is needed to determine priorities for development. This examination should focus on the features of competing sites which affect management's task of producing and sustaining worthwhile recreation values (McCosh 1973; Jubenville 1976).

Assuming that questions of location and convenience of access to potential users have been satisfied, many physical features of the site itself can impinge upon the quality of the recreational experience and, hence, the role of management. For instance, both the size of an area and its configuration are important. It is almost always helpful to have an area somewhat larger than required to allow rotation of use and provision of a buffer zone in order to segregate the site from adjoining developments. In most cases, too, a long narrow site is less efficient in terms of internal arrangement of attractions, facilities and services, than one of more regular configuration (see Chapter 9 for a more detailed discussion of buffer zones).

The nature of the terrain, the degree and direction of slopes, rock types and presence of rock outcrops, soil stability and compactibility, drainage and susceptibility to flooding, and availability of construction materials, can all have engineering implications for site development and maintenance. So, too, can the size, variety and density of the vegetation, and the extent and location of open space.

The importance to the recreation landscape of waterbodies of the right quantity, quality and dimensions has to be considered in site selection. Water is needed for drinking, sanitation and possibly irrigation, so that sources and suitability of water supply need to be determined, along with the costs of pumping, treatment, storage and disposal. Adequate estimates of water quality also require knowledge of groundwater, and of climate and weather patterns (e.g. precipitation, evaporation and snow cover) over an extended period. Other climatic factors which may have a bearing on decisions concerning a recreation site, include aspect, exposure to winds and seasonal conditions (e.g. length of shadows in winter and the incidence of high Spring pollen counts).

Finally, a site could have certain negative or undesirable features which could influence selection. For instance, Jubenville (1976) suggests that a hazard survey be carried out for each potential site, to identify possible hazardous conditions such as avalanches, falling trees, precipices, dangerous waters, poisonous plants and insects, and dangerous animals. Other annoyances, such as noise, dust, fumes, and aquatic weeds and algae can present problems for management; problems which, if foreseen, might be avoided, ameliorated or dealt with in a prepared and more systematic manner. Hamilton-Smith (2003) presented a commonly used set of criteria in the assessment of relatively undisturbed sites (see Table 6.2).

McCosh (1973) stresses the value of prior study and sound judgement in recreation site selection. Poorly chosen sites will become inefficient areas with problems that cannot easily be solved. Of course, some of the negative site characteristics noted may be offset by good planning and design. Use of design in this way as a compensatory device is fine, providing the cost is not excessive. However, it is preferable to implement design measures which are complementary to, and reinforce, the natural features of the site. The idea is:

> ... to utilize the features of the landscape to enhance recreational experience, minimize site maintenance and maintain natural aesthetics. Although fitting the development to the natural lay of the land may be more expensive and require more attention to detail, the resulting site should be more attractive, more able to handle large visitor-use loads and less expensive to maintain.
>
> (Jubenville 1976: 155)

Recreation site planning and design

Effective planning and design of recreation sites involve a combination of attention to objectives, functions and aesthetics. Albert Rutledge (1971) put forward a set of design principles (or 'umbrella principles'), including:

- *design with purpose* – so that the appropriate relationships are established between the various parts of the recreation complex (i.e. natural elements, use areas, structures, people, animals and forces of nature);

Table 6.2 A commonly used pattern in the assessment of relatively undisturbed sites

Geology and land forms	Character of rock, stratigraphy and structure Hydrology Diversity of landforms and landscape
Climatic character	Annual cycles of climate, including temperature, humidity, wind and rainfall patterns Micro-climates occurring within the site
Soils and other surface deposits	Origin and character of soils Existence of rock debris either within the soil or in breakdown heaps Stratigraphy; potential archaeological and palaeontological interest; potential for dating studies
Biodiversity	Inventory of flora and fauna (NB Although actual determination of species may be difficult, this assessment should include invertebrate species) Existence of threatened or endangered species Any special ecological characteristics worthy of attention including unusual or relict ecosystems
Human occupation and land use	Current situation, including modifications of natural environment The historical record Aspects of cultural or historic importance Past or present environmental impacts
Condition and integrity of site	Fallen trees; extent of weeds; water quality
Potential utilisation	For primary intended use Detail multiple intended functions Other secondary (perhaps unintended) functions Probably environmental impacts
Implications for future planning and utilisation	Including advance planning to moderate impacts Provision for future monitoring of change and assessment of impacts for management purposes

Source: Hamilton-Smith (2003: 461).

- *design for people* – rather than to meet some rigid standards, or the impersonal demands of machines, equipment and administrative convenience. More attention to the 'why' of design would go a long way towards structuring outdoor areas to satisfy human behavioural needs;
- *design for both functions and aesthetics* – striking a balance of dollar values and human values with the achievement of efficiency, interwoven with the generation of a satisfying sensory experience.

Rutledge was writing of park design, but the principles and detailed procedures he describes have application in many other situations. For example, Lime (1974) has demonstrated the relevance of good location and design to the effective functioning of campgrounds. Leonard *et al.* (1981, in Hammitt and Cole 1998: 305) developed a linear layout of overnight facilities for densely forested backcountry areas (see Figure 6.2). The idea is to concentrate traffic flows and limit impacts, whilst separating and dispersing visitors among attractive and thoughtfully located sites. Of course, shortcuts between facilities may develop, and if so it is suggested that these be incorporated into the trail system (McEwen and Tocher 1976, in Hammitt and Cole 1998: 305).

While it is probably true to say that aesthetics are often only considered after functional aspects have been satisfied, the two should go together in the design process, because attention to aesthetics can actually strengthen functional efficiency. In practice, functional elements of design tend to receive emphasis because of their more tangible nature – 'it works or it doesn't'. Aesthetics, on the other hand, are like beauty – very much in the eye of the beholder!

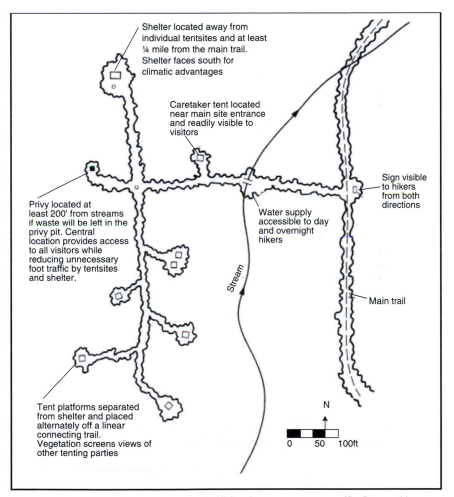

Figure 6.2 A linear layout of overnight facilities that concentrates traffic flow and impacts but separates and disperses visitor groups

Source: Leonard *et al.* (1981, in Hammitt and Cole 1998: 305).

A further source of confusion can arise from overlap between the planning and design phases. In general terms, recreation site planning could be said to be concerned with the broad arrangement of site features, support facilities and circulation patterns necessary for the type of recreation envisaged. Design is related to micro-location and the moulding and fitting of the plans to specific topographic and landscape features of the site, while maintaining the desired positions of the facilities and circulation patterns (Jubenville 1976). For convenience, the criteria to be observed during both phases are considered together in the following discussion.

In the first place, planning and design of the recreation site should conform to known user preferences for given environmental conditions or situations (Christiansen 1977). Merely providing a picnic site is not enough. Service requirements, supporting facilities, equipment and site refinements should reflect the style and characteristics of participants. They should also be located to fit in with normal behaviour patterns, to minimise conflict and confusion, and to facilitate movement within the site.

A basic functional criterion of planning and design is that the recreation site and associated developments satisfy technical requirements (i.e. that they are useable in the sense of meeting standards of size, spacing and quantities). Operational needs and conditions are also important, and apart from meeting health and safety regulations, site developments should provide for the comfort and convenience of users. Rutledge (1971) illustrates the relevance of orientation to natural forces in the layout of recreation sites (e.g. the elevation and path of the sun's rays and the direction of prevailing winds), and stresses a common-sense approach to avoid unnecessary costs, and to provide for ease of supervision.

Recreational use of a site inevitably involves movement. The circulation system adopted can have a pronounced effect on efficiency of use, safety, satisfaction levels and supervision of visitor behaviour. Rutledge points out that the aim should be to get people where they want to go readily, and in doing so, not interfere with other activities. Therefore, the tasks are to anticipate flows, eliminate obstacles and confusion, and to provide unobstructed, well-defined, logical routes. Proper circulation planning and design can become an arm of recreation site management, not only in protecting the natural environment and visitors, but in promoting and facilitating desirable patterns of recreational use.

Sound site planning and design can minimise the task of supervision and the need for restrictive control measures over visitor behaviour. Public welfare should always be a concern, and if provision for visitor health and safety is built into a recreation site, many hazardous situations can be prevented. By definition, accidents are unplanned, but planning and design can go a long way towards eliminating the factors likely to generate emergencies. On-site control of vandalism and other forms of depreciative behaviour is also an important facet of visitor management.

Maintaining law and order at public recreation sites is also a serious problem for management. Depreciative behaviour can reduce or destroy the resource base and facilities, and interfere with the experience and satisfaction of other participants. Vandalism, acts of nuisance, violation of rules and crime, unfortunately, must all be anticipated. The monetary impact is staggering. In the US, the total yearly loss from vandalism was estimated in the 1970s at US$4 billion (Clark 1976). In the City of Boise, Idaho (US), vandalism in parks alone in 2003 cost the city about US$69,000. Almost half the incidents were recorded in five areas and a quarter of the incidents occurred in skate parks (http://www.cityofboise.org/parks/caring/index.aspx?id=vandalism).

The problem can at least be contained by prior attention to planning and design. Weinmayer (1973) believes that proper design can reduce vandalism by 90 per cent, and some observers suggest that much anti-social behaviour actually represents a protest against poor design and management of parks and other recreation sites (Gold 1974). So-called 'vandalism by design' is blamed for providing the opportunity for misuse by equipping recreation sites with objects, facilities and materials which invite disrespect and, ultimately, destruction. The inference is that opportunities for vandalism and other forms of undesirable behaviour can be removed or reduced at the planning and design stage. It is possible, of course, to attempt to devise structures which are vandalproof and virtually indestructible. It is also possible to prepare surfaces to which graffiti paint cannot adhere. It is preferable, and more positive, to provide sturdy, but attractive recreation environments which will be valued and protected by the users themselves. Site developments should be designed for easy maintenance and quick restoration if damaged. Rutledge (1971) suggests that thought be given to the clustering of potentially vandal-prone features, the opening-up of sites to external inspection, more adequate lighting, and the encouragement of higher levels of use, all as deterrents to anti-social acts. Communities can be encouraged to participate in the care and maintenance of local sites and foster land stewardship.

Recreation sites which are properly selected and located, and which have had the benefit of thoughtful planning and design, should almost manage themselves; certainly, the task of management should be made much easier. Unfortunately, it is probably more often the case that managers inherit a poorly selected site, where little attention has been given to adequate development, planning or design. Subsequent problems emerge, either because of overuse, deliberate misuse (above), or unintentional damage through ignorance and inappropriate use. Careful management of resources and visitors then becomes an ongoing concern.

Recreation resource management

Jubenville (1978) saw the managerial role in outdoor recreation as incorporating *resource management*, concerned with the reciprocal relationships between the recreation landscape and the visitor; *visitor management*, enhancing the social environment in order to maximise the recreation experience; and *service management*, involving the provision of necessary and desirable services so that the user can enjoy both the social and resource environment. Whereas each of these managerial roles is an important component of the overall recreation system, Jubenville considered visitor management to be fundamental, since it is the visitor who expresses demand for recreational experiences which require the other two elements. In the ensuing comments, provision of services will be regarded as a complementary, but ancillary, aspect of outdoor recreation management, and discussion will concentrate on resources and people. That said, it will soon become apparent that there is much scope for overlap between the two.

Recreation resource management implies close monitoring of the recreation site, to chart the rate, direction and character of change. It is vital that negative changes be detected early so that appropriate and positive management procedures be taken before site degradation proceeds to the point where the recreational environment becomes a source of dissatisfaction to visitors. Without a systematic monitoring and evaluation programme, management has no basis for comparison to determine change. Indeed, even before environmental deterioration or visitor dissatisfaction become evident, resource management procedures must be monitored and evaluated on a regular basis. 'Management's role, in general, is not to *halt* change within wildland [or indeed other] areas, but to manage for acceptable levels of environmental change' (Hammitt and Cole 1998: 13).

Resource management involves manipulation of elements of the resource base in order to maintain, enhance or even re-create satisfying opportunity settings for various recreational pursuits. In selecting the most appropriate course of action, the recreation site manager needs to balance concern for the resource base against other concerns, such as commercial considerations and the costs involved in loss of patronage. Leaving aside Jubenville's first suggestion – 'cut out and get out' – which is hardly a positive approach to management, other choices, which could be implemented individually or simultaneously, include:

- Select sites that are durable (resilient and resistant) in diverse circumstances.
- Site closure and rejuvenation through natural processes or cultural treatments. Site closure will certainly minimise recovery time and inconvenience, and may be justified for heavily deteriorated sites, especially where alternative opportunities are available.
- Rest and rotation of sites, or perhaps areas within a site, so that some recreation opportunities are always available.
- Leave open and culturally treat the site (i.e. keeping the site operating while implementing rehabilitation measures such as site hardening). If possible, this is clearly the ideal solution, but it can only succeed if treatment begins before site deterioration is well advanced.
- Manage or perhaps regulate the uses to which a site is put.
- Disperse use to avoid concentration of use and impacts (Jubenville 1978; Hammitt 1990).

An extensive range of strategies and tactics for wilderness management was provided by Cole *et al.* (1987) (see Table 6.3). Apart from limiting access and use or modifying user expectations, behaviour and experiences, resource management procedures primarily involve technical and engineering-type modifications to the site and its surrounds, landscaping techniques, and educational strategies. Examples of landscaping modifications include various soil treatments and ground cover

Plate 6.1   Site hardening by use of metal stairs in the Blue Mountains National Park, Australia, also provides easier recreational access to remoter areas

improvements including irrigation; use of fertilisers; re-seeding; replacing or conversion to hardier and more resilient species; and judicious thinning of vegetation and removal of noxious species. These measures are aimed at increasing the durability of the biotic community, as well as inducing its recovery.

Table 6.3 Strategies and tactics for wilderness management

I	*Reduce use of the entire wilderness*

1	Limit number of visitors in the entire wilderness
2	Limit length of stay in the entire wilderness
3	Encourage use of other areas
4	Require certain skills and/or equipment
5	Charge a flat visitor fee
6	Make access more difficult throughout the entire wilderness

II	*Reduce use of problem areas*

7	Inform potential visitors of the disadvantages of problem areas and/or advantages of alternative areas
8	Discourage or prohibit use of problem areas
9	Limit number of visitors in problem areas
10	Encourage or require a length-of-stay limit in problem areas
11	Make access to problem areas more difficult and/or improve access to alternative areas
12	Eliminate facilities or attractions in problem areas and/or improve facilities or attractions in alternative areas
13	Encourage off-trail travel
14	Establish differential skill and/or equipment requirements
15	Charge differential visitor fees

III	*Modify the location of use within problem areas*

16	Discourage or prohibit camping and/or stock use on certain campsites and/or locations
17	Encourage or permit camping and/or stock use only on certain campsites and/or locations
18	Locate facilities on durable sites
19	Concentrate use on sites through facility design and/or information
20	Discourage or prohibit off-trail travel
21	Segregate different types of visitors

IV	*Modify the timing of use*

22	Encourage use outside of peak use periods
23	Discourage or prohibit use when impact potential is high
24	Charge fees during periods of high use and/or high-impact potential

V	*Modify type of use and visitor behaviour*

25	Discourage or prohibit particularly damaging practices and/or equipment
26	Encourage or require certain behaviour, skills, and/or equipment
27	Teach a wilderness ethic
28	Encourage or require a party size and/or stock limit
29	Discourage or prohibit stock
30	Discourage or prohibit pets
31	Discourage or prohibit overnight use

VI	*Modify visitor expectations*

32	Inform visitors about appropriate wilderness uses
33	Inform visitors about conditions they may encounter in the wilderness

VII	Increase the resistance of the resource
34	Shield the site from impact
35	Strengthen the site

VIII	Maintain or rehabilitate the resource
36	Remove problems
37	Maintain or rehabilitate impacted locations

Source: Cole *et al.* (1987: 3).

On-site patterns of recreational use can be influenced in various ways, including channelling the movements of visitors along selected paths (e.g. planting very dense and/or thorny bushes), or discouraging recreationists from entering a particular area (e.g. by fencing or the erection of some barrier designed as a 'people-sifter') (Seabrooke and Miles 1993). The effect may be discriminatory, but obstacles such as ditches and stiles, which prove a deterrent to some classes of visitor, are not insurmountable to all.

Vehicular traffic can be regulated according to mode and route, and many heavily-used sites no longer permit use of private vehicles; shuttle buses and other forms of communal transport are becoming more common in national parks (see Chapter 10). One-way traffic can be made mandatory, especially where parallel routeways exist, and separate trails can be designated for different classes of movement (e.g. skiers and snow mobilers). User fees can be applied.

Such action may be complemented by landscaping in order to enhance carrying capacities. This could involve the hardening or surfacing of intensively used areas such as viewing points and heavily used sections of trails; rotation of site furniture (barbecues or picnic tables) and movable facilities such as kiosks and shelters; rotation of entrances, trails and campsites; and provision of more effective waste disposal systems. As noted earlier, social carrying capacity can also be stretched. This can be achieved by many different management actions, including imaginative plantings to create more 'edges' or borders, or by breaking up the site with artificial mounds and buffers to boost the capacity of the landscape to 'absorb' visitors. By creating more levels or zones, a greater number of users can be accommodated on a beach. Lime and Stankey (1971) also indicate how recreational use can be redistributed, and carrying capacity increased, by improving access to previously under-used areas. Additional roads and trails, the installation of lighting, elevated pathways and bridges, and the elimination of hazards, are effective in redirecting visitor pressure.

With recreational waterbodies, capacity can be enhanced by providing more access points and ancillary facilities, and by manipulating the type and form of landscape features (e.g. addition of sandy beaches). Wildlife capacities, which indirectly impinge upon certain recreational pursuits, can also be built up by provision or improvement of habitats to encourage greater abundance and variety of animals, birds and fish. Wildlife populations will also respond positively to stable food and water supplies, control of diseases and pests (including predators

(a)

(b)

Plates 6.2a and 6.2b   Interpretation provides educational information about the environment and related services (also see Chapter 10)

(c)

(d)

Plates 6.2c and 6.2d   Elevated walkways are one thing, but the tree top walk in the Valley of the Giants (Shire of Denmark, Western Australia) is more than 400 metres in length and reaches up to 40 metres in height

such as feral cats), controlled use of biocides, minimisation of pollution, and reduction of fire and other hazards.

Recreation resource management is directed towards maintaining and enhancing the site as a viable setting for outdoor recreation. Ultimately, however, it is the reaction of the visitor to the site which determines the success of the management programme. Ensuring a satisfying, high-quality recreation experience is the prime reason for developing an outdoor recreation management system. A specific procedure for visitor management which contributes to this aim is the provision of information and interpretation facilities and services (see Chapter 10).

A successful enterprise, public or private, periodically compares its performance against competitors or providers of similar or complementary resources. Benchmarking is a more formal expression of this process and has ready application in the provision of outdoor recreation resources and services. Traditionally, benchmarking is regarded as a continuous learning process whereby a business identifies leading and well regarded providers, compares products, services and practices with reference to external competitors, and then implements procedures to upgrade performance to match or surpass these competitors (Thomas and Neill 1993). Benchmarking is most widespread in the manufacturing industry, but clearly is relevant in the services sector and, hence, in recreation, tourism, and the provision of leisure services.

In the tourism and recreation industry, benchmarking can be an effective mechanism to prompt establishments to relate to, and adapt and adopt, elements of best practice management programmes of market leaders. This has been particularly evident in the development of environmental auditing procedures and targets and the development of self-regulation parameters in adventure, ecotourism and nature-based tourism in protected areas, though not as extensively and rigorously (see Chapter 11) as one would have hoped, given the extraordinary publicity of the National Ecotourism Accreditation Program in Australia and similar programmes (e.g. Green Globe) now launched on an international scale. The process of benchmarking calls for careful selection of features and practices to target and emulate. It also requires identification of appropriate 'partners' against which to compare performance. A national park in a developing country would be better advised to benchmark with examples of successful park management in a comparable developing region, rather than with national parks in the Western world, though it would be foolish to think that lessons could not be learned from parks in developed countries.

The rationale for the application of carrying capacity and the ROS was described in earlier chapters. The following discussion provides a brief overview of recent, innovative management approaches. These approaches seek to address visitor management concerns, including those outlined above, while generally expanding on the principles underpinning carrying capacity and the ROS. The discussion follows the development of these models and frameworks in a chronological manner (see Nilsen and Taylor 1997).

Limits of Acceptable Change

Out of the questioning of the application of the concept of carrying capacity to recreation management evolved a more comprehensive and systematic framework for recreation decision-making, known as the 'Limits of Acceptable Change' (LAC). The planning framework based on LAC is essentially a reformulation of the recreation carrying capacity concept. The emphasis is on the ecological and social attributes sought in an area, rather than on how much use the area can tolerate. Elements of LAC found their way into planning in wilderness areas in the early 1980s (Eagles and McCool 2002). First tested in the Bob Marshall Wilderness Complex in Montana around 1985 (Stankey *et al.* 1984; USDA Forest Service 1985), the system has received widespread endorsement as a rational planning approach to recreation and parks management.

Essentially, the Limits of Acceptable Change approach turns the recreation-environment relationship on its head, transferring the focus from the supposed cause (numbers of visitors) to the desired conditions – the biophysical state of the site and resource base, and the nature of the recreation experience. Moreover, change in nature is seen as the norm, and a certain level of natural variation in the environment is to be expected. 'Thus, there will be diversity in biophysical and social conditions that the LAC process explicitly recognizes. The LAC process then requires that this diversity is examined and a decision about how it should be preserved or modified is made' (Eagles and McCool 2002: 114). It is when the rate of human-induced change accelerates, or the character of change becomes unacceptable, that managerial action may be called for (Figure 6.3).

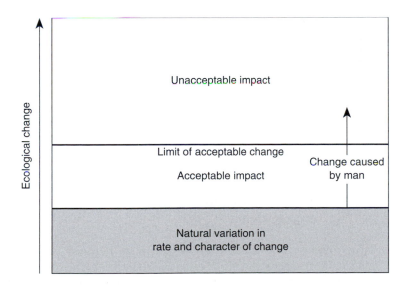

Figure 6.3 Model of acceptable ecological change in wildlands

Source: Hammitt (1990: 26).

The central question for recreation planners then becomes – how much, and what type of change can be accepted? Whereas the response must necessarily be subjective, it needs also to be guided by reference to more than ecological criteria. Socioeconomic and political considerations can also be important elements of the consultative process in setting the Limits of Acceptable Change. A loose analogy can be drawn with the distinction often made between resource 'capability' and 'suitability'. Whereas a parcel of land may be judged *capable* of use, for example as a waste dump, from a biophysical standpoint, other factors, such as economic impacts and social pressures, may render it not *suitable* for such a purpose (also see Chapter 4). It is important, therefore, that a systematic approach be adopted to establishing the Limits of Acceptable Change; one that reflects the natural conditions targeted, as well as economic, social and political realities. Establishing and implementing the Limits of Acceptable Change management framework involves a multistage process, involving nine interrelated steps. The following outline of each stage (Table 6.4) is derived from Stankey *et al.* 1985 and the application of the process at Red River Gorge.

Monitoring is a particularly important part of the Limits of Acceptable Change process. It provides systematic feedback on the effectiveness of the management actions employed, alerting managers to the need to consider more intensive and rigorous efforts, or the use of other measures. It could also point to the need for revision of the standards and indicators specified. This could be the case especially where circumstances external to the recreation site have altered, e.g. changes in access or in contiguous land use.

By applying the Limits of Acceptable Change framework, it is technically possible to establish a rational basis for management intervention. However, Stankey *et al.* (1985) stress the wider context in which decisions have to be made. Recreation planning takes place in a political environment, in which different interests, views and values have to be accommodated. Management techniques based on the Limits of Acceptable Change approach are only part of the recreation planning process. Moreover, the subjectivity and judgement inherent in the identification of acceptable social and environmental conditions, in the setting of standards or thresholds, and in the choice of indicators, need to be balanced by opportunities for ongoing public participation at each stage in the process.

It is important to note that 'the LAC model has a close relation in the tourism planning field in a concept known as the Ultimate Environmental Threshold' (UET) (Mercer 1995: 171; for more detailed discussions of this concept, see Kozlowski *et al.* 1988). This concept refers to:

> The stress limit beyond which a given ecosystem becomes incapable of returning to its original condition and balance. Where these limits are exceeded as a result of the functioning or development of particular activities a chain reaction is generated leading towards irreversible environmental damage of the whole ecosystem or of its essential parts.
>
> (Kozlowski 1985: 148–9, in Mercer 1995: 171)

Table 6.4 The nine steps in the LAC process

Step	Purpose and product
1 Identify issues and concerns	Identify issues and concerns with respect to the site's features and characteristics; unique values and special opportunities identified and featured in management plans, and problems requiring special attention noted.
2 Define and describe opportunity classes or zones	Define hypothetical opportunity classes or zones for the site, including the resources and social dimensions applicable to that class or zone; description of resource, social and managerial conditions for each zone. Close alliance with the ROS.
3 Select indicators of resource and social conditions	Identify the indicators (or specific, measurable/quantifiable variables) that will indicate the condition of the resource. Indicators include water quality, soil erosion, number of trees damaged, vegetation trampled. Social data will include visitor perceptions of resource quality, crowding, interpretation and maps.
4 Inventory existing resource and social conditions	A complete inventory of sites is needed. Baseline data on existing conditions is needed with respect to each resource and social indicator. These are mapped throughout the site.
5 Specify measurable standards for resource and social conditions	Specific standards and measures are assigned to each variable or indicator. These can be developed as a table or matrix.
6 Identify alternative opportunity class/zone allocations	Prescription is called for in deciding levels of resource and social conditions to be maintained. Maps are developed to show these allocations within the broader park or site boundaries. Tables or matrices are used to demonstrate aspects of each class or zone.
7 Identify management actions for each alternative opportunity class/zone	Review baseline data and record differences between existing or current conditions and standards agreed upon. Identify management action to address discrepancies.
8 Evaluation and selection of alternatives	Select preferred alternatives and allocate opportunity classes/zones
9 Implement actions and monitor conditions	Implement the management programme. Monitor zones. Provide feedback and recommend changes needed in areas falling towards or already below minimum agreed standards.

Sources: Graefe (1990, 1991); Vaske *et al.* (1995).

Areas where the UET approach has been adopted, include fragile mountain areas, and single islands and groups of islands (e.g. islands in the Capricornia section of the Great Barrier Reef) (Mercer 1995).

The Limits of Acceptable Change approach: issues and concerns

As a recreation planning framework, the Limits of Acceptable Change approach was originally put forward as a means of rationalising recreation management in wilderness areas. Although it has had application outside this focus (see below), its wider potential is offset by 'lack of understanding of the capabilities of the LAC process … [and] poor or improper execution of the process' (McCool 1990a: 190).

As noted above, a strong element of subjectivity is present in the various stages of assessing acceptable levels of change from recreation use. Turner (1987) singles out the identification of indicators of acceptable conditions as a contentious issue, requiring professional judgement and experience, backed by community consultation. There is little agreement as to what constitutes useful generic indicators of recreation impact, so that it is necessary to derive site-specific indicators for particular environmental attributes at specific locations.

Turner (1987) sets out criteria for selecting environmental indicators (see Table 5.3), and then applies these to a number of measures considered appropriate for the Australian Alps National Park. He notes, however, that few of the indicators adopted are entirely stable, even in the most undisturbed situation. The challenge for managers is to differentiate between recreation impacts and natural variations, and to identify base levels or reference points for particular indicators, outside which environmental values provide an early warning of the need for intervention. This, of course, merely shifts the question for decision to the fixing of a reasonable base level. It might be appropriate to adopt zero as the base level for soil compaction, for example as an indicator of soil condition. However, for other biophysical indicators such as contaminants in streams, 'normal' base readings could be above zero because of natural background levels.

Even then, a good deal of uncertainty prevails with respect to how an ecosystem might react to change in the longer term.

> The best way to handle such uncertainty is to plan on the basis that acceptable conditions will prevail for a certain proportion of the time. It is appropriate to say, for example, that phosphorus levels downstream of a ski village should be less than 40 mg/l for 95 percent of the time, rather than produce a blanket limit which will not be achievable …
>
> (Turner 1987: 10)

The problem, noted earlier, of establishing generally accepted levels for indicators of the social impact of recreation, is equally contentious. Social impacts are obviously important in influencing the quality of the recreation experience, all the more so in remote areas of wilderness, where contact with other humans is usually not sought or welcome. Agreement on indicators such as the number of encounters might be possible, but specifying acceptable levels for such indicators is difficult when interpersonal attitudes and reactions are involved (Turner 1987).

Even the relatively straightforward requirement of monitoring the effectiveness of management actions with reference to the indicators specified raises a number of concerns. Given that the criteria for identifying valid indicators (Table 6.5) have been observed, there remains the question of sampling. Again, Turner (1987) notes as important considerations, the frequency of sampling, the spatial distribution of sampling sites and the need for replication in the interests of consistency. Systematic sampling is basic to monitoring procedures if the cumulative effects of recreation are not to go undetected.

The Limits of Acceptable Change approach is clearly not the panacea for confronting all recreation management challenges, but it does offer the promise of 'more defensible decisions' (McCool 1990a: 191). In reviewing the potential strengths and weaknesses of the process, Knopf (1990) noted twenty possible strengths and only one perceived weakness. The shortcoming that Knopf identified was more concerned with the attitudes of those applying the approach, than with the process itself.

Knopf is concerned with what he considers an element of negativity implicit in the term, Limits of Acceptable Change:

> It seems that the LAC framework has the potential for feeding a certain kind of negative disposition that abounds in outdoor recreation management … that disposition has to do with an attitude that the primary goal of resource management is to arrest the deterioration of environmental quality … people

Table 6.5 Criteria for selecting environmental indicators

Long-term significance	Indicator must detect changes that occur slowly but consistently, and must be able to detect trends over a five-year period.
Short-term significance	Indicator must be able to detect changes in conditions which occur within any particular year.
Responsive	Indicator must detect changes early enough to enable a management response and must reflect changes that are subject to manipulation by management.
Detects amount of change	Indicator should be measurable and allow the amount of change to be assessed quantitatively.
Feasible	Indicator must be reliably measurable by field staff using simple techniques.
Economic	Indicator must produce meaningful information for managers at a minimum cost.

Source: Turner (1987: 20).

being construed as objects that impede quality environmental management
... that litter, form crowds, create noise ... trample vegetation ... pests ...
messing things up.

(Knopf 1990: 207–8)

Some observers may agree with Knopf, and more than one parks manager has
been known to observe that 'parks management would be easy if it wasn't for the
people!'. Knopf's concern is that the Limits of Acceptable Change process has
the potential for encouraging the disposition that people are a problem, rather
than an opportunity, in recreation resource management. He contrasts two possible
statements introducing the process to make his point. The first is *negative*, stressing
the problem of resource degradation and the role of the step-by-step approach in
ameliorating the problems. The second introductory statement emphasises the
positive contribution of outdoor recreation to human growth and development:

Increasing use of our outdoor recreation resource has presented even greater
opportunities for building peak experiences into people's lives. This [LAC]
plan summarizes a step-by-step approach taken to identify new opportunities
for serving our guests ... and expanding our service watershed considerably.

(Knopf 1990: 208)

Perhaps Knopf is right in seeing the Limits of Acceptable Change approach as
reinforcing a tendency towards a 'bunker mentality' on the part of some recreation
managers. However, if applied sensitively and constructively, the process becomes
a valuable recreation tool, offering a framework through which the nature of
recreation management problems can be better understood and more effectively
resolved (McCool 1990b). This is most likely if managers heed Eagles and
McCool's (2002) advice. In summary, managers can modify the LAC process if
there is good reason and, in fact, one is unlikely to move from one discrete step
(say, step 4) to another (say, step 5) without recourse to issues arising that were
not anticipated in earlier steps. In other words, earlier steps may need modification
later in the process. Public participation is essential given that value judgements
are involved and that there are likely diverse interests. Public involvement is
important in learning and consensus building, so that planning is not seen as simply
a technical process. Finally, the LAC is not a comprehensive park management
plan; it 'is primarily a visitor management process that does not necessarily include
other issues, such as fire and wildlife management' (Eagles and McCool 2002:
115–16).

Applications of the Limits of Acceptable Change planning process

As noted earlier, the approach to recreation planning, based on the Limits of
Acceptable Change, was first tested in developing a management framework for
the Bob Marshall Wilderness Complex in Montana (Stankey *et al.* 1984). To

demonstrate further how the system might be applied, a hypothetical case example is described by Stankey *et al.* (1985). The hypothetical area, Imagination Peaks Wilderness, is used to illustrate the flexibility of a management approach based on the Limits of Acceptable Change, rather than restricting and regulating visitors, except when and where necessary. Since this early work, the approach has attracted increasing attention as a decision-making framework for managers of wilderness and similar dispersed recreation settings. The LAC has been considered, developed or applied to varying extents in many locations in the US, Australia and elsewhere (for details, see Brunson 1997; McArthur 2000; Eagles and McCool 2002).

A series of workshops organised by George Stankey and others in Australia, stimulated interest in wider applications of the Limits of Acceptable Change process in that country. Two of these applications will be described briefly, to illustrate the universality and versatility of the approach.

Wild and scenic rivers

In the State of New South Wales, the Department of Water Resources evaluated the applicability of the Limits of Acceptable Change planning system, as a conceptual context for integrated resources management on a catchment basis. In common with water resource managers everywhere, one of the principal challenges for the agency was how to resolve increasing demand for water between competing uses, including recreation. The Limits of Acceptable Change approach was seen as providing a rational, coordinating framework, within which to balance the needs of the aquatic environment with those of other resource users. In particular, the approach was considered relevant to the management of wild and scenic rivers, and wetlands. The following discussion of the Nymboida River draws, with permission, on reports of studies by Don Geering, Project Coordinator, former New South Wales Department of Water Resources, Sydney.

The Nymboida River typifies some of the issues that need to be faced when managing rivers in order to conserve essential environmental values, while allowing their potential for other uses to be realised. The river is part of the Clarence River system on the north coast of New South Wales, and its upper gorge is renowned for its wild and scenic characteristics, providing opportunities for rafting and white water canoeing. Concern continues to grow that the intensity of recreational use is compromising water quality along some reaches of the river; these being the source of urban water supplies, and habitat and breeding areas for threatened species of native fish. Riparian ownership and control are shared between various government agencies and private interests, and an area of wilderness overlaps part of the catchment.

The Limits of Acceptable Change approach was seen as a particular application of management by objectives in a complex resource environment, where a number of competing and conflicting interest groups have to be accommodated. The approach was combined with a procedure known as Adaptive Environmental Assessment and Management (Hollings 1978; Walters 1986) to ensure that all environmental systems and all stakeholders, including recreational interests, were involved in an interactive manner in resource decisions. A modified adaptive procedure, known as Consultative Resource Planning, was developed to allow for even greater public involvement in specifying standards and measurable indicators of resource conditions, and in assessing the predicted impact, over time, of various management options, both positive and negative.

This approach enabled both the agency and the catchment community to apply accepted criteria in order to determine when specified management objectives are being approached. Table 6.6 illustrates how the basic issue of water quality can be clarified, appropriate management action agreed upon, and responsibility allocated. The system linked local knowledge and professional expertise; its flexible nature allowed for fine tuning of levels of acceptable change, indicators and remedial measures as the approach was applied and evolved on site. Acceptable levels of disturbance for defined criteria within certain classes of river settings are set out in Table 6.7, with Class 1 being the least modified. Effective monitoring of these conditions means that the system should be self-correcting, responding to feedback on changes to criteria within acceptable limits for each setting.

Table 6.6 Nymboida River – water quality objectives and management options

Objective	That the water quality of the Nymboida study area remains of a standard that does not affect the recreational, scenic, or urban water consumption potential of the resource.
Desired condition	That water quality be of a standard in its untreated form to conform with urban water supply criteria, during all flow conditions.
Management options	The Soil Conservation Service to investigate the soil erodibility and sediment movement associated with agricultural practices within the Upper Nymboida catchment and recommend a strategy to overcome any identified problems.
	The Department of Water Resources re-schedule catchment to investigate any water quality problems.
	The Forestry Commission to develop operational procedures for logging adjacent to the Wild and Scenic River corridor to ensure a minimum sediment load reaching the river.
	The State Pollution Control Commission to monitor water quality indices, particularly pesticide levels, within the study area.
	The Nymboida Shire to develop standards relating to waste disposal options within the Wild and Scenic River corridor and advise landholders accordingly.
Performance review	The urban authority advise the management committee regularly of changes to water quality, using the monitoring information collected from the urban water supply system.

Source: Geering (1989: 4–5).

Table 6.7 Acceptable levels of disturbance for defined criteria within selected management classes of a river setting*

Criteria	Class 1	Class 2	Class 3
Land-use	95% tree cover in sub-catchment; no settlement evidence or other built intrusions.	80% tree cover in sub-catchment; occasional grazing, clearing or logging; minor.	60% tree cover clearing for agriculture or forestry settlement, in a sub-dominant role.
Scenic value	Outstanding; typically includes deeply incised gorges and waterfalls.	High; including outcrops and rapids.	Moderate; no precipitous slopes, occasional rapids.
Access	No vehicular access within river corridor except for occasional four-wheel drive point sources.	Four-wheel drive access but limited primarily to point sources.	Occasional tracks or minor roads; possible lookouts on valley rim.
Hydrological modification	Essentially free-flowing with no modification.	Occasional minor regulation or water extraction.	Minor modifica-tions including regulation and extraction principally for agricultural uses.
Water	Unmodified; no increase in base load sediment or nutrients.	Partially modified; occasional minor deterioration (typically sediments 5 NIU for part of year and PO4 and NO3 levels around 1 and 6 mg/1 respectively).	Modified; minor permanent deterioration, 25 NTU for short periods and phosphate and nitrate levels generally exceed 1 mg/1 and 6 mg/1 respectively.

Source: Geering (1989: 8).

Note
*Similar tables of criteria can be developed for alternative river and floodplain settings.

Management of public lands

At the local and regional level, government agencies in Australia are frequently charged with preparing management plans for public lands and reserves so as to protect and restore the resource base, while assuring diverse and high-quality opportunities for outdoor recreation. Those objectives underpinned the planning approach for the conservation and development of Wallis Island Crown Reserve, part of a system of estuarine waterways on the central coast of New South Wales (NSW) (Gutteridge *et al.* 1988). The island covers an area of 880 hectares, some two-thirds of which is public land. The surrounding waterways and lake system are heavily used for water-based recreation and commercial fishing; the foreshores are a mix of residential and commercial development, and natural areas (Figure 6.4).

In drafting the management plan for the Island Reserve, the key concern of the resource management agency (NSW Department of Lands) was protecting important environmental features and processes, including wetlands and estuarine water quality. At the same time, it was recognised that increased demand for boating and outdoor recreation would place pressure on the natural environment, and that some low impact modification would be necessary for purposes of access and recreational use.

The planning approach taken was to define and allocate management classes (cf. recreation opportunity settings) for the Reserve, where environmental attributes were specified and certain types of recreation experiences provided for. Two management classes were designated – natural and semi-natural (see Figure 6.5), and within each, management actions were defined to maintain acceptable levels of development, relative to prevailing resource conditions and social considerations.

Thus, the planning process for Wallis Island Crown Reserve represents a further application of the Limits of Acceptable Change approach. The process was driven by the concerns of the public agency and of the local community to develop appropriate classes of land management and use. The two classes were clearly defined by sets of indicators which relate to the resource and social conditions desired, and which specify acceptable limits to recreation use and development.

Selected indicators and standards are set out in Table 6.8 for both of the management classes. The plan of management also specified management actions required to achieve the designated standards.

Incorporation of the Limits of Acceptable Change into the planning strategy for Wallis Island Crown Reserve offers a framework for management of this public land which will provide for a range of purposes appropriate to the island environment. Recreation developments are planned, including revenue-generating accommodation facilities within the semi-natural management class. Fire hazard control, foreshore protection and walking track maintenance are other features of the management strategy to ensure that conditions for each of the management classes are maintained. Field checking and monitoring are also undertaken to assess whether standards and indicators are an accurate reflection of environmental conditions, and whether intrusive actions and associated change are approaching acceptable limits.

These examples from Australia of the application of the Limits of Acceptable Change process are a further indication of the relevance of this approach to outdoor recreation planning and the management of conflict with alternative resource uses.

The Visitor Activity Management Process (VAMP)

Tensions between resources and visitors led to the development of the Visitor Activity Management Process (VAMP) by the Canadian Parks Service (now Environment Canada). VAMP offers a fundamental change in orientation in parks management, from a product or supply basis to an outward-looking market-sensitive

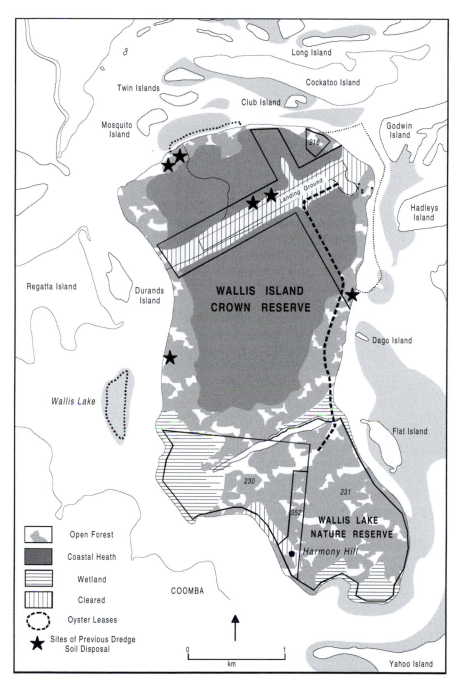

Figure 6.4 Features of Wallis Island

Source: Adapted from Gutteridge *et al.* (1988).

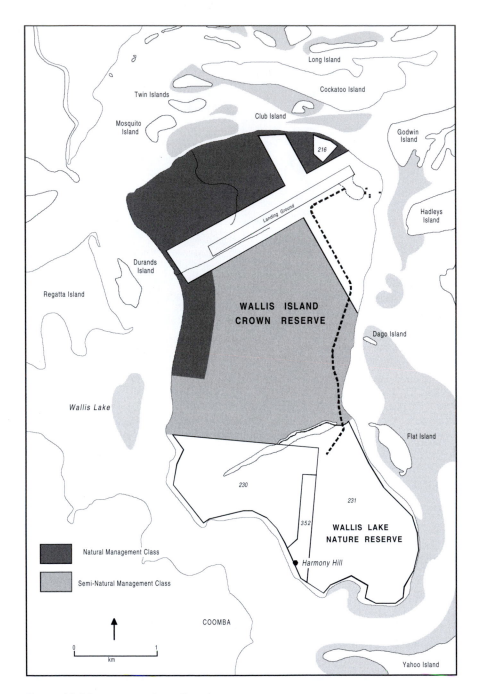

Long Island

Cockatoo Island

Twin Islands

Club Island

Mosquito
Island

Godwin
Island

216

Landing Ground

Hadleys
Island

Durands
Island

Regatta Island

WALLIS   ISLAND
CROWN   RESERVE

Dago Island

Wallis Lake

Flat Island

230

231

352

WALLIS LAKE
NATURE RESERVE

Natural Management Class

Harmony Hill

Semi-Natural Management Class

COOMBA

0   1
km

Yahoo Island

Figure 6.5 Management class allocation

Source: Adapted from Gutteridge *et al.* (1988).

Table 6.8 Wallis Island Crown Reserve: environmental conditions for management classes

	Standards for each indicator	
Indicator	Standard	
	Natural	Semi-natural
Resource		
Percentage of the blocks cleared of natural vegetation cover or not adequately regenerated	Nil cleared	Cleared area less than 5% of any 1 ha block except where necessary at any recreation development site
Presence of exotic species as dominant vegetation form	Minor component of herb layer only	Minor component of understorey only
Presence of seedlings or juvenile plants	All normal growth stages present – natural stand dynamics	Regeneration potential only. Multiple age classes present
Length of shoreline affected by structures, modification (permanent)	No shoreline affected	Less than 10% of shoreline modified
Number of clusters of buildings	No dwellings or structures	Dwelling structures 1 km apart (singly or in clusters)
Presence of services	Services not evident	Services to be not visually intrusive
Length of trafficable road/ presence of sealed roads	Walk trails/well maintained fire trails only	Unsealed tracks/walk trails
Presence of mooring/jetty facilities	No shoreline facilities or disturbance	Limited facilities catering for lowest use only on a share basis. No permanent moorings
Social		
Number of encounters/day (peak period)	Less than 10/day. Only few large groups/day	Less than 50/day at defined use areas. 20/day at other areas
Activities requiring mechanised access	No activities utilising motorised transport	No land-based activities. Only activity powered watercraft
Management personnel presence/day	Occasional field patrols. Less than 1/week	Daily patrols – resident manager
Visibility and enforcement of regulations	No outward or visible regulations	Visible and enforce regulations

Source: Gutteridge *et al.* (1988: 31).

approach (Graham *et al.* 1988). Resource managers are thereby encouraged to be strategic in developing and marketing visitor experiences which will appeal to specific market segments.

> VAMP is a pro-active, flexible, conceptual framework that contributes to decision-building related to the planning, development and operation of park-related services and facilities. It includes an assessment of regional integration of a park or heritage site, systematic identification of visitors, evaluation of visitor market potential, and identification of interpretive and educational opportunities for the public to understand, safely enjoy and appreciate heritage. The framework was developed to contribute to all five park management contexts: park establishment; new park management planning; established park planning and plan review; facility development and operation.
>
> (Graham 1990: 279)

In the same way as carrying capacity, ROS, LAC and VIM (Visitor Impact Management), the Visitor Activity Management Process uses information from both social and natural sciences to facilitate decision-making with respect to access to and use of protected areas (although it has the potential to be applied to a wider range of environments), and incorporates an evaluation requirement to measure effectiveness in outcomes and impacts (Graham 1990). It employs an overt marketing orientation to integrate visitor activity demands with resource opportunities, in order to produce specific recreation opportunities (Lipscombe 1993). A generic version of VAMP (see Figure 6.6) generally involves the following steps:

1   set visitor activity objectives;
2   set terms of reference;
3   identify visitor management issues;
4   analyse visitor management issues;
5   develop options for visitor activities and services;
6   provide recommendations and seek approval of activity/service/facility plan;
7   implement recommended options (Graham *et al.* 1988).

Variations on this initial process have emerged, but nothing too significant (e.g. see Nilsen and Taylor 1997). Quite clearly, VAMP is 'issue-driven' (Hamilton-Smith and Mercer 1991: 58), and is flexible enough to incorporate process, planning and programme monitoring and evaluation. Since the late 1980s, VAMP has been applied to Canada's new park proposals and various park management plans (see Graham and Lawrence 1990). More specifically, VAMP has not been applied widely, save for a limited number of sites in Canada (e.g. Glacier National Park, British Columbia; Cross-country (Nordic) skiing, Ottawa; Mingan Archipelago; Point Pelee National Park; Kejimkujik National Park). VAMP is not a familiar planning approach in such countries as Australia (Lipscombe 1993), where long-run integration of visitor data is lacking, as most visitor management studies are

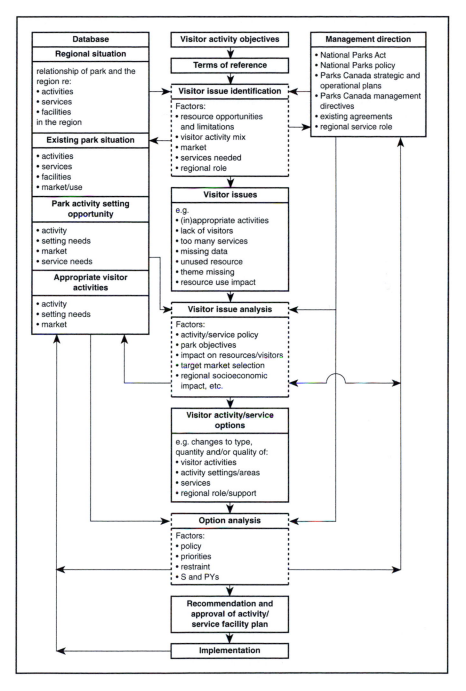

Database	Visitor activity objectives	Management direction

Regional situation

relationship of park and the region re:
• activities
• services
• facilities
in the region

Existing park situation

• activities
• services
• facilities
• market/use

Park activity setting opportunity

• activity
• setting needs
• market
• service needs

Appropriate visitor activities

• activity
• setting needs
• market

Visitor activity objectives

↓

Terms of reference

↓

Visitor issue identification

Factors:
• resource opportunities and limitations
• visitor activity mix
• market
• services needed
• regional role

Visitor issues

e.g.
• (in)appropriate activities
• lack of visitors
• too many services
• missing data
• unused resource
• theme missing
• resource use impact

Visitor issue analysis

Factors:
• activity/service policy
• park objectives
• impact on resources/visitors
• target market selection
• regional socioeconomic impact, etc.

Visitor activity/service options

e.g. changes to type, quantity and/or quality of:
• visitor activities
• activity settings/areas
• services
• regional role/support

Option analysis

Factors:
• policy
• priorities
• restraint
• S and PYs

Recommendation and approval of activity/ service facility plan

Implementation

Management direction

• National Parks Act
• National Parks policy
• Parks Canada strategic and operational plans
• Parks Canada management directives
• existing agreements
• regional service role

Figure 6.6 Visitor impact management/planning process

Source: Adapted from Parks Canada (1985, in Graham 1989: 278).

carried out in isolation (Hamilton-Smith and Mercer 1991), and their findings rarely reported publicly. Nevertheless, according to Graham *et al.* (1988):

> VAMP continues to evolve and be modified as it comes into wider use, and as new concepts and techniques are developed ... VAMP is a flexible management framework, undergoing transition and change. This framework must not become an end in itself, merely justifying design and development of services, programmes and facilities. Case study research which highlights the effective [or otherwise] use of VAMP will be required ... Further development of VAMP will require a supportive organizational environment, a condition not easily found in an era of constraint and pressure for more efficient management.
>
> (Graham *et al.* 1988: 61)

There is a good deal of overlap between VAMP and LAC and other recreation planning theories, models and frameworks. The overlap is not a result of happenstance, but stems from the evaluation of these theories, models and frameworks in particular settings. The lesson is that managers and researchers must be prepared to adapt what is learned elsewhere to the specific circumstances in which they are working.

Visitor Impact Management framework (VIM)

The development of VIM in 1990 demonstrates the increasingly widespread view that recreational management requires scientific and judgemental consideration (e.g. see Hendee *et al.* 1978; Stankey *et al.* 1985; Shelby and Heberlein 1986; Vaske *et al.* 1995: 36), and that effective management of the recreation resource is much more than setting visitor use levels and specific carrying capacities (e.g. see Washburne 1982; Graefe *et al.* 1984; Vaske *et al.* 1995: 36) (also see Chapters 4 and 5).

The Visitor Impact Management framework resulted from a study by the US National Parks and Conservation Association (NPCA), which had two main objectives. The first objective was to review and synthesise the existing literature dealing with recreational carrying capacity and visitor impacts. The second objective was to apply the resulting understanding to the development of a methodology or framework for visitor impact management, that would be applicable across the variety of units within the US National Park System. A number of other goals underpinned the development of the VIM framework:

- to provide information and tools to assist planners and managers in controlling or reducing undesirable visitor behaviour;
- to suggest management approaches that build on scientific understanding of the nature and causes of visitor impacts;
- to consider impacts both to the natural environment and to the quality of recreation experiences, and to develop consistent processes for addressing such impacts.

(Graefe 1991: 74)

The review of the scientific literature relating to carrying capacity and visitor impacts identified five major considerations underpinning the nature of recreation impacts, which should all be incorporated into programmes for managing visitor impacts:

1   *Impact relationships*: impact indicators are interrelated so that there is no single, predictable response of natural environments or individual behaviour to recreational use.
2   *Use–impact relationships*: use–impact relationships vary for different measures of visitor use, and are influenced by a variety of situational factors. The use–impact relationship is non-linear (i.e. it is not simple or uniform).
3   *Varying tolerance to impacts*: not all areas respond in the same way to encounters with visitors. There is inherent variation in tolerance among environments and user groups; for instance, different types of wildlife and user groups have different tolerance levels in their interactions with people.
4   *Activity-specific influences*: the extent and nature of impacts vary among, and even within, recreational activities.
5   *Site-specific influences*: seasonal and site-specific variables influence recreational impacts.

(Graefe 1990: 214, 1991: 74; Vaske et al. 1995: 35)

These five issues represent important considerations for the management of ecological, physical and social impacts (Graefe 1990).

In brief, the VIM framework is designed to deal with the basic issues inherent in impact management, namely: the identification of problem conditions (or unacceptable visitor impacts); the determination of potential causal factors affecting the occurrence and severity of the unacceptable impacts; and the selection of potential management strategies for ameliorating the unacceptable impacts (Graefe 1990: 216). Given these basic issues, the VIM framework comprises eight steps (see Figure 6.7 and Table 6.9). Importantly, Graefe (1990, 1991) and Vaske *et al.* (1995) note that the task of managing visitor impacts is not over when management strategies are implemented, and that continuous monitoring and evaluation are necessary.

The Visitor Impact Management framework is based on high-level natural and social science research. As a planning framework it has the capability to deal with recreation impacts at a site level in a range of environments, and in conjunction with other planning frameworks within the management planning process. VIM has been applied in Australia (e.g. Jenolan Caves), Canada (e.g. Prince Edward Island), and in the US (e.g. Icewater Spring Shelter, Great Smoky Mountains National Parks; Logan Pass/Hidden Lake Trail, Glacier National Park; Florida Keys National Marine Sanctuary, Florida; Buck Island Reef National Monument, Virgin Islands; and the Youghiogheny River, Western Maryland) as well as Argentina, Mexico and The Netherlands (e.g. see Graefe 1990; McArthur 2000; Newsome *et al.* 2002).

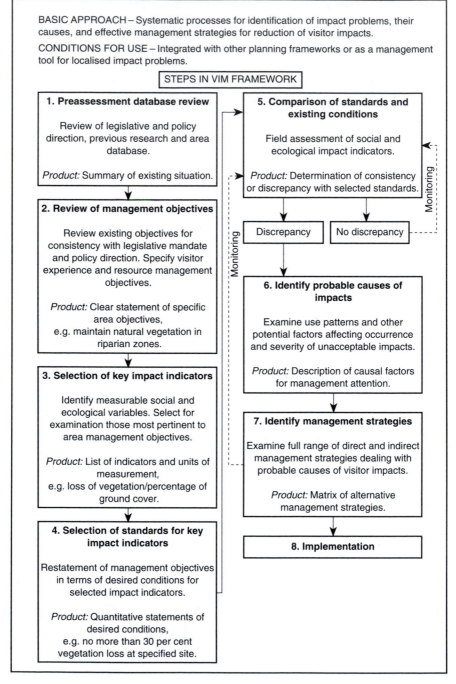

BASIC APPROACH – Systematic processes for identification of impact problems, their causes, and effective management strategies for reduction of visitor impacts.

CONDITIONS FOR USE – Integrated with other planning frameworks or as a management tool for localised impact problems.

STEPS IN VIM FRAMEWORK

1. Preassessment database review

Review of legislative and policy direction, previous research and area database.

Product: Summary of existing situation.

2. Review of management objectives

Review existing objectives for consistency with legislative mandate and policy direction. Specify visitor experience and resource management objectives.

Product: Clear statement of specific area objectives, e.g. maintain natural vegetation in riparian zones.

3. Selection of key impact indicators

Identify measurable social and ecological variables. Select for examination those most pertinent to area management objectives.

Product: List of indicators and units of measurement, e.g. loss of vegetation/percentage of ground cover.

4. Selection of standards for key impact indicators

Restatement of management objectives in terms of desired conditions for selected impact indicators.

Product: Quantitative statements of desired conditions, e.g. no more than 30 per cent vegetation loss at specified site.

5. Comparison of standards and existing conditions

Field assessment of social and ecological impact indicators.

Product: Determination of consistency or discrepancy with selected standards.

Discrepancy   No discrepancy

6. Identify probable causes of impacts

Examine use patterns and other potential factors affecting occurrence and severity of unacceptable impacts.

Product: Description of causal factors for management attention.

7. Identify management strategies

Examine full range of direct and indirect management strategies dealing with probable causes of visitor impacts.

Product: Matrix of alternative management strategies.

8. Implementation

Monitoring

Monitoring

Figure 6.7 The Visitor Activity Management Process (VAMP) – a generic representation of VAMP

Source: Adapted from Graefe (1989: 218).

Table 6.9 Steps in the Visitor Impact Management framework

Step	Actions
Step 1 (The pre-assessment database review)	Compile and review relevant information that is already available in order to gain initial perspectives of problems and issues.
Step 2 (Review of management objectives)	Review current management objectives to ensure that they are stated clearly and specifically, and to define the type of experience to be provided in terms of appropriate ecological and social conditions. 'The objectives should be prioritised, since any single objective may lead to potentially conflicting goals' (Vaske *et al.* 1995: 38).
Step 3 (Selection of key impact indicators)	Select measurable indicators that reflect the management objectives set out in Step 2; i.e. define what variables, that reflect the planning objectives, will be measured. Indicators will vary at different sites. 'Useful indicators include those that are directly observable, relatively easy to measure, directly related to the objectives for the area, sensitive to changing use conditions, and amenable to management' (Vaske *et al.* 1995: 39).
Step 4 (Selection of standards for key impact indicators)	Decide how and when the key indicators identified in Step 3 will be measured, specifying appropriate levels or acceptable limits. 'The selected standards become the basis for evaluating the existing situation. This step serves the function of describing the type of experience to be provided in units of measurement compatible with available measures of the current situation' (Vaske *et al.* 1995: 39).
Step 5 (Comparison of standards and existing conditions)	Compare the existing situation with desired situations. The question to be resolved being: Is the area providing the types of recreational experiences identified in the management objectives, within appropriate maintenance of environmental conditions? Document problem situations.
Step 6 (Identification of probable causes of impacts)	Isolate potential factors (e.g. type of use, length of stay, group sizes, use timing and concentration, behaviour, and site characteristics) that may contribute to impact conditions. Identify the most significant causes of the problems identified in Step 5.
Step 7 (Identification of management strategies)	Identify a range of alternative management strategies given some understanding of how the amount, type, and distribution of people using an area affect given impact indicators. The focus is on dealing with the causes of problems (see Table 6.3 adapted from Hendee *et al.* 1978, in Vaske *et al.* 1995: 41).
Step 8 (Implementation)	Implement the selected management strategies as soon as the necessary resources are available.

Sources: Graefe (1990, 1991); Vaske *et al.* (1995).

VIM is a means of controlling or reducing the undesirable impacts of recreational use (Graefe 1991: 80). It has a sound scientific basis, and presents a systematic process for assessing visitor impacts by way of problem-solving. It is, in addition, a more detailed alternative to the concept of carrying capacity, and has potential for wider application in resource management (i.e. as part of an overall site or regional plan) (Graefe 1990, 1991), perhaps in conjunction with the LAC model.

Visitor Experience and Resource Protection (VERP)

The US National Parks Service is finding it more difficult to fulfil 'its dual mission to provide for the enjoyment of national parks while conserving resources for future generations' (US Department of the Interior, National Parks Service 1997: 4). The Visitor Experience and Resource Protection framework was developed in the early 1990s and first applied in about 1993 by the United States National Parks Service as a response to increasing visitor pressures and impacts in US National Parks and growing concerns about carrying capacities. The 1978 National Parks and Recreation Act (P.L. 95-625) actually requires the National Park Service 'to address the issue of carrying capacity in general management plans' (US Department of the Interior, National Parks Service 1997: 8). VERP evolved from the LAC model in a fashion to suit the National Parks Service's mission, structures, functions and operations. These matters must be emphasised when developing and applying the model under any circumstances.

The VERP framework has close links to both LAC and VIM, with an emphasis on managing visitor use and resources simultaneously and continuously. Hence, VERP is 'a planning and management framework that focuses on visitor use impacts on the visitor experience and the park resources. These impacts are primarily attributable to visitor behavior, use levels, types of use, timing of use, and location of use' (US Department of the Interior, National Parks Service 1997: 8). The VERP framework has a well defined foundation comprising nine integral elements. These elements are described in Table 6.10. While these elements appear discrete, they are in fact integrated not only during the initial development of a plan, but in its ongoing monitoring, evaluation and responses. Adjustments must be made whenever necessary. Like all planning frameworks, VERP is doomed to failure if sites are not monitored and evaluated against core criteria and clearly specified standards. VERP has been applied in countries such as the US and Australia.

Visitor management models – a brief critique

Over the past 25–30 years, a number of interdisciplinary or multidisciplinary planning and management processes and frameworks have been established, trialled, and variously described and evaluated, most on the precondition that they will help resolve, to a greater or lesser extent, visitor management problems. Many of the approaches stemmed from researchers and federal agency staff in the United States and to a lesser extent Canada, and drew on significant work of people such as 'Driver and Brown (1978), and Clark and Stankey (1979) on ROS, and in the

Table 6.10 Framework foundation for VERP

Foci and Elements	Description
Framework foundation	
1 Assemble interdisciplinary project team	Assemble a core team to develop and implement plan. Outside specialist assistance (e.g. consultants) may be needed and should be identified.
2 Develop public involvement strategy	Involve the public and develop a public involvement strategy to promote information exchange. Make explicit the links between public involvement and decision-making. Public involvement is essential when dealing with value laden issues and concepts like carrying capacity.
3 Develop statements – purpose, significance, themes, constraints	Statements of mission, purpose, interpretive themes and significance of the site (i.e. park) are essential and everything that is done must refer back to these. Why does a park exist? How significant are particular resources of the park (are they endangered or do they have some cultural significance)? Statements of significance help set priorities. Stated themes should highlight significance. A constraint may be financial or some obligation required under legislation (e.g. permit fishing or use of motorboats).
Analysis	
4 Analyse park resources and visitor use	This must be well documented. Develop understanding of resource characteristics and how these affect visitor experiences, the resistance and resilience of resources, the appropriateness and current location of existing facilities and infrastructure. From this analysis a range of zones and their possible locations can be identified, and their locations justified. Baseline data is needed for impact analysis or, in other words, assessing the implications of zoning areas in particular ways.
Prescriptions	
5 Describe potential range of visitor experiences and resource conditions	Potential zones are described for application to specific geographical areas in element 6. These zones involve descriptions of visitor experiences and resource characteristics consistent with park mission, purpose and significance. Levels of activity, site development and management are also included in the descriptions.
6 Allocate potential zones to specific locations (zoning)	A zoning scheme is prescribed. Zones are assigned to specific locations within the park.

continued…

Table 6.10 continued

Foci and Elements	Description
7 Select indicators and specify standards for each zone; develop monitoring plan	Each zone requires specific indicators (measurable variables for monitoring) and standards (minimum acceptable conditions). The monitoring plan will include statement of indicators and standards, as well as strategies for monitoring indicators and standards. These strategies need to be prioritised, staffed and funded, and include clear statements of methodologies and analyses.
Monitoring and management action	
8 Monitor resource and social indicators	Resource and social/experiential conditions are monitored and evaluated. Usually, given resource constraints, only the most critical areas are monitored frequently or regularly.
9 Take management action	If social or resource indicators are falling towards minimal acceptable standards, management action will be required.

Source: US Department of the Interior, National Parks Service (1997: 9–11).

mid-1980s with the development of LAC (Stankey and others 1985) and VAMP (Parks Canada)' (Nilsen and Taylor 1998: 49). Each recognises limitations in the traditional concept of carrying capacity and 'was a response to both legislative and policy requirements, as well as to increasing recreation demands, impacts, and conflicts' (Nilsen and Taylor 1998: 49).

These advances in thinking, theory and application have gradually moved agencies towards more integrated approaches to environmental (including visitor) management. Perhaps not surprisingly, no single process or framework has received unanimous support among resource managers and researchers, as a means of solving the problems associated with visitor management. Put simply, a specific process or framework will be more effective depending on the scale at which it is applied (e.g. whether it is applied in a regional or single-site context), or on any number of other possible intervening variables.

Nevertheless, in addressing questions applicable to LAC, ROS, VIM and VAMP, a syndicate of conference representatives in 1990 identified the following characteristics:

- LAC and VIM are mostly resource oriented and 'reactive' (i.e. they are frameworks which are applied after a resource input);
- ROS and VAMP are more oriented to visitor and use/activity, and include an interpretive activities component – a 'recipe' for potential interpretation;
- LAC and VIM applications can provide the opportunity for interpretive services and activities;
- ROS and VAMP applications result in the production of an 'interpretive prospectus-type' document. Including interpretation in the ROS and

VAMP frameworks is automatic. LAC and VIM require a conscious managerial decision to include interpretation;

- none of the frameworks provides a superior timeframe, and none appears to be more efficient than the other;
- improving the timeliness of the interpretive applications of any of the frameworks is primarily dependent on a quality database, management priorities, and commitment to the interpretive element;
- all of the frameworks are information-gathering/decision-building processes rather than decision-making processes;
- inclusion of measurable standards and performance requirements would improve the use of the frameworks and their application to programmes such as interpretation.

(Pugh 1990: 354)

A more recent comparison of these models plus VERP established the following weaknesses and strengths listed in Table 6.11. It is interesting to note how some of the weaknesses were observed in the models even some seven to eight years after their first implementation. Perhaps this is more a symptom of their inherent steps and themes and the fact that 'no model can be everything to every situation' than poor management. What this questioning does emphasise is the need for careful consideration of models employed, their relevance to the task at hand, and the need to adapt models to particular circumstances. The outcomes of applying LAC at the Bob Marshall Wilderness Complex in the US are likely far removed from the circumstances that would be encountered in applying LAC at the Royal National Park, Sydney, Australia or sites in the Lakes District of the UK. Indeed, LAC might not be the best model to use, and McArthur's (2000) research on the evolution of the Tourism Optimisation Management Model (TOMM) on Kangaroo Island, Australia is an excellent case in point.

It would seem, then, that the application of recreation/visitor planning and management frameworks is largely a case of management skill in adapting and integrating components of one or more models or frameworks. For instance, the ROS was used effectively in an urban open-space study, despite its initial development and application in parks. As Hamilton-Smith and Mercer (1991: 57) noted, 'The possibility of using the ROS in any urban study [such as the Newcastle/ Lake Macquarie Open Space Study] should not be dismissed, but it is likely to require adaptation to the specific circumstances of the planning study concerned'. Perhaps, also, much hinges on the vagaries of politics, limited finances, and unresponsive organisational cultures and managerial frameworks, which can all thwart the best-laid intentions of innovative planners and managers.

Summary

The management of recreation resources is inherently difficult because of the extensive range of variables involved in the recreation–environment relationship. Resource managers do not always have the funds to manage a site adequately.

Table 6.11 A comparison of visitor management frameworks

Model	Strengths	Weaknesses
ROS	Practical process Principles require managers to: • protect the resource; • provide opportunities for public use; and • meet preset conditions Links supply and demand Can be integrated with other processes Ensures provision of a range of recreation opportunities	The recreation opportunity spectrum and its underlying principles must be accepted by managers before any options or decisions can be made Does not cope with disagreement among stakeholders ROS zoning and maps need to be related to the physical and biophysical characteristics of each area
VIM	Encourages careful and balanced use of both scientific *and* judgemental considerations Places heavy emphasis on understanding causal factors which are then used to identify resource management strategies Provides a classification of management strategies and a matrix for evaluating them	The process does not make use of ROS, but it could Assesses current conditions of impact but not potential impacts
LAC	Development of a strategic and tactical plan for the area Limits of acceptable change for each opportunity class and indicators of change that can be used to monitor ecological and social change Conditions are clearly defined	Strong focus on present issues and concerns in data collection and analysis Management topics where there are no current issues or concerns may be neglected
VERP	VERP is a thought process Draws on the talents of a team Guided by policy and the park purpose statement Statements of significance and sensitivity guide resource analysis statements defining important elements of the visitor experience guide visitor opportunity Zoning is the focus for management	To pilot the approach in different environments much revision and adaptation may be required 'Experience' is not defined Indicators for experience are often absent The will and ability to conduct monitoring and evaluation are not guaranteed and must be tested
VAMP	Comprehensive decision-making process Based on a hierarchy Structured thinking is used to analyse opportunity and impact Combines social science principles and marketing in developing and managing visitor opportunities	Well-developed at the service planning level, but lacks 'clout' at the management planning level 'Opportunities for experience' definition has not been built into management plans or into the zoning

Source: Derived from Nilsen and Taylor (1998: 50–2).

Some of the most popular tourist sites (e.g. beach and coastal zones; walking tracks, campgrounds and waterways in isolated mountain areas and wilderness) may require extensive restoration work, but costs of materials, their delivery to the site and the labour involved may be beyond the reach of the organisation concerned. Furthermore, information concerning the impact of tourism on the environment is limited (see HaySmith and Hunt 1995).

Despite these problems, several progressive technologies and planning approaches are being used in designing and managing recreation resources (particularly in protected and natural areas) across the globe. Geographic Information Systems (GIS) are being used, among other things, to identify ecologically sensitive areas and to plan tourist development (see Chapter 12). LAC is being utilised to monitor and set standards for acceptable levels of impact. VIM and VAMP call for monitoring and evaluation, the setting of clear and specific objectives, and the establishment of indicators and standards. These techniques, however, are not mutually exclusive. They can be integrated to identify, develop, implement, monitor and evaluate strategies for visitor management. Nor are these frameworks set in concrete. Political circumstances, resource availability and other circumstances will require organisations to adapt to their planning and management environments. This is best achieved by a strategic management approach, a subject discussed in Chapter 12.

Guide to further reading

- The Limits of Acceptable Change concept was introduced by Stankey *et al.* (1984), and reviewed in McCool (1990b) and Knopf (1990); McCool and Cole (1997); McArthur (2000); Eagles and McCool (2002). Also see USDA Forest Service, Red River Gorge, available online at: http://www.fs.fed.us/r8/boone/lac/docs/lacsteps.pdf; applications for bushwalking in World Heritage Areas in Tasmania at http://www.parks.tas.gov.au/manage/batr/WHA_Walking_Opportunities.pdf.
- For discussion and applications of the Ultimate Environmental Threshold concept, see Kozlowski *et al.* (1988) and Mercer (1995).
- For detailed discussions on VIM, see Kruss *et al.* (1990); Graefe (1990, 1991); Graham and Lawrence (1990); Vaske *et al.* (1995); McArthur (2000).
- For VAMP see Graham *et al.* (1988); Ashley (1990); Reynolds (1990); Graham (1990); Lipscombe (1993); McArthur (2000); Eagles and McCool (2002). For application of VAMP to mountain biking in the Canadian Rockies see http://www.mtnforum.org/resources/library/mosej02a.htm.
- Graham and Lawrence (1990), Hamilton-Smith and Mercer (1991), Lipscombe (1993) and McArthur (2000) contain discussions on the frameworks and applications of carrying capacity, ROS, LAC, VAMP and VIM, with McArthur's thesis providing the most comprehensive coverage.
- VERP – US Department of the Interior, National Parks Service (1997) *VERP: The Visitor Experience and Resource Protection Framework: A Handbook for Planners and Managers*, US Department of the Interior, National Parks

Service, Denver, is available online at http://planning.nps.gov/document/ verphandbook.pdf; Visitor Experiences and Resource Protection Implementation Plan for Arches National Park, Utah, US, see http://planning.nps.gov/ document/archverpplan1.pdf; Research to Support the Application of the Visitor Experience and Resource Protection Framework at Zion National Park, see http://www.pwrc.usgs.gov/products/sciencebriefs/SB%2011%20Marion. pdf.

Review questions

1   Discuss the relationships between resource management, outdoor recreation and conservation.
2   What are the respective merits and limitations in applications of each of the following: the Limits of Acceptable Change; the Visitor Impact Management framework; the Visitor Activity Management Process; carrying capacity. Is any particular framework or approach better than the others?
3   What are some of the distinguishing features of the Limits of Acceptable Change (LAC) approach?
4   To what extent do you agree with Knopf's concerns about the LAC process?
5   Visit a nature-based outdoor recreation site (e.g. a national park). Conduct a resource inventory of that site. Make an assessment of the site's present condition. How might the site be improved? Determine what variables should be measured in order to monitor and evaluate recreational use of that site. If there is no current management strategy for that particular site, develop a management approach/framework to manage the site. If there is such a strategy, review that strategy and determine how it might be improved.

7   Outdoor recreation in urban areas

Most of the population of industrialised nations live in urban areas. Human beings are overwhelmingly social creatures and, as such, prefer to live together in communities created to serve individual and collective human needs. These communities can be of varying size and characteristics, but they have one feature in common – the potential to offer a wide range of functions to satisfy the needs of the population. In the UK context, Williams (1995) suggests that the development of recreational opportunities and services moved through three phases (see Table 7.1). He then goes on to point out that 'a great deal of recreation provision in urban areas, particularly public open space, reflects historic accident and/or municipal intervention' (p. 21). Internal reserves (or pocket parks) in Melbourne are a good case in point (see Box 7.1).

Underpinning the attraction of urban places is the presumed availability of a diverse spectrum of recreation opportunities in a relatively limited and accessible

Table 7.1 The phases of recreation provision in the UK

Phase	Characteristics
Phase I: Formation (1800s to early 1900s)	Emergence of town planning Experiments in urban design Improved housing, street environments Spatial design and patterning of urban places
Phase II: Consolidation (1918–39)	Clear establishment of gardens and allotments Provision of playing fields for sports, but used for a wider range of activities
Phase III: Expansion (post-1945)	Growth in demand for leisure Recreation planning and provision is a statutory requirement Extensive recreation space Urban environmental movements have benefited from recreation opportunities (e.g. bush regeneration; greenways; green corridors)

Source: Adapted from Williams (1995: 17–18).

Box 7.1 **Pocket parks**

Internal reserves are an historic form of planned open space, serving as semi-private parkland at the rear of residential allotments and lacking street frontages. They have something in common with the exclusive town gardens and private courtyards, and agricultural allotments of Britain and Europe. In Australia, urban planners saw the internal reserve or pocket park as a ready means of addressing the need in new residential subdivisions for both children's play space and provision for adult outdoor recreation. A shared backyard could serve as a secure area for children and an opportunity for passive or active recreation pursuits, as well as a way of using relict pockets of land unsuitable for building. Not all internal reserves have been successful or survived suburban expansion. The dynamics of urban population change can be an impediment to the fostering of community support and responsibility for maintenance of the reserves. Concern over safety and security and privacy aspects can also act as deterrents to their use as communal recreation space (see below – use and non-use of urban recreation space) and inhibit local residents from acting as stakeholders in ensuring the recreation potential of these valuable but disappearing components of the urban recreation system.

Source: Nichols and Freestone (2003).

spatial context compared with rural settings, ranging from natural areas (e.g. urban bushland) to built environments (e.g. outdoor tennis and basketball courts). Part of the challenge of sustaining a liveable urban environment, is to ensure the maintenance of a choice of quality leisure experiences through the existence of a spectrum of recreation opportunities, with the flexibility to adapt to the dynamics of a changing city landscape and evolving socioeconomic and political relationships. Indeed,

> the extent to which public outdoor recreation will flourish depends very much upon how planners and those in positions of power and influence face up to the challenge of urban restructuring – physical, economic and social – that is being posed by contemporary processes of urban change.
>
> (Williams 1995: 2)

The trend of urbanisation in Western industrial countries is well established. Some three-quarters of the population of the US and Britain live in urban areas and that figure is exceeded in parts of Western Europe. More than 80 per cent of the Australian population lives on about 2 per cent of the country's land area and hence is concentrated in urban areas.

Despite this concentration of humanity, only in the past two to three decades has growing concern for quality-of-life issues, and in particular outdoor recreation, focused attention on the relative deprivation of city dwellers, and the need for more enlightened planning of the urban environment. This recent attention is not surprising. Many of the world's great cities are 'sick' – they are losing people and jobs; they have experienced or are experiencing declining fiscal solvency; they are less convenient, safe and attractive; and they are short on justice, tranquillity and general welfare. Increasingly, they are also seen as short on outdoor recreation opportunities, especially in some high or medium density housing estates.

Castles and Miller (1998: 3, in Sandercock 2003: 20) argued 'The closing years of the twentieth century and the beginning of the twenty-first will be the age of migration'. Migration reshapes cities of all sizes and forces planners (or at least should force them) to rethink the way they shape them. These cities and urban spaces will include places where struggles are familiar. They include 'ethnic neighbourhoods, sub-urban migrant camps, sweatshops, places of worship, and the zones of the new racism ... they are part of the landscape of modernity, which is a landscape marked by difference' (Sandercock 2003: 21). It means planners and decision-makers will need to help find answers to very complex questions about such matters as what places mean, and how to make these places safe and accessible.

Planners regulate space – its production and use; that is, the extent and nature of space and who can use it and when and why (Sandercock 2003: 21), and not always successfully. Although Sandercock's discussion was focused on migration and culture, the four different ways in which multiethnic cities challenge existing planning systems, policies and practices are more broadly relevant. The challenges relate to the values and norms of the dominant culture embedded in planning frameworks and in the attitudes and behaviour of the planning profession. Human attributes like xenophobia and racism can find expression in conflicts over the location of a religious centre, just as they can over emergence of unfamiliar recreation preferences and practices. Outbursts of violence in several US cities have been linked to ethnic rivalries and underlying resentment based on colour or class.

Sandercock raises issues similar to those raised by urban geographers such as Fred Gray some 30 years ago. Gray asked questions concerning whether urban geographers really knew what was going on in the city, and whether their explanatory and analytical frameworks were relevant to anyone other than geographers. Planners work within a capitalist society which provides structural and analytical challenges (see Gray 1975). Does choice rest with individuals and their households? Do people live and play where they do because they freely choose to do so or as a result of various barriers that have limited their opportunities? What are the constraints people face (e.g. see Chapter 3)? For Gray,

> it is wrong to infer that the decisions made by the household unit, in any but the most naïve sense, determine the 'social geographic patterns' of cities ... Instead the process of capitalist economic development and its associated social, political, and ideological relationships is the underlying and missing variable causing the surface of reality which geographers [and indeed planners should] examine.

(Gray 1975: 231)

For all intents and purposes, this initially seems a good point echoed elsewhere. However, Gray's explanation was unorthodox, harsh and to some extent misleading. He certainly highlighted shortcomings in urban geography and urban planning that still exist today and provided compelling reasons to understand macro level issues (e.g. the role of the state). Yet, his decision to downplay the fact that there is some choice and opportunity to express and realise preferences gives a 'lop-sided' account (or as Hamnett 1974, in Gray 1975, put it, amounts to 'throwing out the baby with the bath water') of what affects households' and individuals' decisions and gives priority to supply over demand. Moreover, planners take on experience and past practice and do so in organisational settings (e.g. planning departments; local councils). So perhaps the ways in which organisations develop an organisational culture or a way of doing things is worth considering in terms of the ways in which planners go about their work. 'The concept of culture occupies a powerful place in academic and managerial discourse' (Bolman and Deal 1991: 268).

An organisation's culture is one expression of its 'personality': its characteristic way of doing things (Jacques 1951, in Jans and Frazer-Jans 1991: 336).

> In other words, an organisation's culture is the unique style of a group of people which incorporates the informal coping strategies considered necessary to carry out the tasks and functions required by the organisation. Putting this another way: culture is to the organisation what personality is to the individual. It is a hidden, unifying core that provides meaning, direction and mobilisation in organisations. Within any organisation, then, there is usually some system of shared beliefs or meaning about the organisation and its function which distinguishes it from others. In this respect, organisational culture could be regarded as the glue which binds the parts of the organisation together.
>
> (Britton 1991: 2)

So, an organisation's culture reflects the things its staff value, the goals they collectively pursue, and the way they prefer to operate and to manage themselves (see Jans and Frazer-Jans 1991: 336). The organisational culture or symbolic management perspective 'assumes that many organisational decisions and actions are almost predetermined by the patterns of basic assumptions that are held by members of an organisation' (Shafritz and Ott 1992: 48). Those patterns of assumptions continue to exist and to influence behaviours because they repeatedly lead people to make decisions that 'worked in the past' for the organisation. Ultimately, organisational culture affects organisational behaviour, and 'can block an organization from making changes that are needed to adapt to a changing environment' (Shafritz and Ott 1992: 482). By way of example, Geographic Information Systems (GIS) has not been applied in leisure services provision to any great extent, despite the potential of GIS for park planning and management being highlighted by the early 1990s (Nicholls 2001). In her study of Bryan, Texas, Nicholls used GIS mapping to critically examine accessibility and equity of provision in Bryan's park system. According to Nicholls (2001: 216–17), 'less than 40 per cent of Bryan residents have good access to any form of everyday open space, with only 12 per cent being able to reach a neighbourhood park within the distance specified'. She attributed these problems to 'a lack of sufficient open space rather than its poor distribution relative to the population'. How can this happen?

Parks and recreation are not the highest priorities in urban development, but their roles in social welfare and urban renewal are becoming recognised and cannot be ignored. However, if there are no strong arguments for recreation provision, especially in instances where standards (see below) are not applied, urban spaces may become little more than concrete jungles, with concentrated populations lacking adequate neighbourhood spaces to play. The philosophies and ideologies underpinning urban open space and recreation provision and management are critical and eventually lead to questions about provision – should resources and activities be supplied by government or commercial enterprises?

Williams (1995) sees the provision of urban recreation facilities as drawn from three sectors – the public, the private and the voluntary. The public sector is dominant in that it normally exercises control over key resources and indirect control through regulation. Yet, public provision of recreation services is marked by compromise, trade-offs between what is equitable and what is economically feasible and between socially desirable and cost-effective outcomes.

By contrast the private sector is driven by market forces and the expectation of financial returns. Provision of commercial facilities for leisure outside the home is typically the role of the private sector. The voluntary sector is normally dominated by community-based clubs and groups organised around specific recreation activities. Whereas there should be scope for complementarity between the three sectors, Williams sees only limited evidence of this in Britain and considerable competition and overlap.

Enhancement of recreation opportunities in urban areas is now seen to contribute substantially to the quality of life of local residents, and to assist with the creation of a sustainable urban environment. For instance, increased attention is being given to the historical and cultural significance of park and recreation resources in the US (and elsewhere), with greater commitment to the restoration of historic buildings, facilities and designed landscapes. Julia Sniderman (in Dwyer and Stewart 1995: 607) of the Chicago Park District described her experience with park restoration in Chicago:

> The parks of Chicago are important cultural and historic resources. Documen-
> tation is the key to realizing the full historic value of urban parks. There is a
> vault under Soldier Field that was sealed off and forgotten for decades. When
> it was recently opened, whole archives of architectural plans for city parks
> were discovered. Documents like these are central to historic restoration work.
> Interpretation of restored parks is also important. Our recent restoration of
> Columbus Park is a good example. Seventh graders from the neighbourhood
> were trained as park docents to explain the key features of this Jens Jensen
> design to other kids and adults. One area in Columbus park was designed as
> an open-air theatre; to highlight this feature, we presented a play there during
> opening ceremonies.

Urban restoration projects are occurring in cities around the globe, taking in such areas as remnant ecosystems (see Gobster 1994, in Dwyer and Stewart 1995) and waterfront development (e.g. see Law 1993; Craig-Smith and Fagence 1995; Williams 1995; Hall and Page 2002). Some areas, too, are applying principles of ecosystems management, which emphasise relationships between physical, biological and social elements in the urban landscape (Dwyer and Stewart 1995) to underpin urban plans.

One of the first comprehensive attempts to document concern for urban recreation opportunities in the US, was the National Urban Recreation Study, undertaken in 1978 by the Heritage Conservation and Recreation Service. The primary objectives of the study were:

- to examine perceptions of needs and opportunities held by recreation users and administrators in urban areas across the country from the neighbourhood to the metropolitan level;
- to identify major problems of recreation and open-space providers in meeting needs;
- to explore possible solutions to problems with a wide variety of citizen and governmental interests;
- to identify a variety of open-space areas with potential for protection;
- to define a range of options for all levels of government, with emphasis on Federal alternatives which could assist or facilitate local, State and private efforts.

(US Department of the Interior 1978: 20)

The study concentrated on seventeen of the nation's largest cities, along with smaller towns and countries within their immediate vicinity. The sample field study cities were considered to reflect the dominant recreation issues and problems facing highly populated urban areas in the US.

The 1978 report established that no coherent national policy existed at that time for a balanced system of close-to-home recreation opportunities for all segments of the urban population. The study also found that recreational deprivation was not always a function of lack of facilities. In many cases, existing or potential recreation resources were not being fully utilised because of inappropriate locations or physical characteristics, deteriorating conditions, and poor quality management and programming.

Despite the broad spectrum of urban recreation issues addressed, the report was able to set common guidelines to indicate major directions for public action:

- conserve open space for its natural, cultural and recreational values;
- provide financial support for parks and recreation;
- provide close-to-home recreation opportunities;
- encourage joint use of existing physical resources;
- ensure that recreation facilities are well-managed and well-maintained, with quality recreation programmes available;

- reduce deterrents to full utilisation of existing urban recreation facilities and programmes;
- provide appropriate and responsive recreation services through sound planning;
- make environmental education and management an integral part of urban park and recreation policies and programmes;
- strengthen the role of the cultural arts in urban recreation.

Almost three decades later, and with allowances for scale and local circumstances, the findings of the 1978 report have relevance for other parts of the developed world. In particular, the report recognised the great disparity in the wealth of urban communities and the unevenness in resource endowment, both being factors which make a common strategy for addressing shortcomings difficult.

There are obvious differences in the physical and social geography of individual cities. Sydney, Australia, for example, is 'blessed' with a magnificent harbour and a wealth of accessible sandy surf beaches which provide unparalleled opportunities for water-related recreation. Many other coastal cities across the globe are likewise fortunate, whereas urban concentrations away from the coast typically present a different and more limited recreation environment. The presence of natural features and opportunities for contact with nature, within or close to the built environment, also enhance the potential for outdoor recreation. Sydney, again, is fortunate in being ringed with magnificent national parks only a short distance from the city's central business district (CBD) and periphery.

Climatic conditions also play an important part in the availability of a range of recreation opportunities in urban areas. The snowfields backing the city of Vancouver, Canada, go a long way towards compensating for restrictions on outdoor recreation in the city itself, because of otherwise pervasive rainy conditions. Cities in the tropics and subtropics can usually support more diverse forms of outdoor recreation than those where short summer seasons and severe weather can restrict activities.

A city with attractive natural features and an agreeable climate can take advantage of these for recreation; those not so fortunate may need to compensate by the creation of artificial environments. The provision of extensive facilities for indoor sports and other recreation activities in cities in the higher latitudes is, in part, a reaction to the severe winters of that part of the world.

Social differences between cities can also account for disparities in opportunities for recreation. Where cities are large, long-established and densely populated, diverse cultural features are more likely to exist, and these can be the basis for varied forms of recreation experiences, from participation in ethnic festivals and traditional celebrations, to the sampling of exotic foods and shopping for unusual products. On the other hand, a bland urban environment with an essentially monocultural population and a narrow social focus, can offer a strictly limited, and perhaps, predictable range of outlets for recreation.

Plate 7.1   Sydney, Australia is not only ringed by beautiful beaches, large national parks and an array of recreational opportunities, it also contains beautiful inner urban parks such as Hyde Park, named after Hyde Park, London

The dynamics of urban recreation environments

> The urban domain is complex – towns and cities present an outwardly confusing mosaic of land uses into which recreational provision must fit and over which recreational activity must be superimposed. Patterns of opportunity are in most cases the product of lengthy periods of urban evolution in which physical growth, economic development and social change have combined to produce an environment that is dynamic, competitive and diverse.
>
> (Williams 1995: 14)

Urban areas are complex. They require planning approaches which are integrated (thus recreation and tourism are planned and developed in conjunction with other urban functions), flexible, and focused on 'the complementary function of the city and its region' (Jansen-Verbeke 1992: 33). Understanding the processes influencing urban development, and thus the nature and potential of outdoor recreation and tourism in urban areas, requires some understanding of global and local-scale developments, impacts and issues. Several important processes have shaped urban development, including:

- global economic restructuring;
- the physical expansion of built-up areas from compact, densely settled areas, with low overall populations, to modern post-industrial cities of highly populated but relatively low-density occupation, with residual zones of higher-density occupation in older, inner areas;
- increased social segregation as a result of mobility, preference and powers of different groups, and the ability of some to exert influence over emergent municipal authority;
- increased municipal regulation of development by way of attempts to control suburban sprawl, to plan new towns and to establish green belts;
- contemporary interests in greening the city and in encouraging environmental enhancement;
- urban redevelopment as a result of war damage, the establishment of urban development corporations, and the undertaking of high-profile projects (e.g. in dockland areas) to regenerate areas;
- urban redevelopment and restructuring, stemming from the need to address urban decay.

(Williams 1995: 15–16)

Given such processes and the problems that cities, generally, are encountering, it is not surprising that urban policies are now largely 'concerned with both winning economic growth and regenerating the core areas' (Law 1993: 23). The key elements in current urban policy comprise:

- emphasis on economic policies;
- emphasis on obtaining private investment;
- emphasis on property investment;
- public sector investment in infrastructure;
- public sector 'anchors', e.g. convention and entertainment centres, museums and art galleries;
- focus on the city centre;
- public-private partnerships;
- semi-autonomous agencies such as urban development corporations;
- flagship projects, e.g. Commonwealth and Olympic Games, Formula One Grand Prix, America's Cup;
- image and reimaging strategies (after Law 1993, in Hall *et al.* 1997a).

Tourism as an element of outdoor recreation has become an extremely influential element in urban planning in some areas (see Page 1995). The potential of recreation and tourism as instruments in the policy of urban revitalisation is being increasingly recognised by local authorities and urban managers (Jansen-Verbeke 1992). Indeed, 'the growth of tourism as a form of economic development is having a major impact on the urban landscape of some cities, and reflects a changing attitude toward inner cities as well as a need to diversify repressed economies' (Lew 1989: 15). Since the early 1980s, tourism has emerged as an important element in urban planning for such reasons as:

- economic globalisation and the consequent economic decline and restructuring of heavy industries and manufacturing in Western nations, led to a search for economic and employment alternatives by government at all levels, particularly in service industries;
- changes in transport technology have contributed to a decline of waterfront areas;
- tourism was seen as a way to rejuvenate and redevelop urban areas, often inner-city areas, which had experienced economic decline;
- in order to assist urban regeneration, governments have consciously sought to integrate tourism policy and development with cultural events and festivals, sports and leisure policies, and conservation of heritage, to help develop, market and promote urban regions and thus to attract the tourism and investment dollar.

(Hall *et al.* 1997a: 199)

Of course, while it is possible to generalise to some extent about the patterns and processes of urban development and restructuring, there are some national and regional variations. Apart from the obvious physical, economic, social and political differences between cities which affect recreation potential, intra-urban contrasts develop over time in urban morphology, land-use patterns and socioeconomic characteristics. These can impinge on recreation needs and opportunities. Features such as the decline of the CBD; the establishment of satellite shopping complexes in the suburbs; the gentrification of inner city slums; the alteration of conditions of accessibility by the construction of new transport links; the emergence of ethnic enclaves within the urban system; the effect of political decisions on investment in recreation and sporting facilities; and ongoing changes in the economic, social and age structure of the population, can all have dramatic effects on opportunities for outdoor recreation.

The City of Sydney, Australia, provides ready examples of the dynamics of urban recreation potential. The 'centre of gravity' of the city has now moved westward, with population growth, to the western suburbs of Parramatta and Penrith (among others), some 30 to 50 kilometres (approx. 18 to 30 miles) inland. Moving with it are the sporting and recreation facilities which have long dominated the eastern core of Sydney. These moves were strengthened by the need to provide world-class facilities for the Sydney 2000 Olympic Games. In turn, the availability of large areas of vacant land, once occupied by stockyards and industrial sites, gave further stimulus to the move westward and inland.

Any major city around the world can offer similar examples of the shifting and fluctuating nature of the spectrum of urban recreation opportunities, in response to changing physical, socioeconomic, environmental and political circumstances. Barrett and Hough (1989) point out that many municipalities in Ontario, Canada, have a legacy of parks and recreation space designed to meet the needs of earlier generations, yet, frequently they are inappropriate for the changing preferences, needs and lifestyles of the present population. They note that the social, economic and political context of urban communities continues to change at an increasing rate, but the planning response has not kept pace.

Pressure on the urban recreation landscape comes from a number of directions. Social concerns include the growing proportion of seniors and people with disabilities in the urban population – both groups with special recreation needs. Multicultural diversity in Canada, for example (as in Australia – see below), is placing a different set of demands on parks systems which were designed for a more homogeneous, predominantly white Anglo-Saxon culture. Urban recreation space also attracts the socially disadvantaged – the poor, the homeless, the transients and the unemployed. Apart from averting conflict with other users, the challenge is to develop programmes which encourage the disadvantaged to use recreation space more constructively. Again, the nature of recreation demand in cities is changing, with higher levels of environmental awareness, health and fitness programmes, and the emergence of a more varied array of leisure activities requiring specialised equipment and facilities.

At least some of the shortcomings in the urban recreation environment can be related to the above-mentioned dynamic elements in the character of towns and cities. An evolving pattern of urban growth and development can be recognised, marked typically by inner decay, suburban expansion or peripheral sprawl, and increasingly mobile and sophisticated groups of inhabitants. The implications are that any corrective measures proposed, must be adjusted for a particular geographical setting, area and population. Moreover, a distinction should, at least, be made between the inner city, the suburbs and the urban fringe.

The inner city

There is a good deal of evidence to suggest that the greatest deficiencies in regard to urban recreation space and facilities are to be found in the inner cores of large cities. Serious physical problems exist relating to the age, design and location of components of the recreation system. These are made worse when coupled with emerging recreation demands within rapidly changing urban precincts.

Population dispersion tends to take place from the centre, leaving behind both a diverse ethnic and cultural heritage, but typically, also, less affluent, elderly and otherwise disadvantaged groups. Any reverse movement of population is often representative of dissimilar and incompatible lifestyles, and merely adds a further dimension to the task of recreation provision. With gentrification of old inner neighbourhoods, basic deficiencies are aggravated by a new set of recreation demands from a diverse and rapidly changing clientele.

The inner city suburb of South Sydney, Sydney, for example, is experiencing a changing population. Recent census figures reveal that people born overseas made up a significant proportion of the population of the municipality. Many of these people were from a non-English-speaking background. Major birthplace groups of employed people are shown in Table 7.2. In other inner suburbs of Australian cities, ethnic enclaves are well established (e.g. Vietnamese in Cabramatta, Sydney;

Table 7.2 Birthplace of persons residing in South Sydney

Country	Persons
Australia	44,227
Canada	299
China (excludes SARs and Taiwan Province)(a)	1,382
Croatia	167
Egypt	179
Fiji	237
France	311
Germany	554
Greece	894
Hong Kong (SAR of China)(a)	673
India	484
Indonesia	810
Ireland	680
Italy	500
Korea, Republic of (South)	632
Lebanon	388
Macedonia, FYROM(b)	87
Malaysia	690
Malta	202
Netherlands	233
New Zealand	3,862
Philippines	472
Poland	309
Singapore	490
South Africa	437
Sri Lanka	115
Turkey	235
United Kingdom(c)	5,118
United States of America	713
Viet Nam	891
Yugoslavia, Federal Republic of	342
Born elsewhere overseas(d)	7,572
Not stated	12,772
Overseas visitors	5,292
Total	92,249

Source: Adapted from Australian Bureau of Statistics (ABS) 2001 *Census of Population and Housing*, at: http://www.abs.gov.au/Ausstats/abs@census.nsf/0/9b30e04ea929480cca256bbe00837672? OpenDocument, accessed 25 March 2005.

Notes
(a)   SAR is an abbreviation of 'Special Administrative Region'. SARs comprise 'Hong Kong (SAR of China)' and 'Macau (SAR of China)'.
(b)   FYROM is an abbreviation of 'Former Yugoslav Republic of Macedonia'.
(c)   Includes 'England', 'Scotland', 'Wales', 'Northern Ireland', 'Channel Islands', 'Isle of Man'.
(d)   Includes 'Inadequately described', 'At sea' and 'Not elsewhere classified'.

Koreans in Campsie, Sydney; and Greeks in Coburg, Melbourne, supposedly the largest 'Greek city' outside Athens).

Some older core-city areas are fortunately able to retain a sound financial tax base with an established network of parks and open space. Other declining, fiscally-troubled core cities are forced to allocate a large share of their recreation budgets to operation and maintenance, at the expense of acquiring and developing new facilities and programmes. Those capital funds which are available for investment in recreation in older cities are typically spent on rehabilitation of ageing facilities.

In some of the world's larger cities, further difficulties are encountered in attempting to cater for the recreation needs, not only of residents, but also of commuters and visitors, all within the same inner-city environment. A good example is the City of Westminster, central London, where, apart from permanent residents who number about 222,000 (and of whom only 55.8 per cent were born in the UK), some 547,000 workers commute daily (as London's largest employment centre), and 28.5 million visitors arrive annually from all parts of Britain and the rest of the world (City of Westminster 2005). On any given day there are likely to be more than 1 million people in the City. In this case, the Westminster City Council recognised that its particular responsibility was towards its resident population, especially in providing recreation opportunities close to home for this group, and spends a sizeable proportion of its budget on leisure facilities.

London, in common with most of the world's great cities, developed without the benefit of a comprehensive recreation plan. By the time the need for planning was evident, many options were closed off by the massive social and dollar costs of acquiring recreation space. Skyrocketing land prices and finite funding sources placed any available recreation space beyond the reach of urban authorities. In downtown Atlanta, for example, the excessive valuation placed on a 1.7 acre (approx. 0.6 ha) site sought by the city, meant that its acquisition was only made possible by donation. The result is that traditional recreation activities requiring large expanses of land are now simply not possible in the densely populated neighbourhoods of most inner-city areas. Fortunately, though, for Londoners and visitors, Hyde Park, which was established and opened to the public in the 1630s, and Regent's Park, which was a central feature of John Nash's housing development in the early 1800s, remain as recreational assets. Clearly, historical developments are important, and so it is also worth noting that the number and accessibility of parks increased dramatically in the period between 1850 and 1880, when 111 urban parks were created in Britain, compared with 49 between 1820 and 1849 (Conway 1991). Interestingly, in that same period – and more specifically in 1879 – Australia's first, and the world's second, national park, Sydney Royal (see Chapter 10), was established, mainly to serve recreational functions. It still serves as an important recreational asset within a dominant conservation management perspective as does Golden Gate Park, San Francisco, California (see Box 7.2).

In places where outdoor recreation opportunities are lacking, recreation needs can be met, in part, by providing indoor facilities or by innovative programmes to create additional urban recreation opportunities. Seattle, Washington, for example, transformed the air space over a ten-lane interstate highway into the 3.5 acre

---

Box 7.2 **The new urban park**

The earliest plans for the City of San Francisco, California, made no provision for public parks, so that residents had only private gardens and small urban squares to serve as retreats from noise and crowding. In 1870, Golden Gate Park was created from wind-swept sand dunes, comprising over 1,000 hectares (2,500 acres), extending from the city to the Pacific Ocean shores. It is the second largest urban park in America and is noted for the incorporation of roads, walkways and shelter belts deliberately designed to attract birds and wildlife and for the pleasure of visitors. In more recent years, San Francisco's green space has been expanded into the Golden Gate National Recreation Area, established in 1972 and comprising over 75,000 acres (approximately 30,000 ha) of land and water, from north of the Golden Gate Bridge to the peninsula south of the city. Golden Gate National Recreation Area is rich in natural, historical and cultural resources and is seen as a prototype for the new urban park of the twenty-first century. In many senses it is a hybrid – in its location, its diversity, its close relationship with a range of communities, its ability to involve the public and, at the same time, comply with agency standards. Management of the Golden Gate National Recreation Area presents a challenge for the US National Parks Service in addressing the expectations of conservationists and recreation users about the design and purpose of urban parks.

Source: http://www.nps.gov/goga/.

(approx. 1.4 ha) Central Freeway Park. Spanning the 'concrete canyon' on a bridge structure, the park offers an unusual retreat in downtown Seattle. Sydney, Australia, is another example of a large modern city where hectares of unused space on the rooftops of city buildings have been transformed into sporting and recreational facilities for office workers and residents. New buildings were the prime target of this policy, with incentives for developers to incorporate use of rooftop space into their plans. In the US, Britain and Europe, gardens, swimming pools and recreation areas were established on the rooftops of hotels, private homes and city apartment blocks. More specifically, more than 30 years ago, the US Bureau of Outdoor Recreation (1973) identified several successful space-conversion projects resulting in useful additions to the urban recreation resource base:

- in Baton Rouge, Louisiana, 35 acres (approx. 14 ha) of unproductive land beneath an elevated highway interchange was transformed into Interstate Park as a neighbourhood recreation area;
- in San Francisco, several park areas were developed on top of underground parking facilities (including Union Square), and in downtown Los Angeles two large corporations created a 2.5 acre (approx. 1 ha) rooftop park above a garage, as part of an urban renewal project;

(a)

(b)

(c)

Plate 7.2 Golden Gate Park, San Francisco. It contains open greenspaces, tennis courts, childrens' playgrounds, small pockets of bushland walks, shared cycle and pedestrian routes, a music concourse, Japanese Gardens, an arboretum and a host of other attractions and facilities

- in Albuquerque and Honolulu, airport buffer lands were transformed into a community golf course;
- along the lower Rio Grande, the city of El Paso recognised the potential of the river's floodway, in developing a linear park incorporating recreation activities and facilities capable of withstanding periodic flooding;
- the surface of covered water storage facilities in Denver and San Francisco were developed for public tennis courts and sports facilities;
- in Washington DC, a sanitary landfill site was transformed into a useful and valuable recreation resource;
- in New York City, construction of a 30 acre (approx. 12 ha) park on the roof of a sewage treatment facility, provided a picnic area, baseball diamonds, tennis courts, trails, swimming pools and an ice rink.

In view of such initiatives, it is surprising that not all undeveloped areas of urban land and water are perceived as recreation resources or used as such. Many city authorities apparently lack the imagination and/or the means to capitalise on the potential of neglected areas such as floodplains, water supply reservoirs and catchments, waste treatment facilities, waterfronts, parking lots, service corridors and abandoned rights-of-way and railway lines. Especially valuable are strips of linear open space, where the edge effect promotes greater recreation use.

The inner city remains the focus of intense competition for space, for commercial and industrial premises, for transport and communication, and for high density/high rise residential purposes. It is important that provision for recreation space is not ignored in the redisposition of the land and water resources of the urban heartland.

The suburbs

If it is difficult to make useful comparisons of recreation provision between inner cities, it is impossible to generalise regarding the recreation environment in the suburbs. Clearly, the dispersion of population from the core area referred to earlier, is stimulated by a perceived improvement in the quality of life, part of which is reflected in a better range and standard of recreation opportunities. However, the extent to which this is experienced depends upon the particular local mix of such factors as location, resource base, socioeconomic status, community spirit, and the affluence and initiative of the local government authority.

Suburbia has, or should have, one advantage – relatively newer facilities are likely to mean lower operating and maintenance costs, leaving more funds for investment in capital projects and acquisition of land. However, recently settled suburban communities with small, but rapidly growing populations and limited financial resources, are more often concerned with the availability of basic services than with the 'luxuries' of amenity provision.

Again, the very nature of the suburban environment, typified by a real sprawl of dispersed housing units and dormitory-style subdivisions, with heavy reliance on the motor car, makes it difficult and expensive to provide a full range of recreation facilities. Site design of many early subdivisions precluded the

Plate 7.3   Canberra is the capital city of Australia. Urban living is enhanced by large amounts of greenspace, cycleways and waterways, including Lake Burley Griffin. This picture is taken from Mount Ainslie and shows the proximity of urban housing to greenspace, the Australian War Memorial (bottom left), the Old and New Parliament Houses and the Lake, which is ringed by cycle/pedestrian paths, attractions such as Telstra Tower and the National Museum, and important facilities such as the National Library and Questacon (a place of hands on demonstration of science for people from all walks of life)

Plate 7.4   Unused railway corridors can be fruitful sources of recreational activity. This corridor is in Toronto, New South Wales, Australia. The disused station houses an historical trust

establishment of large open spaces for community recreation. Both private developers and public housing authorities appear to have given only minor consideration to this aspect, apart from labelling a mandatory minimum area as 'recreation reserve'. There seems little evidence of any comprehensive planning of recreation facilities as an integral part of emerging neighbourhood and community development patterns. Where tracts of recreation land are set aside, they are usually developed on an *ad hoc* basis, with scant regard for other than the immediate needs of the existing population.

Large modern cities typically spill over, unimpeded, into the surrounding countryside, in a process aptly termed 'metropolitan scatteration' (Wingo 1964). The rapidly diffusing residential frontier is allowed to outpace provision even for basic service needs, the urgency of which merely reinforces the traditional low priority given to recreation planning. Unimaginative land subdivision perpetuates the conventional grid street system, with associated large-scale alienation of potential recreation space. Little thought is given to the most appropriate size or form of the overall neighbourhood, or its relationship to the rest of the city.

Australian cities, in particular, are much less compact than their older European counterparts, with corresponding lower residential densities. Perhaps this has less to do with the relative scarcity of land than with Australians' preferences for maximising private space at the expense of public amenity.

The resulting featureless sprawl promotes a degree of introspection in urban Australia, or a tendency towards 'privatization' (Mercer 1980b). In the absence of local clubs or pubs, community meeting centres or sports complexes, cinemas, or even shopping centres in some cases, the inhabitants place greater emphasis on the home environment for their leisure pursuits, which are increasingly geared towards technology (television; computer games, email and the internet; DVDs). The sheer lack of public facilities forces households to maximise private space by way of compensation, but in the restoration of homes in many areas there is a tendency to build large 'palaces' which cover almost the entire allotment. The idea of the quarter acre (0.1 ha) block is diminishing, and where it exists in some form, it bears little resemblance to the quarter blocks where backyards, gardens and outdoor facilities such as pools and barbecues were once prominent. In these cases, the provision of public recreation space becomes even more critical.

Moreover, the only practical means of transportation in a highly dispersed, rapidly expanding, low-density suburban area, is the private motor vehicle. Those without a car are severely disadvantaged with respect to leisure options. The scale of metropolitan planning is geared to the car and not the human being. As the metropolis spreads, pressure to construct intra-metropolitan freeways to accommodate the car and overcome traffic congestion, also increases. Such freeways accelerate residential development towards the periphery. Ex-urban recreational opportunities are pushed further and further away from the centre of gravity of the population, to the detriment both of people living in the inner suburbs and of those without access to private transport. Thus, transport improvements, proposed as a solution to one urban problem, merely give rise to other problems, and recreation opportunities decline further.

It appears, then, that the modern city has let its inhabitants down as far as outlets for leisure in any communal sense are concerned. What seem to be lacking are the essential ingredients to create the 'village' atmosphere of earlier times – a setting which will generate a sense of togetherness, belonging and place. Features which once had an important recreational function as part of that setting have no place in present-day suburbs. The town square, the village green, the dance hall-cum-cinema, even the local 'pub' or bar in some cases, have given way to home-based recreation, centred on the television set, perhaps the backyard pool, and all manner of electronic gadgetry. The sterile facilities which often serve for community recreation purposes do nothing to offset urban alienation. It is difficult to identify with a slab of concrete or fibre-glass, and it is little wonder that the potential users attempt to humanise or deface these structures with graffiti. They see nothing wrong with vandalism of incongruous features to which they cannot relate and which apparently cannot satisfy their recreation needs.

Local authorities, which have the prime responsibility for recreation, face a deepening cost-revenue crisis, made worse by a general indifference on the part of the higher tiers of government to the problems of cities. This situation serves to underline the need for fresh initiatives in urban recreation planning. Part of this strategy should be a broader approach to the provision of leisure and outdoor recreation opportunities in the suburbs, with greater emphasis on self-help and community involvement. Out of necessity, planning bodies might come to realise that some of the deficiencies inherent in suburban life and living may be remedied by encouraging fuller utilisation and management of communal recreation resources.

The urban fringe

In 1981, the US Department of Agriculture estimated that some 3 million acres (approx. 1.2 million ha) were being converted each year to urban and built-up uses across North America. More recent observations give no basis for optimism that this trend is in decline. In these urbanising areas, local initiatives to direct development away from critical agricultural, environmental and recreational uses, are often weak or non-existent. The city periphery thus becomes the focus for some of the most urgent programmes for general living and open-space retention purposes.

One of the problems in discussing peri-urban recreation is to decide where suburbia ends and exurbia, or the urban fringe, begins. Yet, it is important to consider recreation opportunities in this transition zone, because mobile city populations readily incorporate nearby fringe areas into their effective recreation space. Despite the growing importance of home-based recreation noted earlier, the neighbouring countryside is increasingly perceived as an extension of life in the city. The tendency for people to seek natural settings to offset the pressures of an urban-industrial existence is well-documented, and is prompted, in part, by the urbanisation process itself. Janiskee (1976) explained the recreation appeal of extra-urban environments in the context of a push–pull model of motivation. Periodically, environmentally undernourished urbanites are 'pushed' from the city because of stresses imposed by their lifestyles. At the same time, they are 'pulled' into the more natural hinterland by the opportunity to experience compensatory alternative surroundings and activities. Apparently, urban dwellers, who have voted with their feet for city living, are not totally adapted to the urban environment. They have a physical and social need to seek novel, irregular and opposite situations, in exchange for routine, boredom and the familiar. This need is reinforced by growing awareness of what the surrounding countryside has to offer, together with enhanced means of making use of its recreation potential. Natural settings offer city dwellers the capacity for self-renewal in a different, specifically outdoor setting, inevitably leading them to the urban fringe and beyond for recreation. That said, the challenge remains of maintaining the essentially undeveloped character of the urban fringe so that its function as recreation space is unimpaired.

The task is given added urgency by the different perceptions held of the urban fringe. To land developers, it could be seen as a speculator's paradise; to urban planners it might represent a useful reserve of land for future urban expansion, or for the location of less compatible elements of urban infrastructure such as motorways, airports, waste disposal sites and noxious industries. Alongside these, is the potential to expand the city's spectrum of recreation opportunities by the creation of active and passive, large-scale and specialised recreation facilities.

Pullen (1977) argued strongly for the establishment of permanent areas of 'Greenspace' beyond the periphery of cities, as a means of guiding and containing urban development. Pullen saw Greenspace as a valuable resource for the provision of important social functions, including recreational activities. Despite the attraction of the concept, experience in major world cities suggests that the protection of a permanent zone of Greenspace is difficult in the face of compelling pressures to maximise economic use of valuable land.

Short of public acquisition and creation of formal park land, the zoning of land in the urban fringe, as reserve for recreation, should act as a further deterrent to development. This step is being enhanced by the establishment of community forests on the outskirts of major cities in Britain (National Community Forest Partnership undated). The inspiration for the new community forests came from Bos Park near Amsterdam and the ancient stands of trees in Epping Forest, close to the heart of London. They were a central plank of commitments to double tree cover in England made by the then central government at the Rio Earth Summit (Hunt 2000: 219). In the early period of this initiative, two parks were established in the North East of England. The Great North Community Forest, for instance, was established in 1990 and covers some 250 square kilometres (96 square miles) of the urban fringe. This 'project' has resulted in the reclamation of over 200 ha (approx. 494 acres) of derelict land, the planting of 800 ha (approx. 1,976 acres) of woodlands, and the creation or improvement of more than 450 ha (approx. 1,112 acres) of wildlife habitats.

There are now twelve such forests, involving 58 local authorities, the Country-side Agency and the Forestry Commission. Despite the desirability of these imaginative projects, in terms of adding to outdoor recreation opportunities in the urban fringe, their viability was initially considered uncertain given the long-term investment involved, along with financial constraints on new plantings (Wilson and Biberbach 1994). Of course, financial and other constraints aside, recreational access is not always guaranteed by private land-owners reluctant to take on added management responsibilities. However, this initiative does seem to hold much promise and the government in this part of the world has a lengthy history in offering incentives to owners, while owners themselves have a more 'relaxed' attitude to recreational access to their properties than landholders in countries such as Australia or the US (see Chapter 8). Demand for the countryside will be deflected from national parks to areas managed for conservation, recreation and education, right in the urban fringes (Evans 2001). These are truly significant developments for recreational provision for urban residents and mirror developments in Australia, particularly with respect to the development of regional parks in Sydney.

This illustrates an important qualification regarding implementation of any plan to utilise the recreation potential of the urban fringe. The plan cannot succeed without the firm commitment of responsible public authorities, both financially and in terms of statutory powers over land use. Cooperation with controllers of private land and resources in the semi-urban countryside is also necessary if this resource is to fulfil its role as an integral part of the urban recreation environment.

Urban open space and recreation space

In the highly urbanised countries of the Western world, the city functions primarily as a place of residence and as a base for work commitments. The growing segment of life given over to leisure appears to find only restricted expression in urban environments. More and more people are looking beyond the city limits to find their 'activity space' for outdoor recreation in rural areas. However, for many urban residents, this alternative is not accessible for reasons such as lack of transport, time or money. These people must turn towards open space within the city for relief from perceived deficiencies in the urban environment.

Much of the dissatisfaction with urban living, and many of the concomitant social problems, can be traced to the apparent inability of the modern city to meet the basic needs of its inhabitants. One of the objectives of urban environmental and recreation planning is to produce a more satisfying array of amenity stimuli and responses. The range and intensity of amenity responses are, in turn, a function of the nature, characteristics and location of what may be called amenity precipitants. In an urban situation, a fundamental component of the amenity response system is again the availability of open space for recreation.

According to Gold (1988), an effective recreation experience in cities calls for opportunities to experience freedom, diversity, self-expression, challenge and enrichment. Servicing such opportunities provides much of the justification for providing open space within cities. In this context, open space is basic to the structure and function of the built environment in meeting human needs. Yet, various factors can affect its role as part of the urban outdoor recreation resource base. In the first place, it is too simplistic to equate open space with recreation space, since not all urban open space is equipped to function as recreation space. By way of example, some modern, planned national capitals are blessed with vast areas of open space, geometrically arranged, trimmed and manicured, yet devoid of any feature which would encourage, facilitate or even permit leisure activities. In many cases, any recreation function, apart from perhaps passive viewing, is specifically excluded by physical barriers, equally forbidding signage or other effective means of discouraging participation. 'Open space it may be; recreation space it is not' (Pigram 1983: 109).

This is not to deny that urban open space *per se* has value, apart from a potential recreation role. Demands for lower residential densities in affluent areas, and for extensive landscaped sites for public buildings and industrial estates, demonstrate a growing social awareness of space as a community asset. Added to this is the acknowledgement of what economists term the 'existence value' of open space and green areas within cities. Nearby residents can develop strong attachments for, even rather ordinary, local parks, which they may rarely use for recreational purposes.

However, satisfaction of the leisure needs of urban dwellers requires more than the existence of open space. In the provision of recreation space in cities, it is not a matter of how much, but how good that space is. In part, this will reflect the characteristics of urban open space in terms of size, range of facilities and accessibility. The importance of the natural setting in contrast to the surrounding built-up environment would also seem to be paramount. With reference to Sydney, Australia, McLoughlin (1997) argues persuasively for the retention of bushland within the urbanised area. Earlier in this chapter, the close proximity of a number of national parks to the City of Sydney was noted. However, despite this, McLoughlin identifies a range of complementary values for bushland, in or near urban areas:

- natural and cultural heritage values;
- habitat for resident and migratory species;
- aesthetic landscape values separating parts of the city, and as a screen for unpleasant urban structures;
- environmental protection values;
- recreational values for a variety of activities;
- scientific and educational values.

(McLoughlin 1997: 166)

McLoughlin also identifies threats to remnant areas of bushland, and the measures which need to be taken to minimise the impacts of city growth and development on this valuable element of urban open space.

In a study of urban parks in Melbourne, Australia, the attractiveness and variety of the vegetation, and the presence of waterbodies, were found to be important factors in accounting for variations in recreational use (Boyle 1983). At some parks, a strong preference was expressed for peace and quiet in relatively natural areas with few facilities. A significant number of respondents at two native eucalypt parks, for example, where minimal equipment has been installed, insisted that more facilities were *not* needed.

A similar preference for nature-dominant environments was revealed in a major study of inner-city parks in the City of Brisbane, Australia (McIntyre *et al.* 1991). Results of the study suggest that the natural setting of inner-city parks and green areas provides a venue for rest, recreation and release from tension for urban residents, as well as an opportunity to appreciate nature. The preference revealed for natural settings 'emphasises the need for the preservation of these "islands of naturalness" within the cityscape' (McIntyre *et al.* 1991: 16).

In the US, corridors of protected open space, known as 'Greenways', are managed for conservation and recreation purposes under a programme established by The Conservation Fund. Greenways often follow natural land or water features, linking nature reserves, parks and cultural and historic resources with each other, and with populated areas. Some are publicly owned, some are privately owned, and some are the result of public/private partnerships. Some are open to visitors, others are not. Some appeal to people, and some attract wildlife. Greenways, linking large natural areas, have also been developed and promoted in rural areas. According to The Conservation Fund, Greenways protect environmentally important lands and native plants and animals, simultaneously linking people with the natural world and outdoor recreational opportunities. Greenways can also preserve biological diversity by maintaining connections between natural communities; soften urban and suburban landscapes; protect the quantity and quality of water; direct development and growth away from important natural resource areas; provide alternative transport routes; and act as outdoor classrooms (for more details see http://www.conservationfund.org/conservation,greenway/htm).

The above discussion adds emphasis to the importance of matching park and protected area settings to the preferences of users. It also raises questions of multiple use, and of non-use or under-use of urban parks.

Multiple use of urban recreation resources

Implicit in several studies of urban open space is the waste involved in setting aside resources for some exclusive use. Public institutions, in or near urban communities, frequently provide opportunities for innovative recreation programmes. Despite additional surveillance costs and possible problems with anti-social behaviour, establishments such as schools, hospitals, child-care centres, health clinics, religious and cultural facilities, fairgrounds, sporting arenas and even military bases, can all have significant potential in multiple use of cost-effective communal recreation space.

School properties, in particular, represent a sizeable part of readily accessible publicly-owned resources. They are usually well distributed within cities, and occupy strategic locations in residential neighbourhoods. Most have playgrounds or playing fields attached, and many have indoor gymnasiums and pools. Yet, aside from their primary role, they are often one of the least utilised public facilities, remaining empty when recreation pressures are greatest – after working hours, at weekends and during vacations. In many areas, opportunities also exist for reclamation or conversion of abandoned public buildings to provide indoor recreation centres. Key elements are diversity and flexibility: the opportunity for a range of recreation opportunities likely to attract a broad cross-section of the community, yet amenable to a change of function and orientation.

Children and play-space

> How do we create an environment that meets a child's need and urge to explore, test and experiment?

Play-space for children is a particularly sensitive issue in urban environments. Play which involves interaction with nature and natural processes is considered important for childhood development (Cunningham and Jones 1987). Whereas natural environments have been shown to have innate appeal for pre-adolescents, the absence of suitable, accessible and safe sites precludes this experience for many children (Cunningham and Jones 1994). In many cases, it is not the decision of the child which dictates the play location so much as the perception of the parent or guardian regarding what constitutes a suitable child-friendly environment.

According to Raymond Unwin (in Williams 1995: 18), the amenities of life have been neglected in that 'we have forgotten that endless rows of brick boxes, looking out upon dreary streets and squalid backyards, are not really homes for people'.

> The types of city Unwin and his followers wished to plan afforded not just those opportunities for leisure that were already associated with parks and gardens, but extended significantly the scope of the home and the suburban streets to support informal recreation and children's play. In time, the utility of the street as a recreational environment would decline in the face of the environmental onslaught of increased road traffic but in, say, the years between 1918 and 1939, the streets of leafy suburbia became an almost unnoticed, but significant recreation resource.
>
> (Williams 1995: 18)

More recently, playing on or near city streets, for example, has been generally frowned upon. Yet, with children, the streetscape tends to be popular for recreation purposes, despite obvious hazards. In fact, it appears that the busier the street may be, the more appealing it is. The unstructured nature of city streets and footpaths, with its clutter and ever-present element of danger, apparently offers an exciting and challenging contrast to conventional playgrounds. Rather than attempting to counteract this appeal directly, it would seem more productive to take advantage of the opportunities at the street-scale for design of imaginative and safe play areas. Bannon (1976) uses the example of Central Harlem, New York, to illustrate the potential for transformation of small blocks of vacant land in built-up areas, into 'vest-pocket parks' and 'tot-lots' as a viable alternative to the streets for play, or for quiet relaxation by older residents. 'Adventure playgrounds', where children are allowed and encouraged to create their own play environment under non-restrictive supervision, provide an unorthodox, but potentially very important, setting for spontaneous enjoyment:

> In urban areas where space of any kind is at a premium ... adventure play-grounds are the closest we have come to emulating some of the mysterious and exciting pleasures of childhood ... The land is left in its original state, with building materials (such as wood, cardboard boxes, logs, planks, bricks and so forth) provided for the children to build almost anything they desire ... Building a house, planting a garden, digging a tunnel, cooking a meal, swinging on ropes from trees, creating a mysterious artefact, anything children enjoy which does not endanger them or others is permitted.
>
> (Bannon 1976: 205–6)

By way of example, the Lenox-Camden playground, Boston, was run from April to October, 1966, and carefully studied by Robin C. Moore (Bengtsson 1972). From Moore's assessment of his experience, a number of observations warrant mention, and are listed in Table 7.3. Whereas many playgrounds are now developed along less creative and flexible lines, the Lenox-Camden playground serves as an important reminder of the need for, and importance of, less 'structured' playgrounds.

Street closures for an hour, or a day, or for longer periods, perhaps with the introduction of mobile recreation programmes, are another means of harnessing and redirecting the attraction of the streetscape as a neighbourhood recreation resource to provide *ad hoc* play-space for children.

Use and non-use of urban recreation space

Failure to recognise urban recreation opportunities is not always confined to city administrators. Potential users, too, seem reluctant at times to avail themselves of the facilities which are provided. Field observations suggest some surprisingly low levels of utilisation of recreation space, especially in the inner core of some cities. An Australian study in the inner suburbs of Melbourne found that a neighbourhood park was not 'a particularly vital part of most residents' perceived environment' (Cole 1977: 93). Although considerable diversity was discovered in user groups and activities, the dislike for the park displayed by children, in particular, was traced to constraints on natural patterns of active child behaviour; possibilities for creative play in the park were virtually non-existent.

The broader issues of non-use and under-use of urban parks were first highlighted by Gold in 1972. Gold concluded that the major constraints could be grouped into three categories – behavioural, environmental and institutional (Table 7.4). Not all of these inhibiting factors are easily countered, but obviously convenience of access, site characteristics, location, level of facilities, safety

Table 7.3 Lenox-Camden playground, Boston

The playground was from April to October 1966 and carefully studied by Robin C. Moore. From the assessment of his experience the following observations have been abstracted:

1 Creative play is an opportunity for children to manipulate their environment to achieve their own ends, and to sense that the world around them can be changed and need not be taken as given.

2 Activity was the initial attraction and reason for coming to the playground; but if the setting allowed people to sit around watching, they did so – talking, singing, joking, flirting, etc., while the more intensive activity of the playhouse, tower and basketball court provided a background interest.

3 … it often appeared that activity on the playground was only one link in a chain of play activities occurring in and around the child's home.

4 The playground was always the first place to 'check out' when looking for friends and/or action. This leads to the conclusion that play-spaces should be incorporated intimately into housing areas.

5 The most important observation in terms of age was that it bore little relation to physical ability, to courage in particular – as well as spills. A six-year-old girl would, for example, climb up the tower without a second thought, while an 11-year-old boy would be scared and unable to take the same route. This observation has many implications for design, such as the non-segregation of different age groups.

6 One aim was to discover the most popular moveable materials. These turned out to be mild crates, large timber cubes 1 ft to each side, 2 in thick timber up to 12 in wide and 5 ft long, sheets of masonite and ship-board, 50 gallon barrels and many other kinds of robust junk.

7 Moveable materials did raise a number of practical problems. The less robust items tended to get smashed and lost their usefulness. They had to be cleaned up and disposed of. After a while, moveable materials became dispersed over the playground tending to reduce their play potential. Stimulation was increased if they were reassembled frequently by the adults.

8 The greatest amount of creative activity, in terms of frequency and span, took place behind and in the playhouse. It is suggested that one of the reasons for this was the sense of enclosure there: spaces of adequate size for constructive activities, cut off psychologically from the surroundings, even though other activities were going on in the immediately adjacent area …

9 A critical difficulty is trying to comprehend the very small-scale environment that children operate in.

10 Materials that would normally appear as 'junk' in other peoples eyes are very relevant to much creative and imaginative play. Useful junk consisted of objects that could be used for building construction or objects that previously had a specific function and could still be used as such: the steering wheel of a car, for instance, became the steering wheel of a 'fire engine'. These materials were often used individually as props to the imagination and many times functioned as the initial stimulus, setting the child's thinking along a particular line.

Source: Bengtsson (1972).

Table 7.4 Major causes of non-use in neighbourhood parks

Behavioural	Environmental	Institutional
User orientation*	Convenient access*	Goal differences*
Social restraints*	Site characteristics*	Personal safety*
Previous conditioning	Weather and climate	Relevant program
Competing activities	Physical location	Management practices
User satisfaction	Facilities and development	Maintenance levels

Source: Gold (1973: 103).

Note
* Most significant in each category relative to all factors.

considerations and management and maintenance are subject to manipulation. Gold's comments support the view that non-planned designation of open space in urban areas, with little thought to effective location, size and quality, will probably ensure that it remains open space – empty and ignored.

Patmore (1983) also pointed to several factors impinging on patterns of use and non-use of recreation facilities. As noted in previous chapters, effective access is not related to convenience of location alone. Patmore categorised four types of barriers to access:

- *physical barriers*, which include personal limitations and the nature of intervening space;
- *financial barriers*, which impose a direct economic constraint through high levels of admission charges or equipment costs;
- *social barriers*, which arise from the association of the images of certain recreational pursuits with social status;
- *transport barriers*, which relate to lack of access to a vehicle and associated time/cost deterrents on participation.

Godbey (1985) translated many of these constraints on participation into a useful model for summarising the reasons why people do not participate in a specific recreational activity (Figure 7.1). The model, when applied to an urban park, can identify options for action by management. Such remedial measures need not be elaborate and can be as simple as relocation of an entrance or better maintenance of grounds. Some of the techniques used to identify and redress causes of, and responses to, under-use of recreation facilities at water storages in the US are set out in Table 7.5.

More generally, solutions to a lack of readily accessible recreation opportunities in cities rest with more enlightened planning of the urban environment to provide adequate recreation space and appropriate recreation facilities to meet the demands of their citizens.

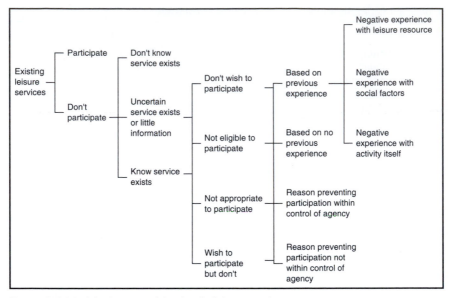

Figure 7.1 Model of non-participation in leisure services

Source: Godbey (1985).

Urban recreation planning

Many approaches to urban recreation planning have evolved in moves away from *ad hoc* responses to outdoor recreation provision. Aside from *ad hoc* approaches, other approaches include standards based, existing use, and investigative approaches. *Ad hoc* approaches have little to commend them except in dealing with urgent provisions of recreation opportunity, and are particularly evident in the facilities that quickly become obsolete (e.g. in Australia, BMX bike tracks), or when inappropriate facilities are provided for small groups.

Whatever approach is adopted, several questions are fundamental with respect to urban recreation space. In no particular order:

- How much is needed?
- What form should it take?
- Where should it be located?
- How should it be managed?
- Who is it for?

The first of these questions concerns measures of quantity, and this inevitably involves reference to space standards – specific numerical indicators of the adequacy of recreation provision.

Frequently around the world, attempts are made to arrive at desirable and practical standards for parks and open space, relative to user populations. At the

Table 7.5 Identifying and solving under-use

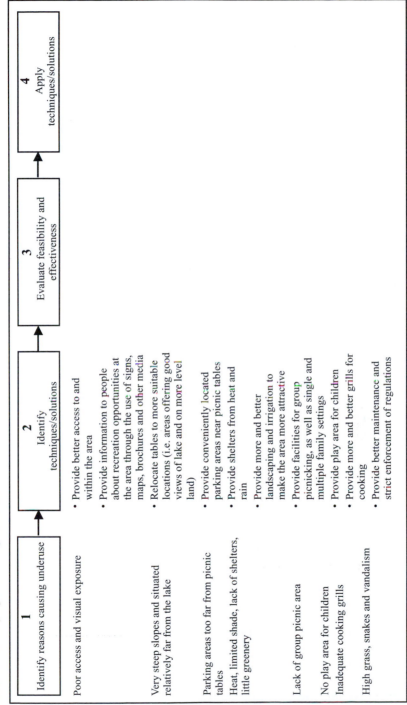

1 Identify reasons causing underuse	**2** Identify techniques/solutions	**3** Evaluate feasibility and effectiveness	**4** Apply techniques/solutions
Poor access and visual exposure	• Provide better access to and within the area • Provide information to people about recreation opportunities at the area through the use of signs, maps, brochures and other media		
Very steep slopes and situated relatively far from the lake	• Relocate tables to more suitable locations (i.e. areas offering good views of lake and on more level land)		
Parking areas too far from picnic tables	• Provide conveniently located parking areas near picnic tables		
Heat, limited shade, lack of shelters, little greenery	• Provide shelters from heat and rain • Provide more and better landscaping and irrigation to make the area more attractive		
Lack of group picnic area	• Provide facilities for group picnicking, as well as single and multiple family settings		
No play area for children	• Provide play area for children		
Inadequate cooking grills	• Provide more and better grills for cooking		
High grass, snakes and vandalism	• Provide better maintenance and strict enforcement of regulations		

national level, the National Recreation and Park Association of the US (Lancaster 1990) listed several reasons for park and recreation standards:

- an indication of minimum acceptable facilities for people in urban and rural communities;
- a guide to land requirements for a range of park and recreation areas and facilities;
- a means to bring into line recreational needs and spatial analysis so as to foster the development of a system of parks and open space areas for a community;
- a guide for regional development;
- as justification for the supply of parks and open space as an important consideration in a region or community.

In urban situations, the most frequently cited standards range from about 3 to 10 acres (approx. 1.2 ha to 4 ha) per 1,000 people, the total encompassing parks and playgrounds under various categories. The space standards for Canberra, Australia (see Figure 7.2), were considered by planners to be appropriate for the particular type and size of population of that city. Cambridge City Council in the UK developed open space standards based on the National Playing Field Association (NPFA) (Earley 1989). Broadly, the standards apply for new residential developments and conversions of buildings to residential use (Cambridge City Council 2004). The Council sets specific standards, applies them to different residential types, and has means of calculating open space requirements (Table 7.6).

A standards approach is adopted to provide open space for recreation by Vancouver-Clark Parks and Recreation, Washington. Park standards are given for two open space systems – the urban park system and the regional park system (see Table 7.7).

Whereas the standards approach specifies a total area of urban space set aside for outdoor recreation, it does little to ensure that such space will:

- be part of an overall scheme to ensure accessibility;
- be designed for specific purposes and community needs;
- complement the regional open space system;
- take into account natural features.

(Schomburgk 1985: 22)

Obviously, standards provide simple and clear instructions on how much to provide. They give some indication as to what is an appropriate or reasonable amount of open space to set aside, ensure some consistency in application and may help to minimise conflict. However, it is unrealistic to attempt to apply common standards across contrasting communities – standards which are inflexible and unrelated to changing socioeconomic profiles of potential users, or to varying space needs for different recreation activities. The fact that supposedly universal norms have not always been attained, reflects the many factors which should

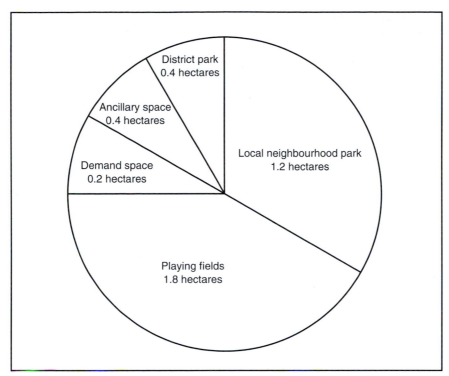

Figure 7.2 Area standards for park planning in Canberra, Australia

Source: National Capital Development Commission (1978: 10).

Notes
1   Ancillary space refers to space for screen and shelter planting, sound-reducing planting, landscape development and easements for overhead powerlines and floodways.
2   Demand space refers to space for tennis courts, swimming pools, bowling greens, squash courts and 'concessional entertainment' for organised and social sport.

influence a more realistic definition of space standards. As the Office of the Deputy Prime Minister, UK, declared under *Planning Policy Guidance 17: Planning for Open Space, Sport and Recreation* (2002), 'The government believes that open space standards are best set locally. National standards cannot cater for local circumstances, such as differing demographic profiles and the extent of existing built development in an area' (Office of the Deputy Prime Minister, UK, http://www.odpm.gov.uk/stellent/groups/odpm_control/documents/contentserver template/odpm_index.hcst?n=3425&l=3).

Clearly, any set of recreation space standards should only be used as a guide, to be modified as required and applied sensibly in the context of the socio-cultural characteristics of the community involved, and the resource attributes of the subject urban environment. In particular, rigid adherence to uniformity should not be allowed to obscure the many possibilities for innovative planning, management and design of leisure opportunities that are less demanding of space. In other

Table 7.6a Cambridge City's open space standards

Type of open space	Local plan standard
Formal open space Organised sports such as playing pitches, courts and greens	1.6–1.8 ha per thousand population, of which 1.12 ha should be for pitch sports (16–18 m² per person)
Informal open space Recreation grounds, parks andcommon land, access corridors and green spaces	1.5–1.8 ha per thousand population (15–18 m² per person)
Children's play areas Equipped play areas and areas provided for youth	0.2–0.3 ha per thousand population (2–3 m² per person)

Table 7.6b Application of Cambridge City's open space standards

Type of residential development	Formal open space (organised sports e.g. fields and bowling greens)	Informal open space (e.g. green spaces, access corridors, casual play space)	Children's play areas (e.g. equipped children's play areas and youth spaces)
Private residential/ Housing Association	Full provision will be met	Full provision will be met	Full provision will be met
Retirement housing	Full provision will be met	Full provision will be met	No provisions
Non-family/ Student housing	Full provision will be met	Full provision will be met	No provisions
Family/student housing	Full provision will be met	Full provision will be met	Full provision will be met

Note

The plan includes several caveats to these provisions. For example, full provision is not sought for informal open space and children's play areas if family student housing is directly linked to a college and it can be shown that adequate provision of formal open space is made by that college.

Table 7.6c Calculation of open space requirements

No. of units	Unit size	No. of people	Formal open space	Informal open space	Children's play area
20 (The formula here is based on the fact that the house sizes are unknown)	2 bedroom	20 × 2 = 40	40 × 16–18 = 640–720 m²	40 × 15–18 = 600–720 m²	40 × 2–3 = 80–120 m²

Source: Adapted from Cambridge City Council (2004).

Table 7.7 Park standards for urban park and regional park systems, Vancouver, Washington

Park system	Acquisition and development standards
Urban park system Includes: neighbourhood parks (3–5 acres); community parks (15–100 acres); open space (forested areas and wetlands)	*Acquisition standard*: 6 acres/1,000 people Population (2000): 251,348 Acres needed to meet standard: 1,508 acres Current acreage: 1,526 acres Current surplus: 18 acres *Development standard*: 4.25 acres/1,000 people Population (2000): 251,348 Acres needed to meet standard: 1,068 acres Current acreage: 653 acres Current deficit: 415 acres
Regional park system Includes: conservation and habitat properties countywide – e.g. regional parks, special facilities (sport facilities), greenway corridors	*Acquisition standard*: 10 acres/1,000 people Population (2000): 340,011 Acres needed to meet standard: 3,400 acres Current acreage: 2,300 acres Current deficit: 1,100 acres *Development standard*: 18% of total site Population (2000): 612 acres Acres needed to meet standard: 417 acres Current acreage: 653 acres Current deficit: 195 acres

Source: Adapted from Vancouver-Clark Parks and Recreation, Washington (http://www.ci.vancouver.wa.us/parks-recreation/parks_trails/planning/standards.htm).

words, a strict standards approach confuses recreation opportunity with area and recreation space *per se*. Standards, originally prescribed as minimums, become maximums and even optimums in some cases.

Again, the pace of modern city development quickly invalidates the setting of inflexible standards. It is not always a lack of conviction on the part of recreation planners regarding the desirability of departing from space standards advocated, so much as the unavailability or cost of land. Although application of standards might be marginally better than a completely *ad hoc* process, it cannot cope with the emergence of 'new' recreation resources, and makes no provision for community input or the involvement of private or commercial enterprises.

It is as well to remember, too, that in fully developed urban areas, it is not generally practicable to redistribute recreation space to match changing needs. However, given sufficient flexibility, the type of facility and the balance between sporting use and informal recreation pursuits, can be adjustable over time.

Moreover, mere figures have little to say about the form, quality and essential characteristics of the recreation space designated under the idealised standards adopted. Too often, the urban recreation system has to make do with 'left-over' or derelict areas, for which no other use can immediately be found. Minute, isolated parcels of low-grade land, devoid of vegetation or other natural features, that are unimaginatively designed and inadequately equipped, may meet the arbitrary space standards set, but do little to meet the recreation requirements of a neighbourhood.

Table 7.8 Planners' checklist for assessing plans of proposed subdivision reserves

1   Reserve function/purpose
(a)   State main function/purpose
(b)   State secondary function(s)
(c)   Is the main function of the reserve;
    (i)   Drainage or Screening (If 'No' go to (ii))
        If 'Yes'   Could alternative sites be used for reserves?
                    Could alternative sites be acquired?
                    Would the monetary contribution be more useful?
        or   Are the proposed drainage/screening locations in accord
                    with a plan of proposed open spaces?
                    Can they contribute to a proposed regional train network?
    Remember: Drainage and Screen Reserves area over-represented in Munno Para.

    (ii)   Recreation?
        Can the reserve provide for active pursuits?
        Can the reserve become a District Community Park?
        Can the reserve provide for two or more categories of recreational activity?
        e.g. walking, picnics/barbecues, informal active pursuits, formal sports,
        bicycle riding tracks, dog walking, jogging, play on playground, horse riding,
        etc.
    Remember: 'Kick-around' areas on reserves and District Community Parks are
                    under-represented in Munno Para.

2   Reserve size
(a)   Is the reserve greater than 0.5 ha, 1.0 ha, or between 5–10 ha? (Yes)
(b)   If the reserve is less than 0.5 ha, can council add to this area, or would the
    monetary contribution option be preferable?

3   Reserve terrain
(a)   Does the reserve have steep slopes or consist of a river valley/gully only? (No)
(b)   Does the terrain provide an alternative to that which prevails in the area? (Yes)

4   Reserve location
(a)   Is the location:
    (i)   central for current and/or future land divisions? (Yes)
    (ii)   providing for through-access? (Yes)
    (iii)   connected to schools? shops? Other reserves (Yes)
    (iv)   encircled with roads (No)
(b)   Are houses/units etc. oriented towards the reserve? (Yes)

5   Focal points
(a)   Does the proposed reserve incorporate:
    (i)   remnant vegetation? (Yes)
    (ii)   cultural artefacts (e.g. ruins, bridges, etc.)? (Yes)
    (iii)   historical sites? (Yes)
    (iv)   encircled with roads? (No)
    (v)   water features? (Yes)

Source: Report on Reserves – Part 1, City of Munno Para, South Australia (Just 1987: 98).

Just (1987) provided a 'planner's checklist' (Table 7.8) to set out guidelines for planning decisions regarding urban recreation space. Size remains important, as research has shown that open space of less than 1 hectare (approx. 2.5 acres) is perceived as too small and is not well patronised. As the checklist suggests, the questions to be addressed are whether the space designated can be enlarged, or whether a monetary contribution would be the preferred option. As noted above, the type of area, including the terrain and configuration, set aside as recreation space in new subdivisions is important for future use. Location is another obvious consideration, not only with respect to potential users, but also in relation to neighbouring land uses (e.g. residential buildings and roadways), and to pedestrian access and bikeways.

Accessibility, generally, is a fundamental concern when decisions are being made about the provision of recreation space (see Chapters 2–4). Access to a diversity of recreation opportunities within urban areas is generally assured for those with automobiles who are willing to travel reasonable distances. As noted earlier, such opportunities are severely limited for people without access to a car – the elderly, the young, the poor and the handicapped. These people, together with those who cannot drive or prefer not to use their cars, rely on public transportation, which is usually commuter-oriented, to work places and shopping centres rather than recreation outlets. Services are often reduced or eliminated during evenings and weekends when recreation demands are heaviest. This means that many city-dwellers are denied access to park and recreation facilities beyond walking distance. In these circumstances, the provision of close-to-home recreation opportunities is even more essential if equity in delivery and performance of recreation services is to be achieved.

According to Cushman and Hamilton-Smith (1980), a spatially equitable distribution of urban recreation facilities would ensure that no person was deprived of access by reason of distance, time, travel cost, or convenience. However, confusion can arise between efficiency and equity in location decisions. A recreation policy based on efficiency-related criteria of minimising costs and aggregate travel, and maximising attendance, would result in the location of a small number of large-scale facilities in high-density residential areas. At the same time, consumers living in lower density areas would be worse off.

Cushman and Hamilton-Smith (1980) advocate a compromise where efficiency is balanced against maximum equality of recreation opportunity. They believe that the degree of equity or inequity can be determined by reference to measurable elements of relative opportunity or relative deprivation (i.e. travel costs, constraints on recreation options arising from facility characteristics, and demographic variations in the population's ability to use services offered).

Equity in location and access within an urban recreation space system must take account of these time/distance constraints and the circulation patterns of user groups. Studies of children's playgrounds, for example, indicate a highly localised service area of up to a quarter of a mile (approx. 0.4 km), and 75 per cent of all visitors to urban parks are said to come from less than a half-mile (approx. 0.8 km) radius. Distance, of course, is only one barrier standing in the way of individuals wishing to make use of a particular facility. Access to neighbourhood parks is often restricted by physical barriers such as highways, railroad tracks or industrial development. Chicago's lakefront parks, for example, have limited pedestrian access from surrounding neighbourhoods, due to the presence of the major thoroughfare of Lake Shore Drive. Yet, these same parks can be easily reached by car. Similarly, a 'tot-lot', separated from its pre-school users by distance or busy streets, can have little role to play in meeting their need for recreation space.

Cushman and Hamilton-Smith (1980) suggest that the first step in reducing inequity is to identify, classify and map the spatial distribution of all recreation facilities in the city and the nature and level of services provided. Deprived residential sectors can then be determined, and deficiencies rectified. For urban parks, for example, the spatial patterns of playgrounds, neighbourhood parks, district parks and large urban parks, can be visually correlated and statistically analysed, according to the degree of dispersion and clustering of parks in each of the park types. In this way, the areas of the city being served and not served by parks in each of the park types may be determined (Cushman and Hamilton-Smith 1980: 171).

In a Canadian study, Smale (1990) went one step further in examining the issue of spatial equity in the provision of urban recreation opportunities, by taking into account variations in the demand for recreation resources, as well as the supply of them. An inventory of urban parks in one part of suburban Toronto was related to household demand indicators for recreation. The study revealed neighbourhoods which were 'supply-rich', in terms of recreation opportunities, and other areas which were 'supply-poor', indicating the need for remedial action.

A somewhat similar approach was used by Mitchell (1968, 1969) to evaluate spatial aspects of Christaller's (1963) central place theory in an urban recreation context. Part of Mitchell's purpose was to seek understanding of the interacting variables and processes which affect the distribution pattern of public recreation sites within the city of Columbia, South Carolina.

Such variables as relative location, distance, time and facilities, appear to be significant to consumers of recreational activities. On the other hand, public demand or pressure, available personnel, budgetary limitations and philosophical orientation, are also factors that seem to be important to producers of recreational services (Mitchell 1969: 104).

Mitchell discussed spacing of recreation facilities within a four-tier hierarchy of recreation units – playgrounds, play fields, parks and large parks – related to the criteria of function, size and service. He proposed a theoretical spatial distribution for each class within the hierarchy, based on uniform hexagonal patterns, equal spacing, regular size and shape of service areas, and standard threshold populations. When simplifying assumptions concerning the strict residential character of the city and its uniform population distribution were relaxed, a more complex distribution pattern emerged. This pattern reflected the overriding significance of population density as the key explanatory variable in understanding the location of public urban recreation sites.

The hierarchy approach, when coupled with the concept of the Recreation Opportunity Spectrum (see Chapter 2), provides a suitable framework on which to base the planning of a functional system of urban recreation space. Figure 7.3a illustrates diagrammatically how positions along the spectrum of recreation opportunities can be related to functional categories in an urban context. Figure 7.3b shows how different opportunity settings are linked with types of recreation activity.

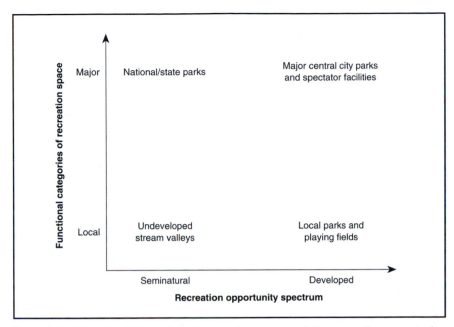

Figure 7.3a Functional hierarchy of recreation space and the recreation opportunity spectrum

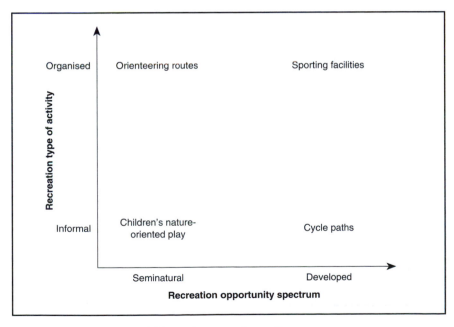

Figure 7.3b Recreation activities and opportunity setting

Source: Adapted from Ministry for Planning and Environment (1989: 7).

The Recreation Opportunity Spectrum has had more limited application in urban areas, but its relevance to 'highly developed' settings is demonstrated in recent work by the US Forest Service (More *et al.* 2003). Highly developed recreation experiences can be supported by settings ranging from large urban parks to small pocket parks or athletics fields. A wide range of settings should ensure that participants can select one to match their preference. An ideal spectrum of urban recreation opportunities should also include linear corridors connecting settings (http://www.parkweb.vic.gov.au/education/tourism_parks/da3.htm).

As with Christaller's original work, the value and practical application of hierarchy techniques in the study of urban amenity provision, rest in discovering, explaining and correcting departures in an existing system from the idealised, theoretical framework. Such remedial action should not be necessary if sufficient regard is given to urban recreation requirements at the planning stage. Further investigative approaches are needed. These approaches include clearly defined settings; inventories of existing resources (i.e. scope and quality of current provision of facilities, programmes and services); attempts to estimate past, present and future demand for open space and other facilities (i.e. trends assessment); comprehensive assessments of community recreation needs in line with sound understanding of the nature and characteristics of community (e.g. shifting or relatively stable population size, socio-demographics and psychographics); diverse community participation strategies and the use of community surveys – quantitative and qualitative; comprehensive plans; clear definitions of roles and responsibilities of government and non-government organisations; cost assessments; and a feasible implementation plan.

Summary

In many cases, the urban recreation planning process does not address the deeper behavioural needs of a leisure-oriented society. More often, it recognises and develops only conventional resources to accommodate present users and uses in stereotypical activities. By positioning a choice of urban recreation opportunities within a flexible hierarchy of recreation space, a functional recreation system can be created to provide for current and future community demands. However, any recreation system must have the capacity to cope with the inevitability of change.

In general, 'the temporal and spatial patterns and processes concerning outdoor recreation in urban areas have not been well conceptualised to date' (see Williams 1995: 20). Due to this neglect, and unless a dynamic element can be injected into the planning process, any recreation development initiative will lose impetus and be unable to respond to changing emphases in leisure behaviour, and associated pressures on resources and management policies. A flexible approach is the key to successful urban recreation planning, one in which priorities rather than rigid programmes are set down, and in which machinery exists for rapid review in the light of changing circumstances. Given this commitment, the recreation planner can make a useful contribution to generating a satisfying leisure environment for city dwellers in both established and emerging urban communities.

Guide to further reading

- For an overview of planning issues in modern cities, see Sandercock (2003).
- Historical perspectives: Clark and Crichter (1985); Williams (1995); Daly (1987).
- Outdoor recreation in urban areas: Stanfield and Rickert (1970); Williams (1995); Cooper and Collins (1998 – several chapters).
- Urban heritage and recreation and tourism: Prentice (1993); Chang *et al.* (1996).
- Urban parks: Hamilton-Smith and Mercer (1991); Nicholls (2001).
- Urban tourism: Ashworth (1989); Ashworth and Tunbridge (1990); Mullins (1991); Law (1992); Law (1993); Page (1995); Hall *et al.* (1997a); Murphy (1997); Law (2002); Hall (2003).
- Waterfront development: Craig-Smith and Fagence (1995); Williams (1995); Hall *et al.* (1997a); Murphy (1997).
- Planning for outdoor recreation and tourism: Mercer and Hamilton-Smith (1980); Jansen-Verbeke (1992).
- Community forest initiatives: http://www.communityforest.org.uk/partnership.html.
- Government or municipalities: Cambridge City Council Open Space Standards: Guidance for Interpretation and Implementation, http://www2.cambridge.gov.uk/planning/reptdocs/open_space_standards_July04.pdf, accessed 6 February 2005. Queensland, http://www.srq.qld.gov.au/zone_files/org_development/sec_4-4.18.pdf; Outdoor Queensland, http://www.qorf.org.au/01_cms/details.asp?k_id=119.

Review questions

1 Why is it difficult to devise a common strategy to address the changing recreation needs of cities?

2 What special problems arise in planning recreation opportunities in the inner core of older cities?

3 Identify the links between a hierarchical approach to provision of urban recreation space and the concept of the recreation opportunity spectrum.

4 How does the physical and human geography of a city affect its potential for outdoor recreation?

5 How might a compromise be reached between spatial equity and efficiency in urban recreation planning?

6 Suggest some specific reasons for under-use of urban parks and how these might be overcome.

7 Differentiate between urban open space and urban recreation space.

8 Discuss the contribution of urban waterfronts or urban bushland to urban recreation opportunities.

9 Examine the role of accessibility in the functioning of urban recreation resources.

10   What do you see as the respective (complementary) roles of the public and private sector, in creating an effective and satisfying spectrum of urban recreation opportunities?

8 Outdoor recreation in rural areas

Explanation of the recreational appeal of extra-urban environments may be found partly in people's reaction to environmental stress (e.g. crowding and noise) associated with everyday urban living. Outdoor activities in a rural setting allow city residents to escape – to exchange the routine, the familiar, and boredom for the recreation opportunities perceived to exist in the surrounding countryside. Even knowledge or cognitive awareness of such outdoor opportunities is considered to act as a psychological safety valve for some in coping with environmentally induced stress (Iso-Ahola 1980).

Recent change in recreation and tourism activities

Rural areas in Western nations have long been used for recreation and tourism. However, since the Second World War, the nature of, and relationships between, the rural setting and the recreational activities engaged therein have changed significantly (Cloke 1993). Recreation and tourism in many areas are no longer regarded as simply passive, minor elements in the rural landscape. They are important agents of change and control of that landscape, and of associated rural communities (Butler *et al.* 1998).

Much recent change in rural areas has been linked to recreation and tourism. Until the 1960s and 1970s, rural recreation was mainly related to the rural character of the setting. Rural recreation comprised, primarily, activities which were different from those undertaken in urban centres, and which could be classified as relaxing, passive, nostalgic, traditional, low technological, and generally non-competitive. Examples include: horse-riding, walking/rambling, picnicking, fishing, sightseeing, boating, visiting historical and cultural sites, attending festivals, viewing nature/scenery, and farm-based visits (Butler *et al.* 1998; Sharpley 2003).

Whereas the above activities are still common, many other quite different activities are now engaged in, which bring new forms of conflict and impact, and require different planning and management responses. These new activities could be characterised as: active, competitive, prestigious or fashionable, highly technological, high-risk, modern, individual and fast. 'They include trail biking, off-road motor vehicle riding, orienteering, survival games, hang gliding, parasailing, jet boating, wind surfing ... snow skiing, and fashionable shopping' (Butler *et al.* 1998: 10). So, a far wider range of recreational activities is now being pursued in rural areas, bringing a requirement for the establishment of specific facilities and settlements (e.g. resorts) to cater to the increasingly more sophisticated demands being placed on resources (e.g. see Sports Council 1991; Butler *et al.* 1998; Roberts and Hall 2001). 'Creation of an appropriate range of settings for rural tourism [and recreation] requires the deliberate selection and manipulation of features of the rural landscape to accommodate different types and styles of visitor use' (Pigram 1993: 163), which have been highlighted in recent studies in the UK (e.g. see Sharpley 2003; Ravenscroft and Curry 2004) (see Table 8.1) and elsewhere (e.g. Sharpley and Sharpley 1997; Butler *et al.* 1998; McCool and Moisey 2001).

The role of rural landscapes in satisfying the recreational needs of a leisure-conscious society has long been recognised in Britain and other countries. For instance, since 1949, the Countryside Commission (now Countryside Agency) has been active in promoting the conservation of the natural beauty and amenity of the English countryside, within the framework of efficient agricultural use. A survey sponsored by the Commission in 1977, found that visiting the countryside was the most popular form of outdoor recreation for the people of England and Wales (Countryside Commission 1979). More recent figures in the UK and other countries provide further evidence of the popularity of the countryside. In the UK, more than 900 million day visits were made to the countryside in 1993 (Countryside Recreation Network 1994). By 2002/3 this had increased to 4.5 billion. In the US, it was estimated more than a decade ago that more than 70 per cent of people participate in rural recreation (OECD 1993) (also see Chapter 1).

Table 8.1 Principal activities undertaken on countryside trips by Surrey residents (2000) and in England as a whole (1998) (% of responses)

Activity	Surrey	England
Short walks	20	22
Walking the dog	11	
Long walks (> 2 miles)	29	17
Bicycle riding	7	4
Horse riding	2	2
Active sports	5	8
General driving/sightseeing	3	6
Water recreation (boats, fishing)	2	3
Pubs and eating out	3	21
Historic sites and gardens	8	4
Organised guided walks	1	0
Camping and caravanning	1	0
Wildlife observation	3	3
Garden centres	1	3
Motor sports	0	1
Other hobbies	4	6

Source: Ravenscroft and Curry (2004: 78).

Note
n for Surrey residents = 1,525.

'For many urban dwellers, it is the rural ambience and the countryside experience which are the main considerations' (Pigram 1993: 161). Recognition of the strong correlation between recreational (and tourist) satisfaction and scenic quality of the recreation environment is an important step towards realising the contribution which rural landscapes, in both public and private hands, can make to the leisure opportunities of the city dweller. Rural recreation and tourism generally require pleasant, aesthetic and peaceful environments. Balancing the demands on rural landscapes, however, is a major challenge (Sharpley and Sharpley 1997).

Unfortunately, public resource-based recreation areas such as national parks and forests are in limited supply, and are not always close to centres of population. Even then their use must be carefully managed and monitored (see Chapters 9 and 10). On those public lands which are accessible, visitation rates at peak periods are often pushed beyond carrying capacity so that fees, permits and other strategies for rationing use become necessary. At the same time, attempts to expand the resource base are frustrated by lack of land of suitable location and quality, by opposition to the creation of recreational settings or establishment of national parks and other forms of reserves, and by budgetary constraints on park management services wishing to undertake further land acquisition programmes or simply manage those lands for which they are already responsible. Therefore, increasing attention has been given to the potential of private land and different land classification categories for the provision of recreation opportunities within reasonable proximity of cities.

Rural recreation space: conflict and multipurpose use

Recreation is just one competitor for the use of rural land and water (Green 1977; Pigram 1986). Many groups have an interest in rural areas, but for different, often competing, reasons. Other uses or interests include primary production (e.g. agricultural, aquatic, horticultural, pastoral and timber production), resource extraction (e.g. uranium and sand mining), conservation or preservation of the natural, cultural and built environments (e.g. national parks, wilderness areas and nature reserves), and transport and communication networks. According to Sharpley and Sharpley (1997):

> In most industrialised nations, up to 80 per cent of rural land is still farmed or forested ... although, significantly, the contribution of agriculture and forestry to income and employment in rural areas has gradually diminished during the twentieth century, as has their relative contribution to GDP in most countries.
> (Sharpley and Sharpley 1997: 23)

The multipurpose character of the countryside represents both an opportunity and a constraint to recreation and tourism development.

> On the one hand, tourism and recreation can be viewed as a valid and valuable form of land use which, if carefully planned and managed, complements other

uses and contributes to the economic and social wellbeing of rural areas; on the other hand, it may be considered that other, more traditional forms of economic exploitation of the land, including farming, mineral extraction and housing, should take precedence over its recreational potential. Thus, it is perhaps inevitable that, given the finite supply of the countryside, conflicts occur between different demands on the rural resource base.

(Pigram 1993: 161)

Australia's large size and low population may suggest fewer constraints on recreation, and lower pressures on rural recreation resources than in Britain or parts of Europe. However, this is not the case. Many groups have an interest in rural Australia, as they do in rural areas in other parts of the world. The perception that many areas are environmentally fragile or unique has encouraged a strong conservation ethic. Parts of the coastal strip are highly urbanised (see Chapter 7), and this places strong pressure on neighbouring rural areas that offer diverse and attractive recreation and lifestyle opportunities. Farming and mining activities predominate over much of the rest. These activities or outlooks are often in conflict, especially as knowledge, perceptions, attitudes and technologies change. People are even beginning to question:

whether a socially-beneficial agriculture can be one which pollutes the land, poisons animal, bird and fish life, and leads to the destruction of the environment. It is becoming clear that ecological diversity and the aesthetic and amenity value of the countryside have become incompatible with modern agriculture and its modern practices.

(Lawrence 1987: 66)

Given recent legislative developments, the expansion of sustainable (including organic) farming practices and the efforts of conservation-minded organisations encouraging replanting schemes and other efforts to revive landscapes in rural areas, some might view Lawrence's comments as dated. However, there is abundant evidence of land degradation, tree felling and land clearing, soil salinisation, and pollution of waterways to suggest we are a long way from changing the attitudes of many rural landholders and extractive industries who fail to act as stewards of land and water.

Recreation and tourism, too, particularly in ecologically sensitive areas, threaten environmental and cultural conservation, while both recreation and conservation pose threats to traditional views (i.e. agricultural, pastoral and mining activities) in rural Australia and the US. The ensuing conflict undoubtedly serves to limit recreational access and tourism development.

The relationships between rural tourism and recreation, and other land and water uses, are largely influenced by landholder attitudes. Landholders include individuals (e.g. farmers), businesses (e.g. agribusiness and tourist resorts) and groups (e.g. recreation clubs) with private ownership rights; leaseholders and licensees (whose land use may be regulated by public agencies); and resource management agencies (e.g. national parks, forestry, nature/wildlife reserves, publicly owned recreational facilities and water reservoirs). Clearly, there is a wide variety of individuals and agencies with different value sets and interests with respect to the rural environment, and with different rights as landholders, according to land tenure and other institutional, legislative or contractual arrangements (see Box 8.1).

The complexity of institutional and ownership arrangements, and the multi-functional character of rural areas, have led to conflict between competing uses and between land managers. Land ownership and the exercise of land-ownership rights are thus critical elements in the supply of tourism and recreation opportunities. Access to land and water in this context is generally contingent upon legislation, public policy interpretations and landholder/management attitudes (Cullington 1981; Pigram 1981; Jenkins and Prin 1998). Effective recreation and tourism planning and conflict resolution are hampered by numerous stakeholders such as government agencies at different levels, conservation groups, developers, recreational groups, and local communities generally.

Box 8.1 **Important agencies for recreation in the countryside and rural areas**

Countryside Agency/Countryside Commission

For more than 50 years the Countryside Commission in Britain was active in promoting the conservation of the natural beauty and amenity of the English countryside, within the framework of efficient agricultural land use. A particular concern has been the provision and upkeep of recreational footpaths and rights-of-way and encouragement of the establishment of farm trails in cooperation with landholders.

In 1999, the Countryside Agency was formed from a merger of the Countryside Commission and the Rural Development Commission. The Countryside Agency is a statutory body aimed at conserving and enhancing the countryside, promoting social equity and opportunity for the people who live there, and helping everyone to enjoy this national asset. Among the priorities of the Countryside Agency are homes, services and opportunities for rural people, access to rural areas for outdoor recreation, reducing the impact of traffic growth on the rural environment, and maintaining farming at the heart of a strong rural economy with attention both to conservation and the production of food and fibre.

The Countryside Agency publishes *Countryside Focus* every two months to provide news and a forum for the exchange of ideas. All Countryside Agency publications can be ordered online through the Agency website (http://www.countryside.gov.uk).

Source: Pigram (2003a: 85).

Agreements and compromise between recreationists, responsible agencies and landholders are often difficult to achieve. On the one hand, as farmers seek to improve productivity (e.g. through more intensive land use practices, including the development of intensive feed lots; or through extensive land clearing and extraction of water from creeks and rivers for irrigation), there are aesthetic and functional changes to the landscape, as well as impacts on the supply of recreational opportunities and visitor experiences and satisfaction. On the other hand, rural recreation activities can become more contentious as their environmental impacts increase (e.g. large numbers of people visiting sensitive sites; the use of potentially destructive recreational technologies, including off-road vehicles). Recreational activities such as hiking, camping, fishing and nature observation may be passive and, therefore, depending on the resilience of the environment and its ability to resist impacts, have less inherent and actual potential to cause conflict between participants and land managers. However, landholder attitudes, perceptions and experiences may be influenced by small numbers of people who fail to consider the relationship between the type and intensity of their activities and the resulting impacts on the environment, including other people. For instance, there are those whose intentions and activities are deliberately environmentally destructive and illegal (e.g. indiscriminate shooting of livestock; stealing from farm households; destruction of gates and fences; lighting of fires in prohibited areas). Thus, any understanding of land use involves both an understanding of the values of the physical, biological, productive, spatial and visual/aesthetic attributes of land, and 'an awareness of the different standpoints from which land use may be considered' (Mather 1986: 6) (also see Chapter 2).

Conflicts

Two features of rural areas in the last half century have been the difficulty of accommodating the structural changes which have occurred, and the much greater range of uses to which rural areas have been subjected. Amidst these developments, it is worth observing that: 'Of all the resource-management conflicts in the countryside, recreation offers perhaps the greatest opportunities for multi-functional land use. Whilst inevitably many forms of recreational activity are compatible with others, some are not, and they have to be specially sited' (Cloke and Park 1985: 187).

Conflicts have arisen between recreation and tourist uses and other forms of land use, and between various forms of recreation and tourism. Conflicts between motorised and non-motorised recreational users of the same area can be severe, and often agreement and compromise is difficult to achieve, as can be seen in the case of disagreements between cross-country skiers and snowmobilers, between non-mechanised trail users and off-road vehicle drivers, and between wind surfers and water skiers. Conflicts also exist between non-mechanised users of the same facilities, as seen in the conflicts between pedestrian trail users and mountain bike riders, between canoeists and anglers, and between hikers and hunters.

Such conflicts will probably become more severe as the overall demand for recreational and tourist use of rural areas increases, and as the range and types of uses widen. Compounding this problem in many countries, is decreasing public access to parts of rural areas because of changing patterns of ownership, and/or the reluctance of many landowners to accept public recreational access to private property. As increasing numbers of people acquire leisure or retirement properties in rural areas, they regard and treat those properties as private preserves; just as the landed élite zealously guarded their own leisure estates in past times (Butler *et al.* 1998). In Australia, such an attitude has a lengthy history dating back to early settlement (see Jenkins 1998).

Recreational use of private land represents multiple resource use and, as such, can generate conflict between recreationists and landholders. The basis for conflict lies in the various functions seen for rural land and the contrasting attitudes associated with these roles. Davidson and Wibberley (1977) suggest a strong polarisation between those whose dominant concern is the efficient production of food and fibre or other economic uses, and those who value more highly the intrinsic character of rural landscapes and wish to preserve this heritage unchanged. Between these two are other groups, for whom different attributes of the countryside are significant. City planners and developers, for example, often view land, especially in the rural fringe, merely as a space and development reserve for urban expansion. For others, the primary role envisaged may be for communications facilities or specific resource uses such as extractive industries or water conservation. Transcending all of these in numbers are those who link the resource function of the countryside with leisure and outdoor recreation.

To some observers, this multiplicity of roles makes conflict almost inevitable if recreationists press their claims to private rural land (Green 1977; Jenkins and Prin 1998). Whether conflict and confrontation are avoided depends essentially upon the goodwill and cooperation of the landholders: their attitudes are of fundamental importance in determining the amount of land available to the public for outdoor recreation. These attitudes, in turn, are a function of the landholder's personal beliefs and experiences, together with legal, economic, social and ecological considerations, national traditions and government policies, and the type and volume of the recreation activity involved (Cullington 1981).

The relationships between these factors are depicted in Figure 8.1. Essentially, the issue is one of balance between incentives and disincentives. Put simply: 'To increase the supply of private land for recreation, it is necessary either to increase the incentives, or to reduce the disincentives, or preferably both' (Cullington 1981: 8).

Incentives may be provided by governments in an effort to encourage wider recreational use of private land, and can include such measures as direct financial support, compensation payments or the sponsoring of access agreements. Disincentives arise from landholders' concerns over such matters as:

• *Legal liability* for injury or damage to recreationists. The degree of the liability and the landholder's responsibility vary between jurisdictions and whether the visitor is an invitee, a licensee, or a trespasser. Several countries have

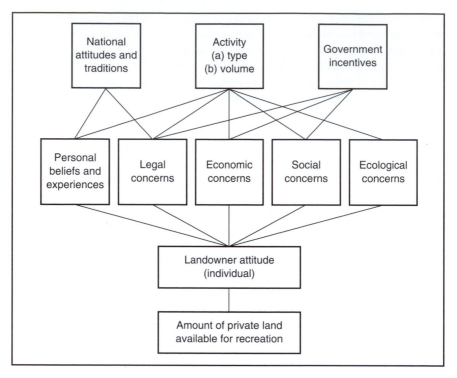

Figure 8.1 Factors influencing availability of private land for recreation

Source: Cullington (1981: 9).

enacted laws which attempt to allay landholders' fears of the liability problems should accidents occur.

- *Economic or financial implications* associated with the costs of providing access and possible dangers of changes in farming practice, set against any compensation payments or opportunities to derive income from recreational activities.
- *Social considerations* such as loss of privacy and problems with trespassing, which may be mitigated to some extent by the landowner's personal satisfaction from providing a community service.
- *Ecological impacts* on the farm environment as a result of recreational activities. These will depend to some extent upon how the farmer's perception of the problem is affected by the particular nature of the recreation activity and its incidence in space and time (Harrington 1975; Cullington 1981; Pigram 1981; Jenkins and Prin 1998).

All of these concerns are interrelated, and the degree to which they influence the landholder's decision to make available or withhold recreational access, is closely linked with the type and volume of recreation activity undertaken and government support and encouragement. The outcome rests very much with the individual landholders and how they perceive the balance between the incentives and disincentives. Undoubtedly, there will remain many who value the economic functions of rural land more highly than any amenity functions it may be deemed to possess. Moreover, conflict between these primary functions would seem most probable in the urban-rural fringe, where the economic value of the countryside is highest, and pressure for amenity and recreation space is greatest. It is here, too, that most problems and disputes over accessibility can be expected to arise.

Accessibility to recreation space

Accessibility has several dimensions, among them, technical, behavioural and socio-cultural aspects. According to Moseley (1979), the concept cannot be divorced from the nature of the desired destination or experience. Certainly, much more than mere mobility is implied. Mobility, or the capacity to overcome space, is a technical and mechanistic condition, derived from such factors as vehicle ownership, travel time and costs, and individual physical attributes. Accessibility, on the other hand, is a broader concept, reflecting the opportunities perceived as available for travel. It is related to the behavioural notion of 'psychic space' or 'movement space'; that restricted area in which potential trip makers react to stimuli within the constraints of their value systems, experience and perceived environmental opportunities (Eliot-Hurst 1972). Moreover, the dimensions of a person's movement space are also, in part, a function of social and legal convention – the institutional 'rules of the game' to which an individual is exposed.

Thus, accessibility has many facets, and use of recreation space can effectively be denied in a variety of circumstances. Examples include: cost of travel, equipment and licence fees; lack of time, especially blocks of suitable time; inadequate information on recreation opportunities; ineligibility to participate on the basis of age, sex, qualifications, membership of group or social class; lack of transport; and special problems for people with disabilities (see Chapter 3). In the recreational use of the countryside, these circumstances may be compounded by the sheer difficulty of physical access; many sites are effectively closed off because of a lack of appropriate vehicles, equipment, stamina or expertise. Consideration of accessibility, too, can be complicated by disputes over property rights and by institutional and legal constraints on movement into and through recreation space.

Property rights

Central to the question of access and availability of private rural land for recreation is the issue of rights to property, and the privileges and responsibilities which ownership and control over land bestow. To some, property ownership, in a legal or economic sense, is the proprietorship of a bundle of rights (Wunderlich 1979). Others go further and question the concept of private property altogether, stressing that property should not be thought of as *things*, but as *rights*, the ownership of which is circumscribed (Dales 1972). In this view, ownership consists of a set of legally defined rights to use property in certain ways and a set of negative rights or prohibitions which prevent its use in other ways; in this sense a proprietor never *owns* physical assets, but only has the rights to their *use*.

In the context of recreational access to the countryside, ownership of the land itself is of no particular relevance. The crucial issues are ownership and exercising of the right to exclude others from use (Thomson and Whitby 1976). Difficulties arise because landholders are only one among several groups with an interest in how the resource is to be utilised and managed. The multiplicity of functions referred to above suggests a number of potential beneficiaries who may value the land for specific purposes. This would include the occupiers and would-be recreationists, but may also cover neighbours, passers-by and conservationists at large (Phillips and Roberts 1973). In economic and legalistic terms, the access issue can be seen as one of allocating among these interested parties the various rights over land in such a way as to maximise social welfare (Thomson and Whitby 1976). It could be that, where a landholder wished to retain exclusive rights to recreational resources (e.g. a stream, a beachfront, or a spectacular scenic view), purchasing such rights, over and above the price of the land, should be mandatory (i.e. the privilege of excluding the public would become taxable). However, others would claim that it is never equitable to permit the holders of land to alienate recreation space to themselves.

A finer definition of rights to property would certainly seem desirable in order to identify those which accrue to the property holder, to the state and to society. It could be held that private ownership rights become merely the residue, after public or communal rights to property are exhausted (Morris 1975). It could further be argued that ownership rights should not apply to the aesthetic component of the resource base, or extend to exclusive access to assets such as wildlife or fish that are found within a property. The landholder, when taking up occupation, also takes up effective control of countryside resources which may be valued by the wider community for recreation. This privilege, in turn, should imply a responsibility for making those resources available to society. It seems that few landholders are prepared to acknowledge this responsibility.

Recreational use of private land

While there is no universally applicable measure of access (Sharpley and Sharpley 1997: 78), and despite the constraints and disincentives noted, much private land is available in the 'old world' for outdoor recreation, either with the tacit or explicit consent of the landholder. In England and Wales, for example, individuals and groups have long enjoyed access to designated parts of the rural environment, although, even there, recreationists in large numbers are unwelcome. In parts of Scandinavia, recreation on private land is accepted and expected under a law known as 'Every Man's Right' (Cullington 1981).

In many parts of Scotland, visitors are allowed the freedom to roam by custom, but not by legal right (Sharpley and Sharpley 1997: 79). In Sweden, subject to

(a)

(b)

Plate 8.1 Hiking in the Lake District, England, is permitted across beautiful countryside under private ownership. Often, paths pass close to private homes. Such access is a right to be respected.

certain conditions, visitors enjoy legal right of access to all public and private land in the countryside (*allemansrett*) (Colby 1988).

> A similar situation exists in Norway, where the great majority of the landscape is undeveloped and available for recreation. Visitors to the Norwegian countryside enjoy the right of Allemansretten, the right of access to most private land, although subject to certain conditions laid down in legislation. Germany, too, has a right of public access (*Betretungsrecht*) enshrined in law, although this is restricted mainly to forests, unenclosed land and along roads and paths.
>
> (Sharpley and Sharpley 1997: 78–9)

The establishment of trails and other means of facilitating recreational use of private rural land, is an indication of some relaxation of the access situation in countries of the 'new world', such as Australia, New Zealand and Canada. However, in many cases, countryside recreation remains inhibited by the prevailing attitude of particular landholders, who fear, with some justification, the consequences of thoughtless negligence or deliberate vandalism by visitors. Their experience suggests that, in many circumstances, recreation is simply incompatible with other uses of the countryside, by virtue of its concentration in time and space, as well as problems of trespass, litter, property damage and general nuisance.

Conflict is most likely to occur closer to towns, where fringe landholders face higher levels of trespass damage, to the extent that some form of boundary protection may become necessary. In extreme cases, the actions of visitors may lead to drastic modification of farming practices or the abandonment of arable farming altogether.

Taking into account impacts of this scale, the negative attitude of rural communities to recreational use of private land can be better understood. More than 30 years ago, Phillips and Roberts (1973) reported that the continuing invasion of the countryside by urban dwellers seeking diversion, set against a background of rapid changes in farming, was leading to a situation in Britain where there was perhaps a greater degree of antipathy between farmers and visitors than ever before.

In Australia, too, where the concept of inviolate rights of property ownership is widespread and generally accepted, the lines are fairly clearly drawn between town and country. The attitudes of landholders to public recreational access are typified by the following statement:

> Access to private land for sport or recreation is a privilege and privilege is not a birthright but something that can be earned by good behaviour and responsibility. This Association will not consent to accept the entry upon private land, without the permission of owners or occupiers, of any persons who are not performing a statutory function, as other than trespass.
>
> (Graziers' Association of New South Wales 1975: 3)

This attitude has not diminished (see below). It was translated into real terms by means of a proliferation of cautionary signs at property boundaries, and warning notices in the rural press advising that all permits to enter land have been cancelled, and that trespassers will be prosecuted or face other dire consequences.

Thus, for many Australians, recreational contact with the countryside remains restricted, and is often confined to illicit and fleeting entry of private land, or viewing from a moving vehicle. Moreover, there seems little prospect of land-holders being willing or able to divert resources voluntarily from what are seen as the land's primary functions – agriculture and the like – to provide recreation space for city dwellers. Yet the potential exists for areas to be opened up by legal means, which governments are reluctant to exercise. Travelling stock routes in Australia, are one mechanism for linear or trail recreational development, while the reserves accompanying the routes provide ideal camping sites (see Box 8.2). These routes and sites have been developed and publicised only to a limited extent.

Box 8.2 Travelling stock routes in New South Wales, Australia

Travelling stock routes (TSRs) are a valuable community land resource in NSW. TSRs were first set aside for the movement of stock into areas of new settlement and generally followed the trails of the explorers along the most accessible routes where water was available. Soon after the initial settlement of an area, the direction of stock movement was, for the most part, reversed, and TSRs were then developed to provide access for stock to market. The area of land dedicated or reserved for TSRs was reduced significantly after the Second World War as the quantity and quality of motorised transport increased and markets became more accessible. Nevertheless, Hibberd (1978) estimated that there remained 2.27 million ha or 2.83 per cent of land in NSW dedicated as TSRs or for similar purposes as at 30 June 1975.

Once a TSR was established, people demanded adequate water supplies along the routes. In response to these demands, colonial laws were enacted which recognised, regulated and maintained stock routes, and provided for the establishment of watering points. The first of these regulations were enacted in the 1860s, to protect the rights of landholders adjacent to TSRs, and to set the minimum rate of progress for travelling stock. The history of the development and administration of TSRs has been well documented by several authors (e.g. King 1957; McKnight 1977; Hibberd 1978; Hampton undated) and the Department of Lands who attributed much of the early popularity of TSRs to several factors that reflected the reasons for the establishment of such routes. However, the traditional use of TSRs as a means of supporting the transportation of stock to markets has declined, and other uses (e.g. recreation) have become increasingly important.

As early as 1951, the recreational value of TSRs was recognised during the Parliamentary Debate of the 1951 Amendment to the NSW Pastures Protection Act, 1934:

continued...

***Box 8.2* continued**

Many Travelling Stock Routes, and particularly those situated near towns and along the banks of rivers, are used by the public as recreation reserves. In that way ... they are serving a useful public purpose and meeting a real need. It would be unwise to withdraw reserves from public use (for disposal) in these circumstances.

(Mr Howarth, in Hampton, undated: 23)

The recreational potential of TSRs and reserves and their potential benefits to conservation if retained in Crown ownership and carefully managed have been highlighted by Breckwoldt (1983: 37), who argued that:

The future ownership, control and management of travelling stock reserves is important ... A policy on the future of travelling stock reserves should consider that some will have high nature conservation values, particularly in the intensively used regions, and should be managed to retain and regenerate existing trees. The cost of managing roadside verges and travelling stock reserves for wildlife conservation benefits many people, particularly when it occurs on public lands, so the cost should be met by the wider community.

TSRs have continued to be sold with little assessment or understanding of their inherent values and despite increasing awareness of their ecological significance and recreational potential.

In the UK, private land has a long history of use for recreation, and 'A central theme to the development of countryside recreation policies has been that of access' (Groome 1993: 5). However, recreational use of the countryside is not

a public prerogative. It constitutes use of a domain, owned mainly (87 per cent) by private individuals, and with public access dependent on certain legal rights or lenient attitudes on the part of landowners. Even the national parks, areas designated specifically for landscape protection and public amenity, are largely in private ownership.

(Shoard 1987, in Glyptis 1992: 156)

British landowners have a long developed tradition of exclusive control of the countryside (Shoard 1996: 13). To make the matter more complex still, there are signs that land ownership is gradually becoming concentrated in the hands of fewer people rather than spreading more widely, while counter-urbanisation is bringing new, affluent and mobile residents, who are accustomed to urban standards of service provision (Glyptis 1992).

Nevertheless, compared with Australia and Canada, the UK has enjoyed relative ease of access to rural lands for recreation, where the broad aim has been to promote and market leisure to as wide a public as possible (Pigram and Jenkins 1994). For instance, recent policy developments in rural recreation and tourism have 'resulted in a change from the perceived need to control the public, to their wholesale encouragement. This has happened quite swiftly and has taken place in tandem with a fundamental reappraisal of the primary role of agriculture in rural areas' (Robinson 1990: 132). Despite initiatives to enhance public recreational access to the countryside in the UK, there are still concerns about landholder attitudes, the ineffectiveness of such initiatives in some cases, and, ultimately, the lack of access (see Jenkins and Prin 1998; Sharpley 2003; Shoard 1996, 1999; Countryside Agency http://www.countryside.gov.uk).

In Britain, a recent policy initiative for access to the countryside is the 'Countryside Stewardship Scheme' (CSS) initiated by the former Countryside Commission. The CSS was launched in 1991, and was undertaken in partnership with English Nature, English Heritage and the Ministry of Agriculture, Fisheries and Food. The Scheme encouraged farmers and landowners to conserve and re-create the beauty of five traditional English landscapes (chalk and limestone grassland, lowland heath, waterside landscapes, coastal land, and uplands) and their wildlife habitats, and to give opportunities for informal public access. The stated 'long-term objective is to develop a basis for a comprehensive scheme to achieve environmental and recreational benefits as an integral part of agricultural support' (Countryside Commission 1991: 1). This initiative certainly broke new ground in environmental management, integrating conservation management and commercial farming, with a view to demonstrating benefits for landscape, wildlife, history/ archaeology, access, or some combination of these objectives. Farmers and landholders were granted annual payments of up to £300 a hectare; a commitment costing the government £13 million for the first three years. Concerns were raised as to whether the access payments secured access, with many open access areas difficult to locate. However, a review of the Scheme (Countryside Commission 1998: unpaginated) stated that:

> Countryside Stewardship delivered significant environmental benefits over and above what would have happened in the absence of the scheme ... An additional survey of public access to land under Stewardship access agreements found that a high percentage were well located and were easily accessible by the public. Sites were used by a wide variety of people of all ages and from all walks of life ... day trippers as well as local people ... The sites met the expectations of two-thirds of visitors. 20 per cent found the sites to be better than they expected. Only 4 per cent were disappointed with the sites.

Many other examples of recent initiatives to secure access to the UK countryside, and in particular private rural land, abound (see Watkins 1996). However, there has been much debate with respect to such initiatives. Shoard (1996) argued:

Farmers now get £247 per mile per year merely for allowing people to walk along access strips ten metres wide, along the sides of, or across fields in ESAs [Environmentally Sensitive Areas] and £145 under the NRSA [Non-Rotational Set-Aside] (MAFF 1994: 6–7). The landowners of the past who established the idea of a right of exclusion would be amazed to learn the size of the potential bounty they have created for their descendants.

The right of exclusion is being increasingly used to turn access into a tradeable asset. The government's endorsement of the right of the landowner to charge others to set foot on his or her land puts the official seal of approval on the notion that access to the countryside is a commodity to be bought from landowners rather than a free public good. This is like bestowing on one group a whole new form of wealth.

(Shoard 1996: 21)

Yet, in the same volume of works (Watkins 1996), Curry (1996: 34) noted that one means of 'improving opportunities for access to closed land … should be pursued through direct payments from the consumer to the farmer and landowner, since, in the absence of non-excludability, this is both more efficient and more equitable'.

The CSS is now a function of the Department of Environment, Food and Rural Affairs' (DEFRA) Rural Development programme which supports another nine schemes such as the Environmentally Sensitive Areas (ESA) scheme and Rural Enterprise Scheme. The aims of the CSS are to:

- sustain the beauty and diversity of the landscape;
- improve and extend wildlife habits;
- conserve archaeological sites and historic features;
- restore neglected land or features;
- create new habitats and landscapes; and
- improve opportunities for countryside enjoyment (DEFRA 2001).

The CSS is regarded by DEFRA as 'The Government's principal scheme for conserving and improving the countryside' (DEFRA 2001: 2) and it is complemented by the Environmentally Sensitive Areas (ESA) scheme under which 22 nationally important areas are conserved. The ESA and the CSS seek to conserve the countryside's natural and cultural heritage. One of the aims of the CSS 'is to improve opportunities for enjoying the countryside by providing new public access to farmland, and for walks, picnics and games' (p. 12). The CSS now provides up to £525 per hectare, depending on the extent and nature of land management undertaken, while £500 million has been made available in the short term. There is an explicit attempt to add hundreds of thousands of hectares to this Scheme in which some 14,000 farmers and land managers participate (DEFRA 2001).

There are no right or wrong answers or simple solutions to the access issue, because of the disparate views on landowners' rights and the complex nature and outcomes of government intervention. This situation is unlikely to change markedly in the near future. The restructuring of rural economies and the difficulties that have been faced by some rural producers and communities, in tandem with an increased demand for recreational and tourist use of the countryside, mean that public sector incentives or disincentives and user fees are likely to become even more attractive to landholders. However, if 'The only honest and effective way of opening the countryside to the people is to require landowners to relinquish some of the rights in their asset, which they are otherwise bound to defend and exploit' (Shoard 1996: 21), then, in the absence of blanket public policy establishing such access, a significant reordering of landholders' values will be needed. More to the point, general and very broad access rights are unlikely to eventuate in Britain in the short to medium term. However, under the Countryside and Rights of Way Act 2000 changes in the policies for and management of recreation opportunities are being introduced throughout England. One of the key changes is that the public has a basic right to walk. So, there is no requirement to stay on particular paths, or trails, or rights of way. Under the Act, open access is being introduced gradually across eight regions of England, and began with the Lower North West and South East on 19 September

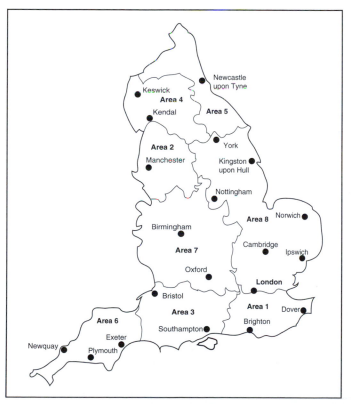

Figure 8.2 Countryside access under the Countryside and Rights of Way Act 2000

Source: Adapted from the Countryside Agency (2004: 4).

2004 (see Figure 8.2). It is expected that the whole of England will be available for open access on designated land by the end of 2005.

Areas being opened up are usually mountain, moor, heath, down and common land with some exclusions, such as gardens, parks and arable land. This will give the public open access to around one million hectares (or approximately 4,000 square miles or 8 per cent of England). Most recreational activities are covered but cycling, horse-riding, using a vehicle, or camping are not included except where these already take place. Fires are not permitted, nor are feeding animals, littering, using a metal detector or causing damage.

The Peak District National Park will be the first national park to benefit from the new access rights with the public's right of access increased from 240 square kilometres to almost 500 square kilometres (92 square miles to 193 square miles). Extensive efforts are being made to ensure that landowners and other authorities are aware of their responsibilities and the implications of the Act. The Forest of Bowland Area of Outstanding Natural Beauty (AONB) led the way for open access in the northwest of England with a pilot study, working with private landowners, to find ways of integrating access, conservation and land management issues. An updated Countryside Code, with the themes 'Respect, Protect and Enjoy' was launched in 2004 by the Countryside Agency and the Countryside Council for Wales. The Code was developed in consultation with key groups, including National Park Authorities, the National Trust and the Forestry Commission (Countryside Agency 2004). Access land and restrictions will be shown on the website, http://www.countrysideaccess.gov.uk and on new OS Explorer Maps.

Changes are also being made in relation to rights to roam in Scotland, via the Land Reform (Scotland) Act 2003 (see Box 8.3).

The initiatives in the UK represent very proactive policy formulation and implementation for multipurpose use in a densely populated country. As mentioned above, these initiatives do have their sceptics, and they also have strong opposition from some farmers' and other landholders' groups. Thus, a much more acute understanding of landowner attitudes is needed. To some extent, such understanding is becoming increasingly evident and detailed in the expansive literature on recreational access in the UK (see Watkins 1996; DEFRA 2001), but is sadly lacking across Canada and Australia.

In the UK, the concept of sharing the countryside is generally well-established in comparison with Australia because use of the countryside for rural recreation has a long history. Legislative provisions in the UK explicitly recognise the importance of recreational access to private lands. Public and private cooperation with landowners is supported by education and formal and informal consultation mechanisms. Furthermore, despite the development of numerous, sometimes very lengthy walking tracks, little progress has been made with respect to public access to private lands in Australia and Canada.

A study of rural landholder attitudes to recreational access to private lands in central western NSW, Australia, was designed to assist in the identification of barriers to recreational access to private lands; in the reconciliation of conflicts in the recreational use of private lands; and in the enhancement of the spectrum of

---

Box 8.3 Recreational access in Scotland

In Scotland, Part 1 of the Land Reform (Scotland) Act 2003 established a statutory right of responsible access over most areas of land and water (http://www.snh.org.uk). Access rights and the Scottish Access Code came into effect on 9 February 2005.

The Bill was debated vigorously both inside and outside the Scottish Parliament where the fears of landowners were discounted, despite strongly held views that the legislation was flawed and inspired by class hatred (of landowners). Some critics suggested that the Act represented an attempt by Parliament to place severe limits on the rights of a landowner to use and dispose of their property (Sellar 2003). In response it was pointed out that the Queen, at her Balmoral Estate, had effectively operated an open access policy for years.

Eventually, the Act was passed, giving 'a right to responsible access to land for recreation and passage'. Relatively few restrictions have been put in place and detailed procedures have been laid down to deal with access disputes. The Act also allows for community purchase of land, including fishing rights and responsibilities under the Act.

Comparable with the Countryside Code of England and Wales, the Scottish Outdoor Access Code has been drawn up by Scottish Natural Heritage (SNH) and explains to countryside users their rights and responsibilities under the Act. It also provides specialist advice, support and guidance for recreation managers – local authorities; national park authorities; and other public agencies which manage outdoor recreation access (http://www. outdooraccess-scotland.com).

---

recreational opportunities in the countryside (Jenkins and Prin 1998). The study's objectives were:

- to identify rural landholder attitudes to recreational access;
- to identify the underlying dimensions which explain those attitudes, and therefore to determine whether attitudes differ because of ownership, land tenure, government incentives, income and other arrangements;
- to review the systems of incentives and disincentives under which landholders are operating;
- to identify means of ameliorating rural land use conflict by way of incentives and removal of disincentives identified by landholders.

The study indicated that much resistance, if not direct opposition, remains to the use of private rural lands for public recreation, at least in this study area. Landholders expressed little or no interest in entering into incentives or agreements to facilitate recreational access to private property, and negative reactions to recreational access were made by some respondents (see Table 8.2). However, the

Table 8.2 Landholders' comments

- The bible says to forgive your trespasses. We don't. We shoot the Bastards.
- Government should provide the land for public recreational activities, for example, State forests … the farmers and landholders have a big enough responsibility just surviving.
- A lot of people have no idea about farming. If they come onto your place they are just as likely to drive over new crops etc. and do damage or get bogged and then you have to pull them out. Irresponsibility with fire is our major fear, dog control also worries us.
- Most people that visit the property are honest and attempt to do the right thing. We are concerned about the small number of less civic minded souls who do on occasion turn up. They ruin the property, owner's confidence and trust and make one wary of strangers in general.
- If we were to adopt a socialist agenda it must apply to all urban and rural lands i.e. open house to all. This must be a decision by all people to affect all people not just a minority i.e. rural landholders.
- We feel that private property should be just that. Access on invitation only.
- Once access is given its very hard to stop people – they think that they have an inalienable right to your place.
- Would you like people camping in your garden without permission?
- Surely there is enough land controlled by crown lands to satisfy the bush walkers etc. without the farmers of Australia having to put up with yet another intrusion into their privacy.
- Have you asked these questions of town or suburban land owners?
- Having been burgled … to the extent of $15,000, I am now very reluctant to draw any attention to our remoteness, as we are not in residence permanently.
- I am entitled to have quiet enjoyment of my own land. This is given by the Law of the Land and must be retained unless I choose to give up those rights.
- Would you like the general public camping and walking around your front yard … we like our privacy as much as city dwellers.

Source: Jenkins and Prin (1998).

study revealed some prospect of endorsing public access, providing specified conditions were met. These conditions related to control of the type of recreational activities permitted, their location and the timing and duration of access, and provisions for legal liability and payment for use.

Clearly, the ability of landholders to regulate recreational activities and the legal liability of farmers for people who use lands for recreational activities, appear to be aspects warranting further research. They also represent potential public policy avenues for increasing recreational access to private lands.

Water-based recreation

Recreating in and around water is a popular activity in most countries. In 1996, it was estimated that more than one-third of adult Canadians participated in water-based recreation (see Table 8.3). The economic significance of water-based recreation in Canada can also be demonstrated economically, as it is estimated that almost $2 billion is spent on recreation fishing (Environment Canada 2001).

In addition to the coastline, inland streams and waterways are extremely popular recreational sites both for Canadians and Australians.

The presence of water is often regarded as a fundamental requirement for outdoor recreation, either as a medium for the activity itself, or to enhance the appeal of a recreational setting. Water provides for a diversity of recreation experiences, some requiring direct use of the water itself (with or without body contact), and others merely requiring the presence of water for passive appreciation and to add to the scenic quality of the surroundings. The more active types of water-based recreation range over boating (sailing, power-boating, rowing and canoeing), fishing in all its different forms, and swimming (including sub-aqua diving, water-skiing and surfing). Some of these are associated more directly with coastal waters, while others are concentrated on rivers and inland waterbodies such as dams. All have experienced a remarkable upsurge in participation during the past two or three decades. In some cases, this upsurge has strained the capacity of the resource base to meet the growth in demand, and, in turn, has generated impacts and conflict between users and uses of water resources.

The impacts of water-related recreation activities were summarised by O'Brien (1983: 65–6, in Pigram 1986: 239–42) (see Table 8.4). The table also illustrates, if

Table 8.3 Participation in water-based activities, Canada, 1996

Participation	Water-based activities (8,532,000)
Total days	134,520,000
Total trips	89,423,000
Same-day	59,239,000
Overnight	30,184,000

Source: Environment Canada (2001).

Table 8.4 Some impacts of selected water-related recreation activities

Activity	Nature of effects
Shoreline swimming	Trampling of vegetation on shoreline; disturbance, litter, erosion of banks and sand dunes; pollution from human excreta; construction of facilities such as toilets, car parks and kiosks
Canoeing	Trampling of vegetation at launch place; littering; disturbance by landing on islands and by penetration into small waterways
Speed boating and skiing	Trampling of vegetation at informal landing sites; increased wave action and shoreline erosion; disturbance and penetration into shallow waters; cars on foreshore
Streambank or shoreline fishing	Trampling of shoreline vegetation; awareness of need for access and preservation of estuaries, waterways and coasts; litter; disturbance of wildlife; pollution from boats if used

Source: Adapted from O'Brien (1983, in Pigram 1986: 239–42).

only superficially, the possibility for conflict due to the incidence of these activities in time and space, and the degree to which they are undertaken.

The quantity and quality of available water can represent major constraints on the location, siting, design and operation of tourism facilities. As pressure grows on increasingly scarce water resources, the potential of areas, otherwise suitable for tourism development, may be compromised by inadequate water supplies. The presence of water serves as an additional dimension to a recreational or tourist facility, enhancing the scenic quality and appeal of the setting, and contributing to the attraction and intrinsic satisfaction derived from the tourist experience. An environment that is rich in water often forms an aesthetically pleasing setting for tourism. Water is essential for recreation and tourism – for drinking purposes, for sanitation and waste disposal, for cooling purposes, for irrigation and landscaping, and for the function of particular forms of water-related activities (e.g. swimming and boating). Water for the making of artificial snow is a major ecological issue in alpine and cool climate areas (Pigram 1995: 211–12; König 1998).

There is ample scope for conflict over use of water for outdoor recreation, and competition can become particularly intense where water resources are in short supply. Conflict can occur between:

- recreation and other resource uses, such as control structures within the river system or agricultural practices and other land uses within a drainage basin;
- incompatible recreation activities, amongst which power-boating and water-skiing probably arouse most opposition from less aggressive forms of recreation such as swimming and fishing;
- recreationists and the environment exposed to use (e.g. the water and shoreline, flora and fauna, and nearby human settlements and communities) (also see Chapter 2).

Conflicts are not confined merely to the water surface, but can occur at access points over ancillary facilities such as boat ramps, parking, campsites, access roads and the like. Even within the one specific recreation activity, excess usage can generate conflict over space at peak periods. Part of the problem is the inability of all waterbodies to satisfy the requirements for particular forms of water-based recreation. At least two aspects are critical (Mattyasovsky 1967).

First, the 'form' or nature of the water, and associated features, is fundamental. Certain wave conditions are an obvious prerequisite for surfing; 'white' water is ideal for wild river-running; and relatively static waterbodies may be preferred for water-skiing, sailing and rowing. Features of the shoreline and the area beneath the water can be important, as are the quantity, permanency and seasonal distribution of the waterbody. Boating enthusiasts who have to carry or drag their craft some distance to the waterline from a poorly sited boat ramp, can vouch for the problems caused by water level fluctuations and drawdown of reservoirs in dry weather, or after large releases of water.

Second, the quality of water (i.e. clarity, purity and temperature) that is appropriate for different recreational uses, needs consideration. Water quality often has to be a compromise, so that minimum criteria are stipulated rather than 'ideal' standards. For some types of recreation, even low levels of pollution can be tolerated, depending upon the pollutants and the activity in question.

Portugal's Algarve region is an example of how a precarious water supply and fierce competition between actual and potential uses can threaten the viability of the tourist industry (Martin *et al.* 1985). In the Algarve, water resources did not keep pace with demands from the intensification of agriculture and rapid urbanisation. At the time, the tourist industry accounted for up to 40,000 users, with peak demand coinciding with the period when water supplies were at their lowest. Groundwater was and still is the primary source of supply, and the major problem is proximity of demand to the coast and the risk of saline intrusion as water levels in wells are depleted. In such circumstances, firm planning control and management of the groundwater resource and alternative sources of supply are necessary in order to avoid contraction of tourist activity and possible abandonment of irrigation agriculture in the region.

Water-based recreation can be affected by contamination which can arise from many sources, including sewage, agricultural fertilisers, urban stormwater run-off, oil and petroleum spills from small boats to large container ships, human and animal faeces, swimmers carrying infections, and people washing with non-biodegradable detergents. So, unfortunately, the pleasure of drinking from the cool waters of even some remote mountain streams has become something of a risk. Two troublesome pathogens have occasionally been detected in streams in varying concentrations. These are the parasites, *Giardia lamblia* and *Crypto-sporidium parvum*. These microscopic, single-celled organisms are generally referred to simply as *Giardia* and *Cryptosporidium*. They can infect people through contaminated and untreated drinking water, or by coming into contact with the faeces of an infected person or animal. Giardia has been present in US natural areas since at least the 1970s (Meyer 1994), and has also been detected in remote areas of New Zealand (in parts where access is predominantly via air!) and is present in the Australia Alps. Australia now faces a situation anticipated by Hecock *et al.* (1976, in Pigram 1986: 216):

> Some streams have become health hazards because of pollution. Others are threatened by accelerated and unregulated shoreline development. Increased recreation use can adversely affect plants, animals and soils along rivers. Erosion of banks, camp sites and landings is a common problem. Growing use has resulted in more littering and vandalism to public and private property along rivers. Problems of maintenance and law enforcement have increased. There is also evidence that crowding, a variety of user-related conflicts and the impacts of recreation use on the environment, have substantially decreased the quality of the experiences for many recreationists.

To avoid the spread and possible contraction of these diseases, the University of Nevada, Cooperative Extension (undated), for example, recommends that people:

1   use toilets when they are available and wherever possible;
2   boil water for 5 and up to 10 minutes;
3   use five drops of iodine or about 2 drops of chlorine bleach for every quart of water (note: *Cryptosporidium* is highly resistant to chlorine);
4   ensure all faecal waste is buried 15–20 centimetres (6–8 inches) deep;
5   bury the toilet paper, mixing it into the soil or burning it;
6   in snow territories dig through snow to the soil and follow steps 4 and 5; and
7   consider carrying out human waste; protect food from contamination from insects and animals.

The incidence of such problems as those described above will not recede as people venture to the countryside and remoter areas, and as increasing population pressure and growing sophistication in water demand generate conflict between users and uses. The availability of water, in sufficient quantity and quality to satisfy demand, has emerged as an important concern in many parts of the world. As competition for water increases, tourism will be forced to justify its claims on the resource, against a range of more conventional uses and priorities. The problem can be clearly illustrated with reference to the water situation in North America and Britain (Pigram 1995).

The low priority given to in-stream uses of water for recreation is apparent in North America. In the US, recreation resource allocation, especially in rural areas, has tended to be *ad hoc*, and provision for tourist opportunities is often a by-product of other major resource developments. The result is resistance to those tourist development initiatives seen to threaten established claims on the resource base.

In Canada, the value of rural water for tourism and recreation is explicitly recognised in the resource appraisal procedures of the Canada Land Inventory. However, public sector initiatives to develop this potential have been intermittent and generally reactive to perceived exploitation of the natural environment by private interests (Butler and Clark 1992). Again, in park development, the emphasis has been mainly on environmental protection. Attempts to implement an integrated approach to the provision of opportunities for water-related tourism have received less attention (Pigram 1995).

In Britain, people have long enjoyed comparative ease of recreational access to rural land and water. The coastline is generally within easy reach (though the water is not always inviting!), and increments to the stock of recreation water space continue to occur from the construction of new reservoirs, restoration of canals and the flooding of disused gravel pits and mineral workings. Since 1974, regional waterbodies have had a statutory obligation to provide for recreation in all new water projects. Yet, few authorities have the personnel or necessary skills to plan and manage facilities in order to satisfy an increasing demand for water-related recreation and tourism (Blenkhorn 1979). Some concern has also been expressed about recreation opportunities for domestic and international tourists at water supply projects, following privatisation. Although legislation provides for public access to water authority land, the requirements are vague and open to differing interpretations (Pigram 1995). The British Waterways Act 1995 (see S 22(2)) requires that the British Waterways Board:

- further the conservation and enhancement of natural beauty and the conservation of flora and fauna on its properties;
- have regard to the desirability of protecting and conserving buildings, sites, objects of scientific or historic interest;
- have regard to the desirability of preserving for the public any freedom of access to towing path and open land and especially to places of natural beauty;
- have regard to the desirability of maintaining the availability to the public of any facility for visiting or inspecting any building, site or object of scientific or historical interest.

More generally, the value of water for leisure and recreation in rural Britain has been recognised by the Countryside Commission and its successor, the Countryside Agency. In its guide to sustainable tourism, the English Tourist Board acknowledged the role of clean waterbodies as an attraction for visitors, as well as the need for adequate water, in quantity and quality, for human and operational needs at tourist destinations. Clearly, the emphasis is on management of water to cater for the many ways in which it can function as a resource for recreation and tourism (Pigram 1995).

With sport-fishing, water quantity and quality are both significant, and for some species, temperature can also be a critical aspect of the fishing environment. It is important to consider fishing conditions for anglers, as well as the fish habitat, in physical and ecological terms. Habitat requirements vary and will almost certainly deteriorate with increased use. Management of the resource may require attention to the form of streams, e.g. construction of fish ladders and remedying pollution and other deficiencies in the condition of waterbodies as well as control of undesirable species. The quality of water is a less important consideration for recreational boating; more important are the size of the waterbody, depth, subsurface features such as rocks, any aquatic vegetation present, and compatibility with other users and uses (Mattyasovsky 1967). Boating of any kind is space-demanding, and power boating, in particular, can cause interference and danger to others, as well as water pollution and bank erosion. In addition, marinas, service facilities and boat launching ramps are often necessary. Provision of sufficient on-water mooring space can be a particular problem in popular, crowded waterways.

Although the primary concern must be provision of an adequate quantity of clean water of suitable quality, modern treatment facilities make many forms of water recreation compatible with this aim. Where recreation is permitted, bank and shoreline activities, as well as fishing and non-powered boating, are usually accepted without question. However, even body-contact forms of recreation could be permitted where water treatment is of a high standard. In any case, often there are many other 'natural' sources of water pollution stemming from agriculture, native birds and animals, and contaminated precipitation, as a study in northern New South Wales demonstrated (Burton 1975). In inland Australia, water for any purpose is generally in short supply, and recreational water space is severely restricted away from perennial streams. In this context, opposition to recreational use of domestic water supply storages is coming under increasing scrutiny, and there are indications that a more reasonable attitude to the issue may eventually emerge.

Summary

The relationships between recreation, tourism, and rural regional development, as in urban areas, are now significant economic, social and political issues warranting attention. Recreation and tourism can contribute greatly to rural development and prosperity, but this is not always the case. Tourism, in particular, has been utilised by governments, industry, regional authorities and other interests, as a means of diversifying and restructuring local economies in response to local and global forces, and more particularly to economic and population decline, and the need to create employment opportunities. Ultimately, the task in utilising tourism is to encourage economic and social development, to maintain or enhance the quality of life of residents and to maintain or enhance the quality of the physical environment.

Planning high-quality and sustainable recreation and tourism developments, which marry rural resources with local and tourist needs and preferences, has not been easy. Perhaps this is related to the lack of sound theories and concepts guiding the role and management of recreation and tourism in rural areas.

Guide to further reading

- Rural recreation and tourism: Simmons (1975); Middleton (1982); Patmore (1983); Cloke and Park (1985); Perdue *et al.* (1987); Wall (1989); Glyptis (1991); Groome (1993); Ibrahim and Cordes (1993); Watkins (1996); Page and Getz (1997); Sharpley and Sharpley (1997); Butler *et al.* (1998); Roberts and Hall (2001); Hall and Page (2002).
- Recreational access to rural lands: Cullington (1981); Pigram (1981); Butler (1984); Ravenscroft (1996); Watkins (1996); Jenkins and Prin (1998); Curry (2000, 2001).
- Constraints: Ravenscroft and Curry (2004).
- History of recreation in rural areas: Towner (1996); Butler *et al.* (1998).
- Rural change: Bowler *et al.* (1992); Cherry (1993); Cloke (1993); Cloke and Goodwin (1992, 1993); Glyptis (1993 – in particular see chapter by Cherry and Cloke); Ilbery (1997); Butler *et al.* (1998); Hall and Muller (2004).

Review questions

1   Define 'rural'.
2   To what extent is 'rural' a geographical term or an experiential concept?
3   What are the main forces influencing rural change?
4   Discuss the relationships between economic change and restructuring, and rural recreation and tourism development.

5 What are 'private' lands?
6 Distinguish between public and private lands.
7 What rights of ownership do public and private lands present for individuals and wider society?
8 Public recreational access to private rural lands is a contentious issue in many industrialised nations. Why is this so?
9 What are your views on public recreational access to private rural lands?
10 Why might there be strong attitudes towards ownership rights in countries like Australia?
11 How might the management of public and private rural lands be better integrated?
12 Discuss the importance of rural recreation as a means of escape for urbanites.

9 Protected areas, national parks and outdoor recreation

In an increasingly complex world, the need to set aside certain areas free of development, where conservation values can be protected, is seen as crucial and of growing importance. A protected area is defined by the IUCN as:

> An area of land and/or sea especially dedicated to the protection and maintenance of biological diversity, and of natural and associated cultural resources, and managed through legal or other effective means.
>
> (see Chape *et al.* 2003: 2)

The International Union for the Conservation of Nature (FNNPE 1993) listed ten protected area categories (scientific reserve; strict nature reserve; national park; natural monument; managed nature reserve or wildlife sanctuary; protected landscape; natural biotic area or philanthropological reserve; multiple use management area/managed resource; biosphere reserve; and World Heritage Site). Since 1994, IUCN has used six categories:

- *Category Ia* – strict nature reserve: protected area mainly for science;
- *Category Ib* – wilderness area: protected area managed mainly for wilderness protection;
- *Category II* – national park: protected area managed mainly for ecosystem protection and recreation;
- *Category III* – natural monument: protected area managed mainly for conservation of specific natural features;
- *Category IV* – habitat/species management area: protected area managed mainly for conservation through management intervention;
- *Category V* – protected landscape/seascape: protected area managed mainly for landscape/seascape conservation with recreation;
- *Category VI* – managed resource protected area: protected area managed mainly for the sustainable use of natural ecosystems.

This chapter focuses on protected areas, a critical factor in the supply of outdoor recreation. Systems of national parks and reserves can now be found in most countries of the world as nations and people recognise the important contribution of protected areas to society.

The urgent need to establish a comprehensive range of protected areas across the globe has been demonstrable. In 1962, the UN published its first list of 1,000 protected areas across the globe. In 2003 it was estimated there were 102,102 protected areas covering about 18.8 million km^2, of which 17.1 million km^2 (11.2 per cent of the world's land surface) were terrestrial biomes. Only 0.5 per cent of the world's oceans are protected (Chape *et al*. 2003: vii). The growth in number and volume of protected areas is globally impressive, but with this extensive development comes the realisation that both existing and new reserves must be managed effectively in the interests of conserving biological diversity. In a world marked by rapid change, economic imbalance and variable access to resources, protected areas face considerable pressure on their integrity and viability.

These concerns are reflected in the Caracas Declaration which emerged from the Fourth World Congress on National Parks and Protected Areas in 1992. The focus of the Congress was on the theme 'Parks for Life', and on the challenges threatening protected areas of the world in making a practical contribution to the health and well-being of humanity. The Caracas Declaration emphasised a number of fundamental principles:

- that nature has intrinsic worth and warrants respect regardless of its usefulness to society;
- that parks protect areas of living richness, natural beauty and cultural significance;
- that such areas are a source of inspiration, as well as places of spiritual, scientific, educational, cultural and recreational value (Lucas 1992).

Similar principles have been embodied in moves to establish national parks and reserves worldwide, but there is by no means unanimity in the philosophy or practice of natural area protection. Even the concept of a national park – the most common type of protected area – has evolved independently over more than 130 years. As a result, and in spite of some common features, there are as many variations on the national park theme as there are park authorities. Add to this the various parts of the environment which come under the description of 'wilderness area', 'marine park', 'nature reserve', 'state/provincial park', 'regional park', 'country park' or 'landscape park', and the picture becomes even less clear-cut.

This is not surprising when the wide-ranging perception of national parks and their role in society is appreciated. Right from their beginnings in the US last century, national parks that were established, were justified, in part, in terms of their potential to generate economic benefits. Much more recently, the same argument has been used to support efforts to expand the US national parks system: '… national parks are good business. They attract tourists and boost economies wherever they are situated. In part, that's why most Californians, including both senators from that state, favour the desert parks … In California, as elsewhere, that's the smart investment' (*USA Today*, 11 April 1994: 7).

Such sentiments are widespread. A survey of visitors to Dorrigo National Park, southeastern Australia, documented the scale and diversity of expenditure associated with park visits. The economic worth of the national park was substantiated by a survey of businesses in the nearby town of Dorrigo, in which the perceived importance to the local economy of park-associated tourism was emphasised. Some respondents went on to suggest that Dorrigo National Park, through exploitation of its tourist potential, could become the engine of regional development in this declining rural area. Other examples of such studies include those conducted in Canada, the UK and the US (e.g. Eagles and McCool 2002).

Whereas considerable efforts have been made to demonstrate the magnitude of the economic benefits of national parks (e.g. McDonald and Wilks 1986a, 1986b; Lambley 1988), there is no doubt that such parks also involve costs. Such costs can be divided into at least two categories – *direct* expenditure on establishment and maintenance, and the *indirect* or opportunity costs of commercial exploitation of resources, usually forgone as a result of the creation of the park. The fact that governments and the community are prepared to accept these costs and support the public funding of national parks, suggests that many people continue to regard them as worthwhile.

Among the range of values claimed for national parks, it would seem that one of the primary justifications lies in the inherent nature of society, and the demands which expanding populations and technological progress place on the natural environment. Without this pressure, large tracts of country would remain under-developed, and there would be less of a reason for national parks and nature reserves. Thus, for many people, the greatest value of the parks lies in their ecological role, in protecting areas and features of outstanding scenic and historical worth, and in preserving distinctive ecosystems, essentially unimpaired, for future generations. For others, provision of recreational opportunities is pre-eminent; the parks being seen as the means of physical and spiritual refreshment in a natural outdoor setting. A more limited segment of the population regards national parks as the vehicle for scientific research, retention of genetic diversity, and the study of natural phenomena in undisturbed surroundings.

Whatever the point of view, there is obviously widespread appreciation of national parks, and considerable support for the development and expansion of parks systems. The concept of a national park, however, is a relatively recent phenomenon.

In Australia, Canada, New Zealand and the US, the establishment of national and state/provincial parks for recreational and tourist opportunities, and the protection and maintenance of representative environments, came about in the nineteenth century. Interestingly, the first national parks in all four countries were set aside for conservation and recreation purposes because the land was considered worthless for such rural activities as intensive agriculture, lumber, mining and grazing (Hall 1992).

Yellowstone National Park, established in the US in 1872, is claimed to be the first national park. It was followed, in 1879, by Royal National Park, on the southern outskirts of the City of Sydney, Australia. This claim has recently been challenged on the grounds that Yellowstone was initially set aside 'as a public park or pleasuring ground', not as a national park. The first time the term 'national park' was used,

(a)

(b)

Plate 9.1   Royal National Park's coastal walk is stunning, yet so close to the Sydney CBD (Australia)

was in the legislation to create Australia's Royal National Park. The first legislative reference to Yellowstone as a 'national park' did not come until 1883.

The merits of the conflicting claims, however, are not the crucial issue here. Rather, since these beginnings and in the space of little more than a century, the modern parks movement has grown to worldwide dimensions. National parks can now be found in all continents, under a variety of economic and political systems.

National park concepts

The evolution of present-day national parks owes much to the American park movement of the nineteenth century, and to the efforts of conservationists such as Olmstead and Muir. The American park movement was motivated by regard for nature, and the revitalising powers of wild landscapes in an increasingly complex society. The dominant themes were the preservation and protection of the resources of nature, and the opening-up of these resources for the recreational needs of the nation. This movement culminated in the reservation of the first extensive area of wild land, primarily for public recreation, in the United States: the Yosemite Grant (in 1864). This was followed by Yellowstone, eight years later, and the Niagara Falls Reservation in 1885.

The International Union for the Conservation of Nature (IUCN) has attempted to clarify the concept of a national park by proposing a standardised definition. For management and planning purposes, a national park is defined under Category II of IUCN Protected Area Management Categories as:

> a natural area of land and/or sea, designated to (a) protect the ecological integrity of one or more ecosystems for present and future generations, (b) exclude exploitation or occupation inimical to the purpose of designation of the area, and (c) provide a foundation for spiritual, scientific, educational, recreational and visitor opportunities, all of which must be environmentally and culturally compatible.
>
> (Chape *et al.* 2003: 12)

As might be expected, the rather restrictive tone of the definition provoked some reaction and is also applied to state and provincial parks (Eagles 2003). The clear bias towards preservation of ecosystems, and the implicit limitations on human use, meant that many so-called national parks in some countries would not qualify as such. Any exploitation of natural resources (including hunting and fishing), all construction (including water impoundments, roads and amenities) and, strictly speaking, all means of transport and communication, could be excluded. In practice, of course, many of these land uses and facilities are permitted, if only to provide the necessary infrastructure to allow the park to function. In most cases, consumptive recreational pursuits (e.g. sport-fishing and even hunting, under certain conditions) are accepted, along with non-consumptive resource uses (e.g. hiking, boating, viewing, mountain climbing and scientific research). Active recreation is provided for and encouraged in many North American parks. Tourist amenities (often concessionaires) are accepted (but controlled), even within the park boundaries, under the operation of the management authority. Many observers would argue, too, that created bodies of water can enhance a park landscape.

Modifications of the IUCN definition have expanded the function of national parks to also include protection of cultural heritage, as well as the conservation of nature. Nonetheless, there is a popular view that at least some facilities for visitors and administration are necessary for the management and enjoyment of a national park. However, the definition probably still applies to parks in Africa, North America, New Zealand and Australia. Few parks would qualify in the 'Old World', where very little unaltered natural landscape remains. Even this qualification could be challenged, depending on how one interprets and understands physical and other kinds of change. The problems of making generalisations about the concept of national parks, can best be illustrated by reference to representative park systems across the world.

National parks in the United States

The US national parks system encompasses more than 380 different protected areas of diverse sizes and types, across some 34 million ha (approx. 84 million acres) (e.g. http://reference.allrefer.com/encyclopedia/N/natlpark.html). National parks are the best-known units within the system, but the Park Service is also responsible for several other areas, with designations such as national monuments and national memorials (only a few of which are actually statuary or historic buildings); national historic sites (especially those associated with American military history); national lakeshores; seashores; parkways; and wild and scenic rivers. In addition, the Service administers a large number of lands and buildings in and around the national capital – Washington, DC. The size and complexity of the American parks system make comparisons with other national systems difficult. Many of its features, however, in particular the approach to national park management, have been adopted by, or have at least influenced, other relatively newly settled countries such as Australia.

National parks within the US system are predominantly large, natural areas, containing a variety of resources, and one or more distinctive attributes or features of such scenic quality and scientific value as to be worthy of special efforts at preservation and protection. In a sense, the American national parks are regarded as 'outdoor museums', displaying geological history and imposing landforms and habitats of interesting and rare fauna and flora. In 1979, there were 37 national parks in the US; most of them in the western states, with a total area of nearly 16 million acres (approx. 6.5 million ha). This figure had grown to 52 by 2005. The better-known national parks (e.g. Yellowstone, Yosemite and Grand Canyon) contain some of the most spectacular scenery in the world, attracting vast numbers of visitors both from North America and foreign countries. Indeed, the sheer numbers of people wishing to visit the parks in peak periods has led to concern for the natural resource base, and has prompted a review of park philosophy and management principles in the US parks system.

Reservation of parkland for recreational purposes was a potent force, if not the primary one, in the early days of the US national parks. This initial viewpoint is interesting in view of the later change in emphasis towards the conservation of nature. In the early decision-making years, the attitude of park authorities was one of active encouragement of visitation by the public (Fitzsimmons 1976). Part of the rationale for these efforts was that exposure to nature would prompt visitors to appreciate and support the parks. Broad popular support was also seen as a means of counteracting political and economic interests hostile to the national park concept. If enough visitors could be attracted, parks would become self-supporting and would provide the income needed for their role in the preservation of natural species and landscapes.

These efforts at 'popularisation' of the parks read a little strangely in view of latter-day problems in North America, stemming from visitor pressure, congestion and fears of deterioration of park landscapes. However, the historical context should be borne in mind. The first parks were remote and difficult to access; transport was relatively slow and primitive, and public funds for park development were very limited. While patronage remained low, there must have seemed little contradiction between use and preservation, nor any need for management plans to maintain ecological values. The major problem, presumably, was how to boost attendance, and justify the viability and continued existence of the parks. Funds generated by publicity programmes 'were in turn used to provide more recreational attractions and visitor services in a spiralling development cycle' (Forster 1973: 17).

Following the end of the Second World War, all the features of the modern outdoor recreation phenomenon emerged and brought unprecedented pressure on national parks and similar resource-based areas. Rapid rises in population, coupled with economic expansion, increased affluence, leisure and mobility, brought new waves of visitors to the parks, seeking more diverse and sophisticated forms of amusement, not all of which were compatible with park values.

During the 1960s, increasing public concern over the impact of rapidly accelerating use and modern technology led to greater awareness and acceptance of the need for positive steps to contain visitor activity and restore park environments. The balance in park management philosophy and practice tipped in favour of restoration and preservation of the resource base. According to Leopold *et al.* (1963), the goal of park management became to preserve and, where necessary, recreate the ecological scene as viewed by the first European visitor. Clearly, the protective function of national parks, and the obligation to maintain the natural heritage 'unimpaired for the enjoyment of future generations'(National Park Service Organic Act 1916 and amendments), was now to receive priority. Although provision for public enjoyment and recreation remained an objective, it became subservient to preservation of natural features and ecological values.

More recent management decisions by the US National Park Service reinforce support for nature conservation as a primary objective. Restrictions on access to national parks are commonplace because of environmental damage and use conflicts. Motor vehicles have been excluded from some park areas, to be replaced

Plate 9.2   Half Dome, Yosemite National Park, provides incredible views. Unfortunately, its popularity and misuse has led to many ecological impacts (also see Chapter 5).

by shuttle buses and mini-trains. Speed restrictions, one-way traffic systems and limited parking facilities have been introduced to dampen visitor use. Although the regulation programme has apparently received general public acceptance, there are some who believe it does not go far enough, and others who oppose the restrictions imposed. Some blame the tour operators and other concession holders, and advocate an increase in fees as part of the answer. A letter to the *National Parks Magazine* targeted tour buses which:

> ... dump 50 or more visitors at a time in one area. The visitor centers, bathrooms, and viewing areas become chaos. The buses are also noisy, spew stinking black exhaust, and take up parking space ... Commercial interests have taken over the parks but very little money spent by park visitors is returned to the Parks Service ... Park fees are ridiculously low... less than the price of a movie ticket ... it's time to return our parks to the taxpayers.
>
> (Linz and Linz 1996: 10)

On the other hand, a proposal to restrict vehicle access in Yosemite National Park, and to require visitors to leave their cars in a parking lot and use a shuttle service, encountered strenuous opposition, although these buses do operate in the eastern end of Yosemite Valley. Whereas some considered the proposal an infringement on their rights, others thought it did not go far enough, given that the ultimate goal of the Park Service Management Plan for Yosemite is to remove all private vehicles from the valley (Nolte 1995: A21).

Despite these initiatives, the strict goal of preservation is obviously unattainable in the absolute sense while *any* level of use is permitted. The implications of this use/preservation dilemma are discussed in Chapter 10. However, it seems that for the present, at least in developed countries, perpetuation of natural and cultural heritage is now recognised as the prime function of national parks. How long the notion of national parks as predominantly nature reserves can be maintained is open to question if community support is alienated in the process and public funding continually reduced. The expectation that national parks will be generally accessible to the community is widely held. The further assumption that they will be, to a degree, self-supporting, is also important in the development and expansion of the national parks system. This situation may prompt renewed support for the involvement of the private sector in national park management.

Canadian national parks

Not surprisingly, the parks system in Canada has features in common with the US; in fact, there is shared responsibility for certain natural and historic sites along their common border (e.g. see Zbicz 2000).

Parks Canada, established in 1911 and perhaps the world's first national parks agency, manages a system of protected areas consisting of 41 terrestrial national parks and 2 national marine conservation areas. Among the 41 national parks, some, such as Banff and Yoho, date from the late 1880s; others, such as Ivvavik and Vuntut (formerly, Northern Yukon), Grasslands and Bruce Peninsula, are relatively recent additions, becoming national parks in the 1980s. The programme to establish national marine conservation areas (and its predecessor, the national marine parks programme) came into being in 1986, with the adoption of the National Marine Parks Policy.

Both programmes depend upon a system of regionalisation, which divides the country into 39 terrestrial and 29 marine natural regions (including natural regions in the Great Lakes) (Canadian Environmental Advisory Council 1991; Parks Canada 2004). Parks Canada has a mandate to establish representative protected areas in each of these natural regions. To date, 27 terrestrial areas are represented in the national parks system. Therefore, the protected area system at the federal level in Canada is far from complete. Parks Canada continues to work to establish protected areas in the natural regions where currently there is no representation. On the other hand, some natural regions in western Canada are over-represented by the establishment of several national parks as tourism destinations in earlier times (Payne and Nilsen 1997). Parks Agency Canada aims to have 34 of the 39 terrestrial regions and eight of the marine regions represented by 2008. Indeed, the Canadian government has given a mandate to Parks Canada to expand three existing national parks and to establish another 10 national parks and five national marine conservation areas by 2008 (Parks Agency Canada 2004).

As with the parks of the western US, recreational opportunities, as well as commercial considerations, were of prime concern in the early years. Interest in the first Canadian national park at Banff (established in 1885) dates from the discovery of hot mineral springs in the 1880s by employees working on the transcontinental railway across the Rocky Mountains. Curious as it may seem, in the light of the magnificent Rocky Mountain scenery in the area, the original reason given for the reservation of land at this site was the 'sanitary advantage' of these waters, and the need to protect them from commercial exploitation and control them for the benefit of the public (Scharff 1972). In 1887, Banff Hot Springs Reserve, when enlarged to an area of 260 square miles (approx. 673 km²), officially became Rocky Mountains National Park. The name was later changed to Banff National Park, and the Canadian government and the railroads combined to develop hotels and facilities for visitors to the area.

It is worth noting that the Rocky Mountains Park Act of 1887, specifically reserved the area as 'a public park and pleasure ground for the benefit, advantage and enjoyment of the people of Canada'. This wording is almost identical with that proclaiming Yellowstone National Park. In addition, the Act went on to spell out the protective aspect, emphasising that no development was to be permitted that could impair the usefulness of the park for the purposes of public enjoyment and recreation.

Canadian Pacific was one railroad group promoting national parks since the 1800s, and they continue to operate hotels and related tourist services. As more and more parks were added to the Canadian system, transport networks were developed, all manner of visitor facilities were provided, and entrepreneurs were encouraged to maintain a high level of service to promote patronage. In some cases, the recreation facilities at sites like Banff and Lake Louise themselves became major tourist attractions to complement the scenic grandeur in the surrounding park landscape. Tourism development was seen by early directors of Canada's national parks (and particularly by J.B. Harden) as a means of getting people to support conservation and national parks. Harden was an advocate of tourism development and the need for access and was involved in the construction of the Canadian Rockies' Banff–Jasper Highway (Nelson and Butler 1974; Nelson 2000). As with the American parks, concern for nature preservation was to come later as visitor pressure mounted on the park environments, and the depredations brought about by indiscriminate hunting, mining and timber-getting became obvious. According to Nelson and Butler (1974), it was only in the period after the Second World War that a strong preservationist movement emerged in Canada. The traditional view of tourism and recreation as fundamental underpinnings for parks was increasingly brought into question and, ultimately, the preservation and protection of park landscapes came to be regarded as first priority.

Park pressures in Canada are enormous and growing. In 1994, Parks Canada revised its national parks programme policy to designate ecological integrity as a prime agency goal, and ecosystem management as the prime means of achieving it. Central to this new policy direction is the acceptance that ecosystem management must address the full range of human issues in establishing and managing parks, including the impacts of human use on natural systems and the impacts of park establishment and operation on human use systems. Although Parks Agency Canada is directed to consider human use issues, such as opportunities for public understanding, appreciation and enjoyment, in establishing national parks, and while the agency possesses a range of tools (such as the Visitor Activity Management Process – see Chapter 6, as well as the associated Appropriate Visitor Activity Assessment and Risk Management processes), it was previously argued such human considerations have not yet figured in new park establishment in any major way (Payne and Nilsen 1997). Yet, now that Canada's system of protected areas receives about 26 million visitors overall, and more than a quarter of Canadians visit national parks each year, Parks Canada is making improved education services a priority. It has undertaken to 'increase our investment in education services supplemented by partnerships with the tourism industry and non-governmental organizations to achieve that end' (Parks Agency Canada 2004: 4). Strategic objectives, planned results and performance expectations have been set, and three particular aspects of visitor services are being targeted: visitor information, recapitalisation of visitor facilities, and public safety (p. 32). A recent audit highlighted the decline in the quality of many of Parks Canada's assets, which range from national monuments to canal locks to national parks. Indeed, the Agency itself states that:

> Many visitor facilities, including campgrounds, visitor reception centres, trails and exhibits were constructed between the late 1950's and early 1970's and have now reached the end of their normal life cycles. Some no longer meet the health and safety standards of the day. Parks Canada budgets are not adequate to maintain all of these facilities and the related services. Potential strategies to address this shortfall include the closure of facilities and the reduction of services and increase to the fees paid by visitors to access and enjoy the national parks and national historic sites.
>
> (Parks Agency Canada 2004: 33)

Parks Agency Canada has been innovative in its approaches to addressing visitor management issues, privatising park services, entering into partnerships with recreation and tourism organisations and altering the status of Parks Canada (it is now a Special Operating Agency – SOA). However, like so many countries, national parks agencies, while confronted with increasing recreational demand and visitor expectations, are met with shrinking budgets, conflicting land use (mining) and shifts in political ideology. Recently, however, the CEO of Parks Agency Canada, Alan LaTourelle, announced a budget increase of $315 million over the next five years. This additional funding is targeted at restoring damaged sites and Canadian Historic Places, and improving public park infrastructure.

A particular overarching and ongoing concern is the question of prior human habitation in areas designated as national parks, and the problem of accommodating traditional resource uses within park management programmes. Proposals for new national parks and reserves in the more remote regions of Canada, such as the Yukon and the Northwest Territories, are examples. Special attention is being paid to protecting wilderness values, while maintaining the rights of native peoples to continue traditional extractive activities, such as hunting, fishing and trapping, in areas like Baffin Island. In the Canadian Rockies, there has been research conducted into the reintroduction of bison into national parks. Research indicates free-ranging plains bison once frequented the Canadian Rockies and were hunted by native people. Moreover, the ecosystems in these parks 'were structured from the top-down by carnivore and human predation – a factor that must be taken into consideration if free-ranging plains bison are to be reintroduced to Banff and other Canadian national parks' (Kay and White 2001: 148). As a condition of reintroduction, it has been suggested that 'hunting by First Nations may be required to maintain appropriate herd sizes and ecological integrity' (Kay and White 2001: 148). According to Parks Canada's Ecological Integrity Review Panel,

> humans have been present for thousands of years on the lands that now constitute Canada. Their association with the land and their traditional activities were part of the ecosystems and, to a certain extent, made the landscape what it was when Europeans first arrived ... [Moreover] the influence of Aboriginal peoples is fully consistent with ... [the] definition of ecological integrity. [In fact] ... this traditional human role is an important element of the ecological integrity of the ecosystems that Parks Canada is mandated to preserve or restore ...
>
> (Parks Agency Canada 2000b: 7-2, in Kay and White 2001: 148)

In many respects, these problems resemble those encountered in tribal territories in developing countries (see below). Prior human habitation also represents a problem, but of a different kind, in the older, more densely settled countries of Europe.

National parks in Britain

It is clear that the IUCN definition of national parks is inappropriate, and largely irrelevant, for a country like Britain, with a long history of human settlement and no great reserves of unoccupied lands in which to create national parks in the North American mould. Moreover, by the time the first moves were made to establish national parks in Britain at the end of the Second World War, widespread acquisition of private land was prohibitively expensive and politically unacceptable. However, the amount of land under national parks status in the UK is not insubstantial. About 8 per cent of the land area in England is national park, 20 per cent in Wales, and about 7 per cent in Scotland.

The result is that areas designated as 'national parks' remain almost entirely in private ownership and productive use. Agricultural holdings, fenced pastures, forestry plantations, quarries, farm structures, transport routeways, and even villages and towns are all found inside the park boundaries. Management plans endeavour to reconcile conflicting interests between landholders and park visitors. At the same time, attempts are made to maintain and enhance the scenic quality and appearance of the landscape by controls over the location and nature of new facilities and proposals to alter existing structures. This can be made more difficult by the intrusion of extractive activities and, more recently, 'wind farms' as energy resources.

A National Parks Commission (later Countryside Commission) was set up in Britain in 1949, and the first park, the Peak District National Park, became a reality in 1951. Since then, another eleven national parks have been created, the most recent being the New Forest National Park in 2004 (see O'Brien 2004). Attention is now being directed to securing national park status for the South Downs. A public enquiry has examined the designation by the Countryside Agency for a national park in this attractive area, under threat from urbanisation and intensive agriculture (Crane 2003). The twelve national parks in England and Wales are operated by their own national parks authorities. Each authority has two statutory purposes:

1   To conserve and enhance the natural beauty, wildlife and cultural heritage of the area; and
2   To promote opportunities for the understanding and enjoyment of the park's special qualities by the public.

Tourism is a major feature of the national parks of England and Wales, and in some of the national scenic areas in Scotland. In 1991, it was estimated that the national parks of England and Wales attracted 103 million visitor days a year, with the greatest number visiting the Lake District and Peak District national parks, some 20 million visitor days a year, each (National Parks Review Panel 1991). More recently, it was reported that in 1994, a minimum of 76 million visitor days were spent in the parks as a whole, but that this figure was likely to be a substantial underestimate of the actual totals, because survey methods did not cover all categories of visitor. Average daily expenditure, excluding accommodation, was estimated at £9.78 per person. These figures, though underestimates, clearly demonstrate the economic contribution of parks to local economies (Countryside Commission 1998). Visitation among the parks varies greatly, with the Lake District and Peak District each receiving more than 12 million visitors, whereas Exmoor and Northumberland each receive far fewer than 2 million visitors.

Until recently, there were no national parks in Scotland. Although proposals were made as early as 1945 for five national parks, pressures on the countryside were much less than those in England and the idea lapsed. The Countryside Commission for Scotland, set up in 1967, established some quite small country parks for intensive recreational use, and proposed a new parks system to encompass urban parks, country parks, regional parks, special parks and national scenic areas. National parks were felt inappropriate in a Scottish context because, under internationally accepted standards, 'conservation must always take precedence over recreation and other land uses' (Foster 1979: 4). Such an approach was seen as lacking flexibility and inhibiting retention of desired characteristics in 'a living, in-use way rather than in a museum sense'.

In 1990, the Countryside Commission for Scotland recommended the establishment of four national parks, the first parks north of the border. Heritage landscapes such as the Cairngorms, Ben Nevis and Loch Lomond were proposed for protection, using a system of zoning for core areas, surrounded by management buffer zones and a transitional community development zone. On 5 July 2000, the Bill for the National Parks (Scotland) Act 2000 was passed and subsequently received Royal Assent on 9 August 2000. This Act aims:

(a) to conserve and enhance the natural and cultural heritage of the area,
(b) to promote sustainable use of the natural resources of the area,
(c) to promote understanding and enjoyment (including enjoyment in the form of recreation) of the special qualities of the area by the public, and
(d) to promote sustainable economic and social development of the area's communities.

Two national parks have been established – Cairngorms National Park, and Loch Lomond and the Trossachs National Park. Both national parks are administered by a national park authority as for parks in England and Wales.

Reviews of the concept and purpose of national parks in Britain have led to more emphasis being given to management procedures to ensure that recreational use does not threaten the scenic beauty and wildlife, and that forestry and agriculture within the parks does not detract from the appearance of the landscape. Concern has also been expressed about quarrying, the design and construction of reservoirs, housing and recreational facilities, and visitor pressure on roads not designed for heavy traffic.

Plate 9.3   Ben Nevis is Scotland's highest peak at just over 1,300 metres or around 4,400 feet. It provides stunning views, but visitor use is high and erosion is evident on tracks leading up the mountain

Plate 9.4   Loch Lomond is part of one of the first two national parks designated in Scotland

Plate 9.5   The Munros are Scotland's highest mountains, which are an attraction for the Scots and walkers around the world. Stunning views can be had from any of the 284 mountains more than 3,000 feet or 914 metres in height. The first list of Munros was compiled by Sir Hugh Munro in 1891. Many people seek to 'bag' or 'climb' Munros, often bagging more than one in a day's hike.

One of the most contentious issues surrounding the management of Britain's national parks is their use for military purposes. Table 9.1 shows the widespread nature of military activities in the parks. Not surprisingly, frequent protests occur, as new proposals for development of training facilities are put forward.

The Otterburn Training Area was established in 1911 as an artillery range and occupies 58,000 acres (approx. 23,500 ha) in Northumberland National Park. This represents 23 per cent of the park's area, with further expansion being planned. A new army training camp has recently been built in Dartmoor National Park, where the Ministry of Defence refuses to abandon its training programmes, which include live firing. The military vigorously defends its need for training facilities, and points to the success of its conservation and restoration programmes. However, the Countryside Agency and the Council for National Parks have long argued that military use of national parks is inconsistent with national park purposes and frequently leads to irreparable damage, and have therefore opposed any extensions or intensification of military activity in the parks. Recent developments were approved, however, for the construction of new roads and infrastructure. Clearly, problems will always exist where privately owned resources play the major role in providing recreational opportunities for park users, and where private interests may conflict with national priorities in conserving the natural beauty and amenity of the countryside.

The national parks of England and Wales were a product of the circumstances prevalent at the time. As these circumstances change, management has to adjust. Few would argue for the abolition of the parks, but their character may change and different solutions may have to be found in order to attain the objectives for which they were established. The designation of Areas of Outstanding Natural Beauty (AONBs) Country Parks have proved to be valuable steps in this direction.

Table 9.1 Military use of national parks

	LF	D/A	A	Air	RPA
Dartmoor	•	•	•	•	•
Exmoor		•		•	
Brecon Beacons		•	•	•	
Pembrokeshire Coast	•	•	•	•	•
Snowdonia	•	•		•	
Peak District	•	•		•	•
Yorkshire Dales		•		•	
North Yorkshire Moors		•	•	•	•
Lake District		•	•	•	
Northumberland	•	•	•	•	•

Where: LF – live firing
D/A – dry/adventure training
A – army camps and bases
Air – low flying aircraft
RPA – restricted public access

Source: Adapted from Lunn (1986: 6).

The main purpose of Country Parks, which arose out of the Countryside Act 1968, is the provision of recreational facilities in an outdoor setting, and in many ways they are the antithesis of national parks. More than 270 Country Parks have been recognised: these act as 'honeypots', providing readily accessible recreation outlets for large numbers of rural users, where existing, more natural areas are under threat from overuse. Nonetheless, some country parks are in decline and have been for years. In this way, pressure on the national parks might well be relieved by provision of a greater range of alternative rural recreation opportunities, accessible to large centres of population. Park boundaries and features also need to be reassessed in order to identify areas and sites where management controls may be eased, or in other cases, tightened.

When the first national parks were being established in England, it was realised that large areas of beautiful countryside were neither big enough or wild enough to meet the definition put forward for national park status. For such landscapes, the description – Areas of Outstanding Natural Beauty – came into use. There are now 41 AONBs in England and Wales covering 15 per cent of the land, a total of 8,200 square miles (approx. 21,237 km^2). The primary purpose of these areas is to conserve and enhance the natural beauty of the landscape. Their diversity and distinctiveness is remarkable, ranging from the Isles of Scilly (16 km^2 or 6 square miles) up to the largest, the Cotswolds, totalling 2,038 km^2 (787 square miles). Since 2004, they are to be subject to statutory management plans under the responsibility of conservation boards (Cecil 2005).

Whenever people are intimately involved, as they are in the British national parks, concern must be shown for their attitudes and welfare. The continued support and endorsement of the park concept by the inhabitants are vital for their continued success.

National parks in Ireland

Few people appreciate that Ireland had a national park long before Britain, or that Irish national parks conform to the strict guidelines laid down by the IUCN.

(Dillon 1993: 5)

Nature conservation in the Republic of Ireland is the responsibility of the National Parks and Wildlife Service, part of the Department of the Environment, Heritage and Local Government. Fauna and flora are protected by refuges and nature reserves, and by 1,200 Natural Heritage Areas. Size apart, Irish national parks are similar to those in North America and Australia, in contrast to those in Britain. Ireland's first national park was established in 1932, near Killarney in the southwest, from the gift of a 4,000 ha (9,883 acre) estate. It has since been extended to just over 10,000 ha (24,700 acres), and includes the lakes of Killarney and surrounding mountains. The small Connemara National Park in Galway (almost 3,000 ha or 7,412 acres) was opened in 1980, followed in 1986 by Glenveagh National Park in Donegal (the largest – nearly 17,000 ha or 42,000 acres), and Wicklow Mountains National Park, near Dublin, in 1990. There are now six national parks in the

Republic (Figure 9.1), the smallest of which is The Burren (1,673 ha or 4,133 acres). In the management of national parks in Ireland, the key objectives are:

1   To protect the natural heritage allowing for, and if necessary, managing for the continuation and restoration of natural processes.
2   To protect and where appropriate to restore and develop other heritage and aesthetic qualities.
3   To provide for public access and to encourage public appreciation and enjoyment of the parks under conditions compatible with the above objectives.
4   To develop a harmonious relationship between National Parks and the surrounding communities, taking into account the social and economic needs of local communities (National Parks and Wildlife Service [Ireland], http://www.npws.ie/en/contactus/, accessed 20 January 2005).

While there appears to be general support for national parks in Ireland, controversy has arisen over a number of localised management issues. The stark beauty of the limestone pavements in the Burren National Park, in the central

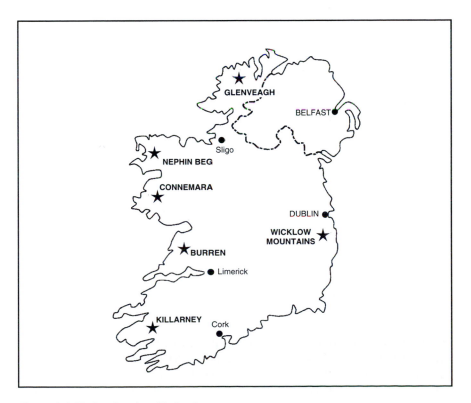

Figure 9.1 National parks of Ireland

Source: Adapted from http://homepage.tinet.ie/~knp/duchas/; also see http://www.npws.ie/en/NationalParks/.

west, has been threatened by a proposal to build a modern visitor centre at Mullaghmore within the park. Concern was expressed over increased numbers of vehicles, pressure on roads and facilities, and the sustainability of the site and degradation of the park's wilderness character (Don 1997). Money and jobs were also a consideration due to the belief that if roads were widened to take coachloads of visitors '... the cash rather tends to flow (out) with them' (Dillon 1993: 5). The visitor centre was put on hold, pending design and location of more environmentally compatible facilities (http://www.iol.ie/~burrenag/ hist72000.html).

As in many national parks in the Western world, conservation of endangered species, and eradication of noxious exotic species and feral animals are ongoing concerns in Irish national parks. An example from Killarney is control of infestations of rhododendrons, an attractive flowering shrub, but one that represents a significant management problem for regeneration of native species.

National parks in Europe

The British Isles share, with much of the rest of Europe, the problem of developing a functioning park system within a landscape which has evolved over centuries of human use. In a country like the Netherlands, the task is made even more difficult; it has one of the highest population densities in the world, and a good proportion of the countryside is the direct product of human efforts to reclaim land from the sea. Yet, even there, 13 per cent of the country is still said to be in a more or less natural state – dunes, wetlands, woods or uncultivated areas – and a number of national parks have been created.

A national park in the Netherlands has been defined as:

> An uninterrupted stretch of land of at least 1,000 hectares (approx. 2,500 acres) consisting of areas of natural beauty, lakes, ponds and water courses and/or forest, with a special character as regards nature and landscape, and a special plant and animal life.
>
> (Netherlands Ministry for Cultural Affairs 1976: 2)

The Dutch Government is endeavouring to meet the requirements laid down in the IUCN definition, and to date 22 areas have been selected which meet the criteria of size, quality and integrity of area and management. It is worth noting that in the Netherlands, 1,000 ha is considered sufficiently large for a national park. This hardly compares favourably with around 4.5 million ha (approx. 173,000 square miles) in the world's largest national park, Wood Buffalo, in Canada, while the smallest parks in Ireland are less than 2,000 hectares (4,941 acres) (see above).

In addition to national parks, the Netherlands is developing an experimental system of National Landscape Parks, which are closely related to the National Reserves in the USA, the National Parks of Britain, the Regional Parks of France and the *Naturparken* in Germany. The concept has much in common, too, with the idea of 'countryside parks' proposed for Australia (see below).

With this type of park, the concern is not with purely natural areas, but with areas shaped by humans and nature in combination over the course of many centuries. National Landscape Parks include villages and towns, agriculture, typical architecture, and other features of human activity characteristic of the Netherlands landscape. The concept envisages that landholders, in addition to working their land, should assist in the management of the landscape park and receive payment for activities concerned with its care, as well as compensation for loss of income as a result of any limitations on farming practice. Thus, farmers will no longer supply only grain, potatoes, dairy produce and meat, but will also provide the community with an attractive landscape in a healthy living environment. Moreover, they will get paid for it. In a country like the Netherlands, in particular, National Landscape Parks are seen as complementary to the national parks system, and as an appropriate way of encouraging people living and working in settled rural areas to maintain the natural and cultural values of the countryside.

The reunification of Germany and the changing geopolitical scene have led to the opening-up of former well-known national parks in central and eastern Europe. These include the Hochharz National Park in the former East Germany, the Ojcowski National Park in Poland and the Tatra National Park near Zakopane, in the south of that country.

It has been predicted that the emergence of the European Union and the operation of the Channel Tunnel will lead to a new era of partnerships and linkages between Europe's protected areas (Simpson 1995: 5). Twinning arrangements and staff exchanges are already helping to disseminate best management practice, and to promote common policies and strategies between national parks in Britain and Europe.

National parks in Australia

'National' parks in Australia, have, until relatively recently, been the sole responsibility of the six State and two Territory governments. Technically, therefore, they did not qualify under the strict requirements of the earlier 1969 IUCN definition, that national parks be under the jurisdiction of the nation's 'highest competent authority'. However, from most other standpoints, they do meet the international guidelines – national parks typically consisting of sizable areas of predominantly unspoiled landscape, with the emphasis on nature conservation. Only since 1975 has an Australian National Parks and Wildlife Service functioned, with specifically 'national' parks being established alongside the State systems.

Originally, the provision of public recreational opportunities (as in North America) was the primary objective of national parks in Australia. The first park established, Royal National Park, near Sydney, provided holiday accommodation, sporting facilities and picnic areas, with the emphasis clearly on human pleasure and amusement. Since the early years, the concept of a national park has broadened beyond this recreational theme.

A gradual increase in the number and area of national parks and reserves followed the growth in environmental awareness which occurred in the 1970s and

1980s. New South Wales (NSW), the most populous State, has more than 600 parks and reserves covering 5.9 million hectares (or more than 7 per cent) of that State. These areas (also see Table 9.2), administered by the State National Parks and Wildlife Service, which is part of the Department of Environment and Conservation, included:

- 169 national parks,
- 376 nature reserves,
- 15 historic sites,
- 10 Aboriginal areas,
- 18 state conservation areas,

Table 9.2 Lands managed for conservation by NSWNPWS, Department of Environment and Conservation

Under the NSW National Parks and Wildlife Act there are seven categories of land managed for conservation: national parks, nature reserves, historic sites, Aboriginal areas, karst conservation reserves, state conservation areas and regional parks.

National parks are relatively large areas of land set aside to protect and conserve areas containing outstanding or representative ecosystems, natural or cultural features that provide opportunities for public appreciation and inspiration and sustainable visitor use and enjoyment. They are permanently reserved for conservation and for public education and recreation and apart from essential management are preserved in their natural state.

Nature reserves are areas of special scientific interest for their outstanding, unique or representative ecosystems, species, communities or natural features. Management practices aim at maximising the value of the area for scientific investigation and educational purposes.

Historic sites are areas of national importance that are preserved and include buildings, objects, monuments or landscapes.

Aboriginal areas are places of natural or cultural significance to Aboriginal people, or of importance in improving public understanding of Aboriginal culture and its development and transitions.

Karst conservation reserves are areas managed to protect significant surface and underground land formations in karst regions.

State conservation areas are permanently reserved areas that contain significant or representative ecosystems, landforms or natural phenomena or places of cultural significance, and that are capable of providing opportunities for sustainable visitor use and enjoyment.

Regional parks are areas in a natural or modified landscape that are suitable for public recreation and enjoyment.

National parks and historic sites are managed in such a way that their natural and cultural features are conserved while still allowing visitors to use and enjoy them. State conservation areas and regional parks are managed to maximise their recreational potential while preserving and protecting their natural features. Because nature reserves and Aboriginal areas are conserved for scientific and cultural values and are small in area, public access is often limited.

Source: NSWNPWS (2003: 10–11).

- 10 regional parks, and
- 4 karst conservation areas (NSWNPWS 2003).

Protected areas in NSW receive more than 20 million visitors annually, while about 50 per cent of international visitors go to a national park when they visit Australia. The largest national park unit is Kosciuszko National Park, which occupies 640,000 ha (approx. 1.6 million acres) southwest of Canberra in the Australian Alps. More parks and reserves have been added recently.

The Great Barrier Reef off the coast of Queensland also has a strong attraction for visitors. Following extensive lobbying by preservation-minded pressure groups, the Great Barrier Reef Marine Park Authority (GBRMPA) was set up, and sections of the reef region are being successively incorporated into the marine park. Such parks, by their very nature, present unusual problems in park management. Contentious issues in this case were the question of oil exploration on the reef, commercial and recreational fishing rights, and the clash between the Queensland State Government and Australian Federal Government over administration of the resources of the region.

The federal government has become increasingly involved in park management since the formation of the Australian National Parks and Wildlife Service in 1975 (now part of the Department of Environment and Heritage). The Service works in collaboration with the States, and has sole responsibility for certain areas of nature conservation interest such as Norfolk Island and Christmas Island, as well as national parks such as Uluru and Kakadu, near Darwin, in the Northern Territory. The Service believes that the plans of management drawn up for Kakadu may well prove a model for the development of similar parks in 'frontier' areas. Certainly, the park has had to contend with some major problems. Apart from preservation of the park environment and providing for appropriate use by visitors to a remote area, protection of Aboriginal interests, regulation of mining (uranium) in and near the park, and control of feral animals, all create difficulties for park management.

Australia's largest city, Sydney, with a population approaching four million people, is fortunate in that a large number of national parks are located within a radius of 160 kilometres (approx. 100 miles). To a great extent, this situation results from the reservation of land for parks in areas where the soil and terrain were considered unsuitable for agriculture and too rugged for housing. Royal National Park, south of the city, Kuringai Chase National Park, immediately to the north, and Blue Mountains National Park, to the west, are located on dissected sandstone plateaus, for which no economic use was perceived at the end of last century. Melbourne, on the other hand, in the State of Victoria, was ringed by good agricultural land, and there is now a serious dearth of parks and reserves close to the city.

The apparent contradiction and conflicts between outdoor recreation and nature preservation, and between different types and intensities of recreational use of national parks in Australia, suggest that the complexity of recreational demands should be matched by an appropriate array of recreational opportunities outside the parks. In the State of New South Wales this need has been met, in part, by the development of state conservation areas (SRAs), state parks and regional parks. SRAs, some of them quite large, comprise both natural areas and man-made features of scenic, historic and recreational importance. Several SRAs adjoin inland waterbodies, and others occupy coastal sites. SRAs were designed to take pressure off the national parks, and while haphazard destruction of their environment is obviously not allowed, they do cater for more intensive forms of outdoor recreation; even trail-bike riding and hunting may be permitted. In some parks, accommodation facilities have been provided, while others are designed for day-visitor needs only. Provided that the recreational emphasis continues, state recreation areas are a useful and popular complement to national parks, and an important additional unit in an integrated system of outdoor recreation opportunities.

The introduction of more recreation-orientated 'people's' parks, came at an important stage in the development of the parks system in Australia. For many years, national parks have made a significant contribution to the recreation resource base. Now, in many parts of the country, and for a variety of reasons, opportunities for further expansion of the national park system are becoming limited. Indeed, it could be said that Australia has entered a 'mature' phase of park development, in which the initial stage of large-scale land acquisition is closing, to be replaced by careful appraisal, development and management of park resources already acquired. At the same time, consideration can now be given to alternative means of expanding the recreation opportunities of urban-based populations, by provision of a range of different park options in accessible rural settings.

Moreover, a good deal of questioning of the role of national parks has emerged in Australia in recent years. To some, parks and wilderness areas appear as enclaves of unproductive land, and havens for noxious plants and animals. Proposals to enlarge national parks on the north coast of New South Wales, and to establish additional national parks in the New England region, further inland, have generated significant local opposition, but the size of the national parks estate (and other conservation reserves) continues to grow. The continued relevance of the North American park model to Australia has been challenged, and the adoption of the British/European style park has been put forward as an alternative (Pigram 1981). The proposed new style of 'living' park environment, if developed along the lines of the National Reserves in the United States, would encompass farming communities and rural settlements within distinctive scenic landscapes. The creation of 'countryside parks' in this way should be seen as a complement for national parks (rather than as a replacement), and as an extension of the land management systems, providing for a dual network of inhabited parks developed in tandem with the traditional parks. Regional parks fulfil some of this need, but Australia is a far cry from the living parks of the UK.

Much progress has been made in the conservation of biodiversity beyond the formal system of parks and protected areas. The Voluntary Conservation Agreements in place in New South Wales are an example, as are the regional parks of Perth, Western Australia (Moir 1995), some of which remain in private ownership. The establishment of new, long-distance walking tracks has witnessed much opposition from potentially affected landholders, but some have agreed for tracks to be negotiated through their properties. As with Canada, controversy has emerged over the extent to which traditional indigenous hunting practices can be accommodated in Australia's national parks. A draft discussion paper has been released on 'Wild Resource Use by Aboriginal People on National Parks and Reserves in New South Wales' (Woodford 2005). Several issues would need to be addressed, including implements to be permitted, endangered species and public liability.

Until recently, park authorities in Australia have generally shown a decided reluctance to depart from the existing national parks system. Brisbane Forest Park, near the capital city of Queensland, was a step in the right direction, although all land within that park is publicly owned. In the state of Victoria, the Regional Strategy Plan for the Upper Yarra Valley and Dandenong Ranges on the northeastern outskirts of Melbourne, encompasses many of the features of the National Reserve concept. Non-urban land in private hands comprises 23 per cent (approx. 175,000 acres or 70,000 ha) of the total land area, with a further 3 per cent classified as urban. Approximately 105,000 people live within the region, and the Strategy Plan provides for protection of the special features and rural character of the area and the maintenance of recreation opportunities on both public and private land. In Sydney, eight areas have been designated as Metropolitan regional parks. These parks were conceived by the New South Wales Government as a way of providing the people of Sydney with green lungs; a similar justification as for the establishment of Royal National Park in 1879. Regional parks are areas of open space for recreation and for the conservation of fragile ecosystems. They vary in size (from 4,000 hectares to less than 50 ha) and in the activities to which they cater, and represent a promising public sector initiative (for details see http://www.npws.nsw.gov.au/parks/regprks.htm).

Apart from these initiatives, a generally negative attitude prevails at official levels towards the introduction of European-style national parks, or US National Reserves, in Australia. This reaction, coupled with growing resistance by rural landholders (also see Chapter 8) to any further acquisition of park lands, means that progress towards establishment of 'countryside parks' is likely to be slow. Clearly, the need for innovations in park planning should be obvious. In a developed country like Australia, with all the pressures on resources for greater output and more efficient production methods, the transition to large-scale stereotyped forms of land use can be very rapid. Therefore, there is an urgent need to adopt an alternative approach to allocating land for parks because the changing nature of agriculture and rural life acts as a disincentive for landholders to maintain the character and quality of the countryside in their keeping.

The countryside park concept could play a useful role in Australia, alongside national parks. If it can be shown by successful pilot projects that the economic and amenity functions of the countryside can be compatible, then a range of park types can be created, as and where appropriate. Given time and enlightened management, such parks have the potential to demonstrate the benefits of sharing the countryside, both for ongoing productive purposes and for outdoor recreation.

National parks in New Zealand

New Zealand was one of the first nations to establish a national park after the creation of Yellowstone. Tongariro National Park came into being in 1887, as the result of a gift from the Maori people of an area of volcanic peaks in the central north island. Since that time, 13 additional national parks, three maritime and two marine parks, 20 forest parks, and over 1,000 scenic and special reserves have been added. In all, about one-third of New Zealand's land area is protected in parks or reserves; this is in a country which has a total landmass only roughly the size of the US State of Colorado (Figure 9.2).

New Zealand has an environment that is unique in the world; its geographic isolation has resulted in the evolution of diverse fauna and flora. This, coupled with spectacular scenery, provides many opportunities for the creation of parks and protected areas. Many of the large reserves are focused on lakeshores and on the higher mountainous country and fiords of the South Island. Fiordland National Park, covering 1.26 million ha (more than 3 million acres), is one of the largest in the world. The park is a World Heritage Site and takes in areas of outstanding natural beauty such as Milford Sound. Another World Heritage Site, Te

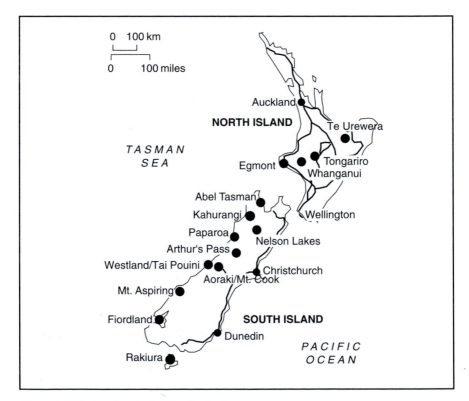

Figure 9.2 National parks in New Zealand

Source: Adapted from Department of Conservation (http://www.doc.govt.nz/Explore/001~National-Parks/index.asp, accessed 14 March 2005).

Waipounamu, covers 2.6 million ha (approx. 6.4 million acres) and several national parks in the southwest, and is a focus for many forms of nature tourism (see Box 9.1). Some of the management problems concerning such protected areas are dealt with in the following chapter.

Box 9.1 National parks in New Zealand

Visitor facilities and opportunities in New Zealand national parks are managed by conservancies. Management is primarily a place-based exercise where decision-making takes account of the contribution of the facilities within a wider geographic context. A range of planning tools is used for categorising the nature of the land, the types of visitors, the types of tracks and the types of huts within the network. Facilities include picnic areas, toilets, walking tracks, huts, signage, wharves, bridges and viewing platforms. The Recreation Opportunity Spectrum is a key planning tool to ensure, as far as possible, a mix of recreation settings for park visitors. One shortcoming of this approach is the over-representation in the ROS of 'back country walk-in' settings, so that the concept of a range of recreation opportunities is not always available. This has been offset to some extent by the development of a great number and diversity of huts and tracks within the parks to contribute to a satisfying range of opportunities. In this way, the hut and track network combine in providing facilities that support the visitor opportunities.

That said, the network of huts used by hikers and climbers can become a source of contention between visitors and park management. Clearly, the huts are an important part of the back country experience which many visitors seek. Some were constructed in the past in conjunction with culling programmes to eradicate introduced species of deer, goats and pigs which can cause extensive environmental damage in the parks. Others were erected as simple and rather primitive survival shelters against the risk of dangerous weather conditions in the more remote mountainous sections of the parks. An ongoing debate has focused on the advantages and disadvantages of the huts. Those that advocate their removal argue that their very presence negates the risk and challenge associated with a 'true' back-country experience. Similar claims have been made in regard to survival huts in Australia's high country national parks. It has also been noted that some huts are in poor condition and are sited in hazardous locations, thus giving a misleading impression of safety. Hut sites are also said to have become the focus of environmental degradation. Proponents of huts maintain that, apart from saving lives, providing shelter, and a means of informal social contact between park users, they can be a useful management tool. Properly located and maintained, the huts can play a role as an aid to conservation and help in directing visitors away from fragile areas. Thus, while a 'no hut' policy applies in wilderness areas of New Zealand, the continued existence of a limited network of huts and tracks in the national parks can assist in the control and management of visitors.

National parks in developing countries

Despite the existence of large areas suitable for designation as national parks, problems can arise with park establishment and management in developing countries. Although ecological considerations and the desirability of preserving unique ecosystems may certainly be recognised in the selection of environments and landscapes for inclusion in the parks system, park proposals are often assessed primarily against potential economic and social benefits. This means that, in negotiating land acquisition and planning the future operation and management of a park, it becomes critically important for the government authority to be able to demonstrate specific benefits, especially for the local people, by way of commercial opportunities and employment. Thus, economic factors may overshadow ecological considerations, to the detriment of the park environment.

National parks are now a reality in all corners of the developing world. Some of these parks and reserves reflect attempts to protect natural landscapes and wildlife for conservation and scientific purposes. In other situations, potentially large returns from tourism appear to have influenced their creation. Much of the stimulus for this tourist activity comes from worldwide interest in viewing nature, and from the diversity of animal and bird life to be found in the national parks. In the less developed countries of Africa, for example, most park visitors come from abroad. Whereas some newly-emerging nations may regard the parks as unwelcome vestiges of previous foreign dominance, parks are tolerated and even encouraged because of their role in providing local employment, and attracting tourists and foreign currency.

Large and varied species of wildlife can be found in national parks such as Tsavo and Nairobi in Kenya; Kilimanjaro and Serengeti in Tanzania; Matopos in Zimbabwe and Whangie Kruger in South Africa. Kruger National Park covers nearly 5 million acres (approx. 2 million ha), and is visited by almost 500,000 people annually, 25 per cent of them from overseas.

This also applies to Ras Mohammed National Park, Egypt's only national park (Egyptian Environmental Affairs Agency undated). Established in 1983, the park covers an area of 480 km^2 (approx. 120,000 acres) in the South Sinai peninsula. The park includes land and marine areas and shorelines along the eastern coast of the peninsula. Two other Managed Resource Protection Areas – Nabq and Abu Galum – have also been created further north on the Gulf of Aqaba. The national park and protected areas take in some of the world's best coral reef ecosystems and fossil coral platforms, as well as spectacular granite mountains and desert landscapes. Visitors attracted to those areas and to the rapidly developing tourist resorts at Sharm el Sheik and Dahab, are an essential feature of the economic development of South Sinai.

Although a number of national parks have been established in Southeast Asia and the Pacific Islands, considerable difficulties still have to be overcome. Countries like Indonesia, Papua New Guinea, the Philippines, Malaysia and Vietnam, have apparently endorsed the concept of national parks and appear convinced of the role they can play in nature conservation. However, such conviction cannot always lead to action in societies where wilderness is still considered an obstacle to progress and the value of conservation is not universally appreciated. There may well be difficulty in diverting money and manpower to the development of parks, and a reluctance to take land out of what is considered to be more productive use. Even in circumstances where the authorities do display enthusiasm and an awareness of the value of parks, obstacles may still surface in attempting to translate the concept into action.

Specific problems can occur in areas of prior human habitation, especially where land is in communal ownership and land use practices, such as shifting agriculture, timber getting and hunting, are destructive of the environment. Problems can be countered, in part, by raising standards of living above the subsistence levels that contribute to these rapacious forms of land use. Moreover, if the local population can receive some tangible benefit from the establishment of a national park, people may be more prepared to respect and maintain the integrity of the park environment. This calls for a fine balance between the creation of a strict nature reserve on the one hand, and a commercially orientated nature-based tourism enterprise on the other. If this is not achieved, there may be resentment and non-cooperation, where a more environmentally compatible, but less rewarding and beneficial type of park system, is imposed on local communities. In practice, as Cochrane (1996: 242) has argued, it is extremely difficult to achieve the aims of ecotourism and to improve the welfare of local people, simultaneously.

A major concern in these circumstances is the extent to which new or existing national parks and nature reserves may intrude upon the lives of local residents, leading to disruption of established patterns of land use and of the social fabric. Trekking in the Nepalese Himalayas has brought profound change to local communities of the region. Tourism development since the early 1960s has been nothing short of phenomenal. In peak seasons, tourists outnumber local residents by a ratio of 5:1, with Everest attracting about 20,000 trekkers per annum and daily densities of trekkers ranging between 1 to 2 people per kilometre to almost 30 per kilometre. As Nepal (2000) explained:

> Declaration of the Everest region as a national park in 1976 and a World Heritage Site in 1980 may have saved this region from further environmental degradation, as compared to other parts of the country. However, the rapid development of tourism has transformed the region's economy, environment and culture in an unprecedented way. As a result of tourism-related problems, Everest has been labelled 'the world's highest junkyard', and the trail to base camp 'the garbage trail'. Namche Baza is called a 'lodge city' where Sherpa life revolves around tourists and a small-scale, locally controlled form of capitalism (Well 1994).
>
> (Nepal 2000: 79)

A related issue for park establishment and management in developing countries, and one shared with the developed world, is the dilemma of promoting national parks as an engine of tourism, while maintaining the biophysical integrity of the park environment. In the absence of sound appreciation of park values, emphasis may be misguidedly placed on maximising visitor numbers in the interests of economic returns, to the detriment of the park itself (Pigram *et al.* 1997).

Wilderness

No discussion of protected areas would be complete without reference to what many regard as the ultimate in natural environments – wilderness. For much of history, wilderness has held a negative connotation; either as wasteland, or some vast, hostile and dangerous place to be avoided if at all possible, or else to be tamed, controlled and exploited.

Today, both people and governments have come to think of wilderness in more positive terms, as something to be valued and preserved for a future world, in which it could become an increasingly rare phenomenon. Many people now perceive wilderness as a large natural area where animals and plants can live undisturbed, and where visitors can enjoy recreational activities of a primitive and unconfined nature. Hiking and canoeing are often given as examples of the types of recreation envisaged – those for which a minimum of mechanical aids is required.

The main benefits of a wilderness experience are often said to be the spiritual and psychological satisfactions gained. Other advantages of wilderness recreation are physical and mental stimulation, the aesthetic appreciation of beautiful scenery, and the experience of conditions similar to those encountered by the first settlers of a region. Wilderness serves as a sanctuary, either temporarily or permanently, for renewal of mind and spirit. In modern jargon, it has become a refuge for those who wish to 'drop out', momentarily, into a simpler, less complicated world; a place where self-confidence can be re-established through physical challenge and reliance on self-sufficiency and subsistence skills.

Wilderness areas are also valued because of their role in nature conservation and scientific research. The size, remoteness and variety of ecosystems represented in wilderness are important for wildlife preservation and the maintenance of ecological stability and genetic diversity. Apart from being a potential source of a wide variety of useful plants and insects, wilderness also provides a reference point against which to measure changes in settled areas, and in crops, forests and animal populations. Some proponents of wilderness argue that these areas also provide a buffer, or safety valve, against long-term disturbance of the global ecosystem, resulting from large-scale human interference. While this may be the case, it is a nebulous argument to use in trying to persuade decision-makers to close off public lands for exclusive use in scientific research. This argument has provoked a reaction in some quarters that wilderness is a selfish concept and the pursuit of a small and vocal élite. The restricted numbers and specialised forms of recreation associated with wilderness do little to destroy this impression (Sax 1980).

A further qualification concerns the degree to which conditions in wilderness areas can remain pristine. Conditions of total naturalness are impossible to find, even in Antarctica. Therefore, wilderness has to be a compromise, taking in areas where there remain no permanent traces of people (e.g. roads, buildings and modified vegetation).

The really large remaining areas of 'true' wilderness can be found, like the big national parks of the world, in North and South America, Australia, New Zealand, parts of Africa, and, of course, the Arctic and Antarctic. In general, these areas have not experienced heavy population pressure on their land and water resources; but, even in polar regions it is growing, although in a necessarily limited way because of climate, accessibility, safety and environmental concerns (Marsh 2000: 134). Impacts in these very sensitive areas need to be very carefully monitored, because they may be especially fragile and vulnerable to damage. Permafrost soils are particularly vulnerable to vehicle use but also people's feet; cold climates mean disposal and decomposition of human waste are inhibited as is use of conventional methods; fishing is a threat in polar waters where productivity is limited (Marsh 2000). Indeed, in these remote, 'wilderness' areas, such as the Canadian Arctic, there are concerns about lack of employment for local people in park agencies and tour companies. The government of the Northwest Territories subsequently 'introduced a programme of training and certification for northerners, especially native people, to learn about guiding and interpretation …' (Marsh 2000: 133).

Wilderness is not a concept generally applicable in Europe; some limited examples of quasi-wilderness might be found, but as in Britain, Western Europe and Scandinavia, potential areas have, with few exceptions, been extensively used by humans.

Wilderness is land which retains its natural character and is without improvements or human habitation. Simple, non-mechanised forms of recreation are envisaged; to preserve wilderness values, it is necessary to protect the natural ecosystems present, and to maintain the topography and plant and animal populations in an undisturbed state. Thus, a prime purpose of wilderness is to keep the area as natural as possible by only allowing levels of use that are consistent with both ecological and perceptual carrying capacities.

Zoning is a common strategy employed in wilderness management. One approach to zoning is the core/buffer concept (Figure 9.3). Here, the wilderness core is surrounded by a wilderness management zone or protective buffer. The protective function is two-way – to protect the wilderness, and to protect adjacent land from disturbances such as wildfires which might originate in the wilderness. Subzones are determined for access and minor facilities, while separate scientific reference areas, with more restricted access, are set aside within the wilderness complex.

The designation of land as wilderness is a contentious issue. A long legacy of resource exploitation brands as strange and unacceptable the sterilisation of land with economic potential. Even among wilderness supporters, the formulation of management policies satisfying those advocating strict ecological preservation and those seeking a 'wilderness experience' is a difficult challenge to meet. Until wilderness is accepted as a legitimate form of land use, and its benefits, both for

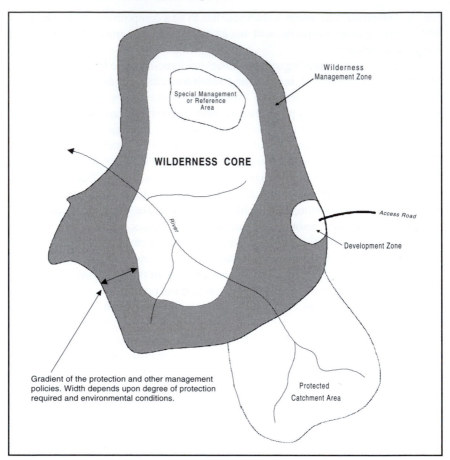

Figure 9.3 Wilderness management zoning

Source: Helman *et al.* (1976: 45).

outdoor recreation and nature conservation, are more generally recognised, controversy will continue to surround the wilderness concept.

Guide to further reading

A number of useful references for further reading are cited below. However, readers are encouraged to access government documents, in particular those containing details concerning the management of specific protected areas. In addition, several other useful sources are cited in Chapter 10.

- General discussions about the establishment and management of protected areas, particularly national parks: Sax (1980); Hall (1992); Green (1992); IUCN (1994); Payne and Nilsen (1997); Jenkins (1998); Butler and Boyd (2000); Eagles and McCool (2002).

- Recreation and tourism in protected areas, including World Heritage Areas and national parks: Forster (1973); FNNPE (1993); Eagles (1996); Shackley (1998); Ryan (2003); Buckley, R. (2004).
- Tourism in protected areas in developing countries: Shackley's (1996) book makes frequent reference to protected areas in developing countries; Pigram *et al.* (1997) present a case study of Cat Ba Island, Vietnam.
- Regional impacts of tourism and recreation in parks: McDonald and Wilks (1986a, 1986b).

Review questions

1 What is a national park? What should be the main goals and objectives of national parks authorities? To what extent should recreation be 'tolerated' and accepted in national parks?
2 Discuss important aspects of the early history of the establishment of national parks globally, nationally or regionally.
3 Select a national park or wilderness area. What are the main sources of recreational pressures on that park or area? What planning and management initiatives have been devised by that park's authorities to deal with such pressures? Have these measures been evaluated? How and to what extent have they been evaluated?
4 Protected places are not for people! Discuss.

10 National parks management

The previous chapter presented a broad overview of protected areas in various environments, with particular reference to the establishment of national parks in countries around the world. This chapter develops a more focused and critical assessment of the management of national parks for nature conservation and outdoor recreation. Competing values, interests and priorities highlight the need for planning procedures that take account of environmental concerns, as well as recreational and tourism use of national parks, in an integrated and sensitive way.

The brief canvass of park systems in the previous chapter illustrated the many ways in which the national park concept has been interpreted. This diversity gives rise to an equally complex range of park problems and approaches to the management of national parks. Despite these variations, a recurring theme with all park environments is the need to strike a balance between conservation, or preservation, and use; that is, how to accommodate appropriate levels of human activity, while maintaining the quality of the natural environment for which the park was established. This is the 'dilemma of development' (Fitzsimmons 1976).

Resource use is primarily concerned with the present generation; conservation and preservation are linked more closely with generations to come. Expediency demands the satisfaction of current wants, whereas prudence suggests limitations on use in order to preserve park values for future generations. Conflict and compromise are inevitable since *any* use involves some disturbance to the park landscape and ecosystem. The aim of management should be, first, to exclude activities which are clearly inappropriate – power boating, organised sports and entertainment centres come to mind as ready examples. Second, care must be exercised to keep unavoidable disturbance to a minimum and to recognise and correct environmental deterioration before it becomes irreversible.

Given the popularity of national parks as a focus for outdoor recreation, it could be assumed that moves to expand the parks system would not be questioned. However, in countries like the US and Australia, proposals to establish new national parks frequently generate strong opposition. Part of this reaction stems from a conviction in some quarters that resources and personnel should be directed towards upgrading the facilities and management of existing parks before acquiring more land. Added to this is concern expressed by neighbouring landholders, who question the record of certain park management practices (e.g. with respect to fire and pest control) deemed incompatible with surrounding land uses. Part of the problem stems from inadequate attention to the setting of park boundaries.

National park boundaries

In the past, the process of establishing national parks might have been described as trying to put boundaries that don't exist around areas that do not matter (Kimble 1951). Kimble was referring to regions rather than parks, but the point is well made. In many cases, park boundaries appear to have been determined with more regard for administrative and managerial convenience than for ecological and other relevant criteria. Examples include national parks which end abruptly at state borders or follow shire or county lines and similar cadastral features. The boundaries of Yellowstone National Park, for example, are mostly straight lines, and the park itself is contained almost within a square (see Figure 10.1). In many other countries

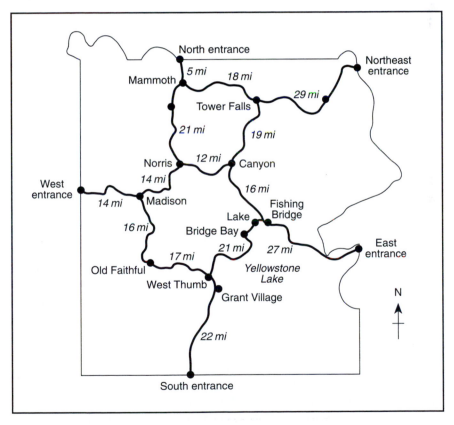

Figure 10.1 The boundaries of Yellowstone National Park

Source: Adapted from http://www.travelsmarter.com/mapynp.htm, moved to http://www.nps.gov/yell/planvisit/orientation/index.htm, accessed 1 March 2005.

such as Australia, where many parks and reserves, especially in isolated areas, are typically regular polygons, the boundaries rarely follow natural features, or bear a close relationship to the 'shapes' of the ecosystems they are designed to protect.

The question of boundaries is of particular relevance to the management of Australia's national parks, whose boundaries often have been (1) determined qualitatively, and (2) influenced by political considerations and opportunism (Pressey *et al.* 1990). This situation gains added significance when it is realised that park boundaries, once established, are generally 'set in stone', with proposals for even minor changes to modify the park area requiring legislation and political contests among political parties, government ministers, resource management agencies, and local communities and indigenous peoples (Howard 1997).

Yet, the allocation of land use, for any purpose, cannot be rigid. The process of allocation should be dynamic in keeping with new information and changes in technology and social preferences (Walker and Nix 1993). Certainly, experience in countries outside Australia does not suggest reluctance to review park boundaries. In Canada, delineation of park boundaries has also come under scrutiny (Theberge 1989, 1992). According to Theberge (1989: 22): '... it is ecologically indefensible to establish parks without provision for boundary or other management adjustments'.

In the US, following the 1980 State of the Parks Report to Congress, an increasing amount of research has been directed towards the formulation of credible and practical guidelines for establishing ecologically and managerially sound boundaries for national parks and reserves. This research culminated in the development of a comprehensive Park Resource Boundary Model for selecting critical park resource variables and determining their spatial extent (Sundell 1991).

In Britain, a decision was taken in 1991 to reorganise the administration of national parks and review their boundaries. Definition of national parks boundaries has been addressed by the Countryside Agency. The Agency's decisions on whether the land is suitable are defined in legislation (S 5(2) of the 1949 Act). The criteria for defining the boundary were defined as the Hobhouse criteria (see Box 10.1). Recently, it was recommended that these criteria have generally stood the test of time but nevertheless require updating (see Box 10.2).

Defining and indicating boundaries in marine environments is perhaps more difficult than terrestrial. In Victoria, Australia, Marine National Parks and Marine Sanctuaries are defined in legislation. There are 24 such areas which have been surveyed and which range in size from 10–20 ha for Sanctuaries to more than 10,000 ha for Marine Parks. These boundaries are being developed so that they are accurately portrayed in marine charts while the markers to indicate their existence or presence include 'Yellow On-Shore Triangles' which by aligning two indicate the boundary on the shoreline and towards the Marine Sanctuary or National Park. The 'In Water Special Mark' is found on special buoys and piles. They are used in calm water and mark channel boundaries and other boundary features (Parks Victoria 2002).

Apart from establishing the spatial extent of a park, the determination of workable park boundaries is fundamental to maintaining the viability of the reserved

***Box 10.1* The Hobhouse criteria**

The Hobhouse Report of 1947 (Report of the National Parks Committee to Parliament Cmd 7121) identified the following criteria for identifying National Park boundaries.

The following considerations should, in our opinion, be taken into account in the precise definition of National Park boundaries.

(a) The first criterion should be the inclusion of areas of high landscape quality.

(b) Wherever possible, an easily distinguishable physical boundary should be chosen, both for administrative reasons and for the convenience of the visiting public. Roads and railways frequently provide such a boundary.

(c) Where County, District and Borough boundaries follow suitable lines, it may be administratively convenient to adopt them. In the majority of cases, however, they are unsuitable, since they follow no defined physical feature, may be subject to alteration and seldom conform for any considerable distance to the limits of landscape value.

(d) Towns or villages should not normally be cut in two by a National Park boundary. The inclusion or exclusion of a marginal town or village should be dependent on its character and beauty and its present or potential value for the accommodation of visitors.

(e) Unsightly development on the edge of a National Park should generally be excluded, but the possibility of its modification or screening should not be overlooked where the immediately surrounding country claims inclusion.

(f) Quarrying and mining of important deposits on the margins of a National Park, which could not, in the national interest, be strictly controlled, should normally be excluded from the Park, except where the deposits are likely to be worked out within a reasonable time and surface restoration seems practicable and desirable.

(g) Features of scientific, historic or architectural value (e.g. Nature Reserves, important archaeological sites and Ancient Monuments) which are situated on the margins of a National Park should be included where practicable.

In general, boundaries should include, as far as possible, any features which are part of the rural economy and community life within the Park, and should normally exclude areas where the needs of urban or industrial development conflict with, or outweigh, the essential values of the Park. The boundary of a National Park should not, however, be regarded as a sharp barrier between amenity and recreational values within, and disregard of such values outside.

Source: The Countryside Agency (2000).

Box 10.2 **The Countryside Agency's approach to defining National Park boundaries**

1. The Countryside Agency shall first determine in broad terms that an area of land meets the statutory criteria for designation as defined in S 5(2) of the National Parks and Access to the Countryside Act 1949, in respect to the Agency's policy in applying.

2. It shall then in drawing a National Park boundary take account together of the following considerations.

    (a) Areas of high landscape quality coherent with the character of the area of land identified by the Agency for designation should be included. Landscape quality includes visual and intangible features and values and embraces natural beauty, wildlife and cultural heritage. It is interpreted as the extent to which the landscape demonstrates the presence of key characteristics and the absence of atypical or incongruous ones, and by its state of repair and integrity.

    (b) Areas to be included may be of differing landscape character, as quality will be the key determinant rather than homogeneity. A variety of landscape character can be an important factor in the overall amenity of the Park. Usually, however, there will be some unifying factors, such as land use, ecosystems, historical or cultural links which bring differing character areas together to be included into a National Park.

    (c) Areas which provide or are capable of providing a markedly superior recreational experience should be included. Recreation in this context means countryside recreation related to the character of the area: that which allows people to enjoy and understand the special qualities of the Park, without damaging it or conflicting with its purposes. It does not mean recreation which will materially diminish enjoyment of the Park by others. Therefore it will generally be quiet recreation, including for example walking, riding and cycling; enjoying natural history; visiting features of historical interest and countryside sports.

    (d) Boundaries should include land and settlements which contribute to the rural economy and community life within the Park, in so far as that economic or social activity contributes to the special qualities of the Park and its purposes. Areas should be excluded where activities, in particular urban or industrial development, conflict with or outweigh the essential values of the Park.

    (e) Wherever possible, an easily distinguishable physical boundary should be chosen, both for administrative reasons and for the convenience of the visiting public. Roads and railways frequently provide such a boundary.

(f) Where County, District, Borough and Parish boundaries follow suitable lines, it may be administratively convenient to adopt them. In the majority of cases, however, they are unsuitable, since they follow no defined physical feature, may be subject to alteration and seldom conform for any considerable distance to the limits of landscape value.

(g) Towns or villages should not normally be cut in two by a National Park boundary. The inclusion or exclusion of a town or village should depend on whether it contributes to the character of the Park, including its economy and community life, and its present or potential value for visitors; e.g. provision of accommodation, access to public transport, information or other services. Large settlements should generally be excluded as a National Park Authority should not be overburdened with responsibility for planning in urban areas.

(h) Unsightly development on the edge of a National Park should generally be excluded, but the possibility of its modification or screening should not be overlooked where the immediately surrounding country claims inclusion.

(i) Land shown in the adopted minerals and local plan as to be worked for the quarrying and mining of important deposits on the margins of a National Park should normally be excluded from the Park unless there is a realistic expectation of restoration to a land use and quality which contributes to Park purposes. This approach should also extend to major industrial and commercial developments shown in adopted local plans at the time of designation.

(j) Features of scientific, historic or architectural value (e.g. Nature Reserves, important archaeological sites and Ancient Monuments) which are situated on the margins of a National Park should be included where practicable.

3. The statutory criteria (S 5(2) of the 1949 Act) point to the duty to designate where both high landscape quality and recreational opportunity exist. This does not mean that all land within the park must necessarily satisfy both criteria (a) and (c). But there should be a high degree of concurrence, in view of the Agency's policy for designation (established in February 2000).

4. The definitive boundary of a Park should not be regarded as a sharp barrier between areas of differing quality. It should be recognised that in most situations there will be a transition of amenity and recreational quality across a sweep of land: the boundary chosen should be an easily identifiable feature within this transition.

Source: The Countryside Agency (2000).

area, its ecological integrity and biological diversity. An accepted part of the rationale for the establishment of parks and reserves is protection of areas and features of special natural or cultural interest. This task is often more acute along the edges of designated areas, where conservation values may conflict with neighbouring land uses. The need to integrate the area harmoniously into its surrounding environment is essential for the effective functioning of natural checks and balances with neighbouring ecosystems. This has been articulated at IUCN Congresses in 1982 and 1992 and in the Durban Accord arising from the 2003 Fifth World Parks Congress (World Commission on Protected Areas 2003). Zube and Busch (1990) indicated that relationships between parks and their neighbours have grown in importance as:

*   ideas and systems for the protection of ecosystems spread;
*   traditional farming and other land uses are challenged;
*   tourism grows in association with the establishment of protected areas;
*   access to traditional food and other resources for indigenous people is affected.

Cross-cultural approaches to management of national parks are receiving attention around the world, and have received impetus from different sources (Zube and Busch 1990; Notzke 1995; Noble 2000; Wearing and Huyskens 2001; Slocombe and Dearden 2002). In Australia, joint management regimes have arisen from evaluations of uranium mining in Kakadu national park, Northern Territory. Additional impetus came from the Millstream Recommendation which encouraged:

*   joint management arrangements between Aboriginal people and national park management agencies;
*   the involvement of Aboriginal people in the development of management plans for national parks;
*   the excision of areas within national parks as living areas;
*   the granting of access to protected areas for subsistence hunting;
*   fishing and gathering rights; and
*   facilitating the control of cultural heritage information by Aboriginal people.

(Lawrence, 1997, in Wearing and Huyskens 2001: 190)

Despite such developments, in few cases have indigenous cultures' values systems been adequately accommodated (Veal 1994, in Wearing and Huyskens 2001). Some of the most significant problems in national park management occur in relation to park boundaries. Further, it can be argued that many of these problems could be avoided if more care was exercised in setting those boundaries. When inappropriate boundaries are established, the very resources which a reserve was intended to protect may become vulnerable to environmental stress from a number of internal and external threats. Thorough assessment of critical resource attributes is fundamental to the delineation of workable boundaries for a viable park or reserve.

Park relationships in Australia

In Australia, the National Parks agencies take seriously the challenge of neighbour–community relationships. In New South Wales, for example, there are more than 30,000 adjoining 'neighbours' to parks in urban areas, and some 8,000 in rural areas (Howard 1997). These interests cover a number of groups, including private and public sector landholders, and authorities managing utilities, services and transport. Among key issues of concern to these neighbours are fire management, control of noxious species and access.

Fire management

Growth in urban areas adjoining parks and reserves in NSW and the Australian Capital Territory has increased dramatically in recent years. In some high-risk areas, residential development has continued to the edge of park boundaries as was the case with the severe fires which penetrated parts of the City of Canberra in January 2003. Over 500 suburban homes were destroyed and some people lost their lives. Prior to the recent spate of fires, the Parks Service performance in fire management has remained at a high level (Table 10.1) and has continued to do so when battling natural elements to reduce fuel and in managing and controlling fires that begin in isolated places. Despite assertions to the contrary, the parks system has to cope with many more fires from outside protected areas than those which escape from the parks into adjoining land.

Whatever the origin, the disastrous bushfires of the 1993/4 summer, especially those close to the City of Sydney, tragically demonstrated the potential for loss of lives and property, and serious damage to the natural environment. There is wide diversity in the way Australian native vegetation responds to fire, and some species depend on fire for reproduction. Others may be wiped out completely. A surprising number of native animals also survive fire by evading the flames or taking refuge in safe sites. Despite this resilience, long-term damage and decline in ecosystem diversity can occur as a result of a sequence of fires (as has recently occurred in parts of the Blue Mountains National Park west of Sydney), especially if closely followed by drought or lengthy dry spells.

Table 10.1 Fire origin and progress, New South Wales

Year	Started on park, controlled on park	Started on park, moved off park	Started off park, moved on park	Total fires	Area burnt (ha)
1989–90	142	8	99	249	66,464
1990–1	303	29	93	425	125,469
1991–2	266	21	109	396	66,409
1992–3	167	6	40	213	21,772
1993–4	216	25	50	300	382,897
1994–5	173	15	62	250	89,112

Source: Howard (1997: 399).

Strategies for fire prevention and control, therefore, are important elements in park management planning (see Worboys *et al.* 2001: 259–64). Unfortunately, however, it is difficult to eliminate fires because of the range of potential causes. In the latest 2001 severe or 'hot' fires, which burned out 60 per cent of Royal National Park's 16,000 ha, power lines clashed and caused a grass fire. But these and other causes of fires are in fact quite rare. Some of the major causes of fire in Royal National Park are the burning out of stolen vehicles, arson attacks, cigarette butts and camp fires. It is perhaps no coincidence that most fires begin near the urbanised western boundaries of Royal National Park. When the risk of fire is a major concern to park agencies (e.g. during extremely high temperatures with prevailing hot, westerly winds), public notices may be issued regarding closures of parks at risk.

Research into fire management in national parks has received considerable attention in the United States. A partnership to develop a National Fire Plan, designed to reduce fire risks and impacts now exists between the USDA Forest Service, the Bureau of Land Management, the National Park Service, the Fish and Wildlife Service and the Bureau of Indian Affairs. Knowledge of the impacts of fires on tourists' attitudes, perceptions of risk and visitation to parks (e.g. McCool and Stankey 1986), on hiking (Loomis *et al.* 2001), and on visitation numbers after very intense or large fires (Snepenger and Karahan 1991), are three areas of research where Australian studies are lacking.

Control of noxious species

Noxious species of plants and animals can cause great damage to rural lands, affect livestock quality, and spread disease. They can also contribute to soil degradation and compete with native animals and plants. The New South Wales National Parks and Wildlife Service (NSWNPWS) control measures concentrate on introduced species and feral animals such as goats, foxes, rabbits, wild pigs, cats and wild dogs, including dingoes. In some parks, wild horses or brumbies, and introduced fish, in particular, European carp, are also a problem.

In the past, the Parks Service has been criticised for an apparently ineffective approach to control of unwanted animal species. However, a wide range of control methods are now in use, including physical, chemical and biological means. One controversial aspect of control of pest animals is the targeting of honey bees, which are said to compete with native bees, birds and small animals for nectar. For this reason, no new beehive sites are being allowed on park and reserve land, and current licences are being phased out.

Controversy also surrounds the Parks Service's attitude to wildlife management. Many landholders regard kangaroos, for example, as pests, whereas the Service has a responsibility to conserve and protect these native animals. Similarly, dingoes are native animals and the Parks Service aims to maintain existing dingo populations within park and reserve lands. Both these species, along with fruit bats or flying foxes, and wild ducks and other waterbirds, are considered 'bad news', particularly by adjoining landowners, who are critical of what they see as inadequate methods to control these 'pests' and contain them within park lands. Similar criticism is aimed at measures taken to eradicate weeds and noxious plant species. Without being too cynical, we can hardly expect animals to respect human boundaries. If people wish to build and live close to the boundaries of national parks, they should surely accept some risks as does a person who resides in alpine or hazardous areas or on major highways.

The Parks Service is working hard to build bridges with neighbouring rural communities, and to avoid the parks and reserves being turned into 'islands' with little relationship to surrounding environments. Research on this issue is underway in other countries such as New Zealand and Canada (e.g. McCleave *et al.* 2004). An immediate risk when creating a park is the loss of opportunity to interact with adjoining ecosystems, and for natural checks and balances to regulate population expansion and changes. In ideal circumstances, parks are most viable when buffered by transition zones of extensive land use (e.g. forestry). Controlled zones, for hunting around parks with large wild animal populations, have also been advocated to regulate the growth and movement of herds.

Park management problems: external and internal

A different type of conflict can arise when incompatible resource uses are imposed *inside* a park from *external* sources. A park can be seen as relatively unused space, and, as public land, involves no resumption or compensation when claims are made on that space for public purposes. Near-urban parks, in particular, are regarded as 'fair game' for waste disposal facilities, motorways, airports, service corridors and water control structures.

Externally derived pollutants such as oil spills, chemicals, sewage, industrial effluents and biocides that are introduced into park drainage systems, can also affect park ecosystems and food chains. Pollution from pesticides is most likely in parks established near zones of intensive land use. Detergent pollution may occur in parks close to the urban fringe, and marine parks are obvious targets for oil pollution, siltation and sewage overflows and other forms of pollution. Even when the source of pollution can be pinpointed, control may be difficult, and would probably only succeed as part of a wider programme to combat pollution generally.

Within a park, the principal recreation-related concerns are with access and the impact of recreation and associated human activities on the park environment. In part, the problem is a function of visitor numbers and the extent and nature of visitor management (e.g. marketing strategies; level and sophistication of facilities and services; educational programmes). Roads, parking, toilets, accommodation, food outlets, refuse and litter, and off-road vehicles, are just some of the ramifications of outdoor recreation that can place pressure on ecological quality and park resources. Much recreation activity is seasonal, and the problems are worsened during peak periods. In Australia's Kosciuszko National Park, for example, an ongoing debate continues over the place of snow sports in the park, and on proposals for development of additional resort accommodation, facilities and extension of access into more remote areas.

The question of access

While it is as well to recall that outdoor recreation is only one of the roles set down for national parks (Chapter 9), these areas provide a diverse range of opportunities for the enjoyment and appreciation of visitors. Access to parks and reserves for recreation purposes requires careful management to ensure that the activities of visitors co-exist harmoniously with the conservation of natural and cultural values, and with the educational and spiritual experiences parks have to offer. It is important also to recognise that national parks are but one element in a system of protected areas and public lands. Recreational access to national parks should not be looked at in isolation, but from a regional perspective, taking into consideration the complementary access opportunities available in public land (e.g. State Forests and State Recreation Areas in Australia) as well as private land. Planning public access to national parks on a regional basis has the potential to increase economic, social and environmental benefits, and to maintain balance in the overall spectrum of recreation opportunities. It also mitigates against pressure to have national parks respond to unreal and inappropriate demands for outdoor recreation, and helps achieve compatibility between environmental protection and visitor use.

That said, community expectations remain high that access to national parks will remain open for a diversity of recreational pursuits. The New South Wales National Parks and Wildlife Service produced a *Draft National Parks Public Access Strategy* (1997b), which provided a basis for offering a wide range of outdoor recreation opportunities, from pleasure driving, bushwalking, horse-riding and biking, to various forms of boating, fishing and water-related activities. At the same time, a number of qualifications and limitations were indicated which constrain unfettered participation in these forms of recreation. Particular sources of friction identified were the restrictions placed on the use of four-wheel drive and other off-road vehicles; such vehicles permit access even to trackless areas of parks. The Parks Service has been sufficiently concerned at their potential for degradation of the environment to declare the more remote parts of national parks and wilderness 'off-limits' to such forms of transport. Likewise, snow-mobiles are discouraged in alpine parks, and horses are banned from several near-urban parks, because of fears of erosion, spread of weeds and other impacts.

Pets, especially dogs, and firearms, are not permitted to be brought into parks; even fishing is subject to controls regarding which areas may be fished, the number and species of fish which may be caught, and the methods used. Although progress is being made in providing access for people with disabilities, many national parks remain effectively closed to these groups, as well as to seniors and young families. These limitations, along with the imposition of visitor fees, mean that there is still some way to go before opportunities for public access to national parks and reserves satisfy appropriate equity standards.

Whereas national parks in Australia have always attracted large numbers of visitors, and indeed, in earlier days, were justified on the basis of the recreation opportunities they provided (see Chapter 9), it is only now that the NSWNPWS is developing a policy and management strategy to promote the parks as a focus for nature tourism. The *1997 Draft Nature Tourism and Recreation Strategy* sought to achieve ecologically sustainable visitor use of protected areas in the State. The strategy points to problems such as disturbance to wildlife, introduction of unwanted species, soil erosion, damage to vegetation, and escaped fires, associated with thoughtless or deliberate acts by visitors. The strategy aimed to balance the protection of natural and cultural values with management of visitor use (Figure 10.2).

In putting forward the draft strategy for public comment, the Parks Service drew attention to the economic benefits flowing from tourism in the State's national parks and protected areas. Focusing first on the City of Sydney – 'City of National Parks' – the strategy identified a select number of parks and nature reserves in strategic locations within regional New South Wales, to be promoted as key destinations to international and domestic visitors (Figure 10.3). Themes ranging from local diversity of the natural environment to the cultural heritage of Aboriginal people, are among the varied experiences represented in the programme. The concept of the Recreation Opportunity Spectrum was to be used to develop an array of settings appealing both to mainstream tourists and to niche markets. It was estimated that by the year 2005, some 28 million visitors would be attracted to the key regional park destinations identified.

This level of visitation has not been reached, with recent estimations suggesting close to 21 million visitors annually. *Living Parks: A Draft Sustainable Visitation Strategy for NSW National Parks* is a recent draft strategic policy framework linked

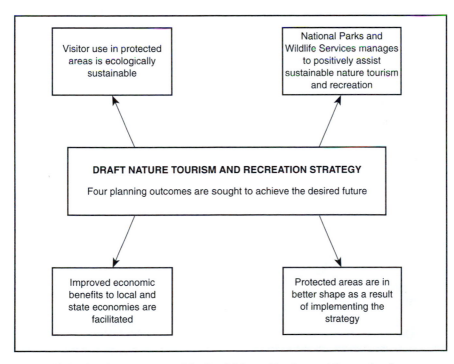

Figure 10.2 Planning outcomes of the Draft Nature Tourism and Recreation Strategy

Source: NSW National Parks and Wildlife Service (1997a: 8).

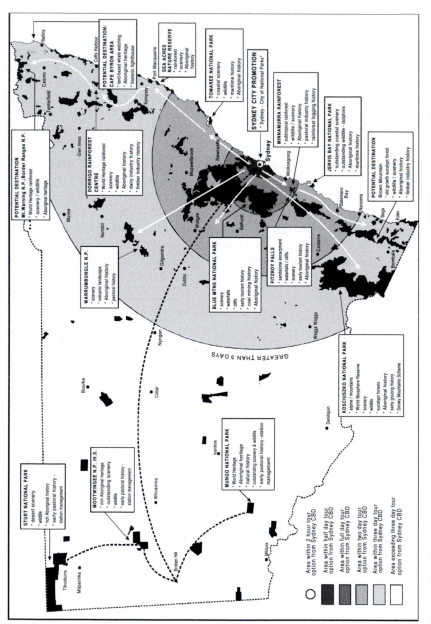

Figure 10.3 Key regional destinations: key nature tourism destinations in regional NSW to be promoted nationally and internationally

Source: NSW National Parks and Wildlife Service (1997b).

to Tourism NSW's *Towards 2020 NSW Tourism Masterplan*. *Living Parks* will provide broad strategic direction and establish principles for visitor management. It is also informed by several NSW State strategies and policies, such as other Government agency open space and recreation plans, the NSW Biodiversity Strategy, strategies relating to cultural heritage management and pest management, and State level park management policies (NSWNPWS 2004) (see Box 10.3).

Box 10.3 **Living Parks**

The message underpinning the Living Parks Strategy is to demonstrate that the New South Wales parks are, indeed, 'living parks' and part of the local community. The strategy sets out an invitation for the people to 'Conserve, Explore and Experience', so as to appreciate the natural and cultural heritage values of the parks, as they build on their awareness of the park environment. Living Parks represents a sustainable visitation strategy aimed at improved visitor management and high quality visitor services, while fostering public appreciation, understanding and enjoyment of the natural habitats, eco-systems, biodiversity, unique landscapes and cultural values present within the park settings. The strategy is based on a set of Guiding Principles linked to the key features of park management:

- Protection and Conservation;
- Enjoyment, Understanding and Appreciation;
- Partnerships;
- Best Practice;
- Communication and Marketing.

Each principle supports a series of outcomes and actions. Thus, the principle of Protection and Conservation seeks to ensure that visitor experiences are enjoyable and memorable while remaining compatible with the conservation and management of the parks' natural and cultural values; and that commercial operations and activities are consistent with that dual purpose.

The principle Enjoyment, Understanding and Appreciation, seeks to promote diversity, equity and informed choice in opportunities for park visitation and the experience of park values and their conservation.

The Partnerships principle recognises the significant contribution that parks make to regional economies and local communities, and the particular rights and responsibilities of Aboriginal people in the provision and care of culturally appropriate park facilities and services.

Best Practice is a guiding principle emphasising excellence in visitor management and support facilities to achieve targeted conservation outcomes and ecological sustainability. The principle Communication and Marketing

continued...

***Box 10.3* continued**

encompasses visitor expectations, satisfaction, and understanding of park management objectives and the role of parks in fostering nature conservation and providing culturally sensitive and responsible experiences.

The Guiding Principles lead logically into a progressive Action Plan, marked by a series of measures orientated to producing coordinated and integrated outcomes, ensuring high standards of visitor experience, the conservation of park values and an ecologically sustainable park environment. The Strategy culminates in an Implementation Plan to operationalise the Living Parks principles and actions and to identify implementation time-frames and responsibilities. Key components are park management plans to focus on visitation needs, opportunities, priorities and potential conflicts, and ongoing monitoring and evaluations of the management initiatives. The draft Living Parks sustainable visitation strategy was released for community review and comment in 2005.

Tourism and national parks

Promotion of tourism in national parks would seem to make good economic sense and, if planned as is envisaged in the strategies mentioned above, should be compatible with ecological objectives of park management. Without such planning, encouragement of visitor use can bring with it the risk of degradation of the park environment and, ultimately, loss of its appeal for nature tourism. The ongoing debate over the merits of tourism in national parks takes on added significance, due to the growing recognition of the economic contributions associated with tourist use of parks and protected areas, especially in developing countries. These benefits can be considerable and include:

- greater environmental awareness and understanding, perhaps leading to increased concerns about parks and stronger public action or motivation to establish and maintain parks;
- income for parks, which may be used to enhance conservation measures within the parks system;
- local employment, directly in the tourist sector and in support areas;
- stimulation of profitable domestic industries in accommodation, restaurants, transport, guide services and artefacts;
- foreign exchange;
- diversification of the local economy to meet new and growing demands for local products;
- improvements to local infrastructure and intercultural communication and understanding.

Against these positive aspects, must be considered a range of potential negative impacts from overuse of protected areas by visitors. Some of these are listed in

Table 10.2, and the consequences are likely to be added to in fragile park ecosystems favoured by ecotourists (Chapter 11). Less obvious impacts can also ensue from resentment generated among local people, who may perceive the parks as being provided primarily for the benefit of foreign visitors or outsiders.

In an Australian study, Buckley and Pannell (1990) documented the kind of problems associated with visitor use of national parks for tourism and recreation. Table 10.3 is especially useful, in that it also locates parks where problems have occurred and includes references to support the incidence of impacts.

Table 10.2 Potential environmental effects of tourism in protected areas: the types of negative visitor impacts that must be controlled

Factor involved	Impact on natural quality	Comment
Overcrowding	Environmental stress, animals show changes in behaviour	Irritation, reduction in quality, need for carrying-capacity limits or better regulation
Over-development	Development of rural slums, excessive manmade structures	Unsightly urban-like development
Recreation		
Powerboats	Disturbance of wildlife	Vulnerability during nesting seasons, noise pollution
Fishing	None	Competition with natural predators
Foot safaris	Disturbance of wildlife	Overuse and trail erosion
Pollution		
Noise (radios, etc.)	Disturbance of natural sounds	Irritation to wildlife and other visitors
Litter	Impairment of natural scene, habituation of wildlife to garbage	Aesthetic and health hazard
Vandalism destruction	Mutilation and facility damage	Removal of natural features
Feeding of wildlife	Behavioural changes, danger to tourists	Removal of habituated animals
Vehicles		
Speeding	Wildlife mortality	Ecological changes, dust
Off-road driving	Soil and vegetation damage	Disturbance to wildlife
Miscellaneous		
Souvenir collection	Removal of natural attractions, disruptions of natural processes	Shells, coral, horns, trophies, rare plants
Firewood	Small wildlife mortality, habitat destruction	Interference with natural energy flow
Roads and excavations	Habitat loss, drainage	Aesthetic scars
Power lines	Destruction of vegetation	Aesthetic impacts
Artificial water holes and salt provision	Unnatural wildlife concentrations, vegetation damage	Replacement of soil required
Introduction of exotic plants and animals	Competition with wild species	Public confusion

Source: WTO (1992: 14, adapted from Thorsell, 1984).

Table 10.3 Environmental impacts identified in Australian National Parks

National Park	Tracks and ORVs	Trampling (human or horse)	Weeds and fungi	Boats damage bank	Firewood collection	Human wastes	Camp sites	Water pollution
Western Australia								
Cape Range								
Stirling Range	*	*			*			
Northern Territory								
Kakadu	*			*				
Uluru	*	*						
South Australia								
Simpson	*	*	*		*			
Coongie	*				*		*	*
Flinders	*		*					
Coorong	*			*	*			
Queensland								
Carnarvon						*		*
New South Wales								
Mount Warning		*			*			
Kanangra Boyd	*						*	*
Blue Mountains	*	*				*	*	*
Ku-ring-gai								*
Kosciusko		*	*		*	*	*	*
Victoria								
Croajingolong	*							
Wilson's Promontory	*	*			*	*	*	
Tasmania								
Cradle Mt St Clair		*			*	*	*	*
Gordon River				*				
Southwest			*					

Changed water course	Water depletion	Disturb-ance to wildlife	Damage to archaeo-logical sites	'Cultural vandalism'	Litter	Visual impacts: roads and buildings	Noise	Reference (see Buckley and Pannell 1990)
	*							Peerless n.d
						*		Brandis & Batini 1985
			*		*			ANPWS 1986b
*			*	*		*	*	Ovington *et al.* 1973
								ANPWS 86a
		*					*	SANPWS 1984
		*	*	*	*		*	Gillen 1988
	*				*	*	*	Williams *et al.* 1988
			*		*		*	SADEP 1984
			*	*				Pitts 1982
					*	*		NSWNPWS 1985
*					*			Brown 1988
					*	*		Brown 1988
*								Snelson n.d
				*	*	*		
								VNPWS 1985
	*				*	*	*	VNPWS 1987
					*			O'Loughlin 1988
								TDLPW 1985
								Cook 1985, Bayly-Stark 1985
								Neyland 1986

* = recorded in reference cited = observed by RB

Source: Adapted from Buckley and Pannell (1990).

A constructive approach towards a positive relationship between tourism and national parks calls for management of both the park resources and people. Agencies in Canada, the US, the UK, New Zealand and Australia are all working towards greater recognition of this in the strategic and operational plans. As noted in Chapter 6, resource management implies close monitoring of features of the environment in order to detect the rate, direction and character of change. In this way, remedial action can be taken before degradation reaches the point where the park environment becomes a source of dissatisfaction to visitors. Management, then, implies the maintenance or even enhancement of the park's resource base to provide satisfying recreation opportunity settings for visitors. These measures need to be complemented by efforts directed towards visitor management. A classification of visitor management strategies is provided in Table 10.4.

As noted in Chapter 6, some cynics have suggested that park management would be easy if it wasn't for the people. Certainly, the physical attributes of parks lend themselves to relatively straightforward procedures and technical and engineering-type techniques. With visitor management, a much more sensitive approach is required in coping with the many sources of conflict and manifestations of overuse. A good balance needs to be struck between regulation and modification of visitor behaviour, otherwise the benefits of tourism may be traded off through lost patronage resulting from regimentation of people in parks.

> The guiding principle for tourism development in national parks is to manage the natural and human resources so as to maximise visitor enjoyment while minimising negative aspects of tourism development.
>
> (World Tourism Organisation 1992: 12)

Manning (1979) distinguishes between *strategies* and *tactics* in the management of recreational use of national parks. Strategies are defined as basic conceptual approaches to management, setting out different paths to preservation of environmental and recreational quality. Tactics are defined as tools to carry out various management strategies. To clarify the distinction, a strategy, for example, might be a decision to limit recreational use. Within this basic strategy a number of tactics or tools might be pursued, including the imposition of fees, the requirement of permits or the erection of physical barriers. Manning classifies available strategies for park management into four basic approaches, each with a number of distinct sub-strategies. These are summarised in Figure 10.4; two of the strategies deal with supply and demand aspects of recreation and park lands, and two focus on modifying either the character of existing use to reduce adverse impacts, or the resource itself to increase its durability or resilience.

Typically, efforts are made to overcome 'people problems' by spreading the load in space and time. Manning's strategies offer four possibilities which are not meant to be mutually exclusive; nor are these approaches exclusive to parks management, but have application in other forms of recreation activity and tourism. Increasing the supply of park space may not be feasible in all situations, so that making better use of available space is the alternative. Redistribution of visitor

Table 10.4 Classification of visitor management strategies

Indirect strategies	Direct strategies
Physical alterations • Improve or neglect access • Improve or neglect campsites	*Enforcement* • Increase surveillance • Impose fines
Information dispersal • Advertise area attributes • Identify surrounding opportunities • Provide minimum impact education	*Zoning* • Separate users by experience level • Separate incompatible uses
Economic constraints • Charge constant fees • Charge differential prices	*Rationing use intensity* • Limit use via access point • Limit use via campsite • Rotate use • Require reservations
	Restricting activities • Restrict type of use • Limit size of group • Limit length of stay • Restrict camping practices • Prohibit use at certain times

Source: Adapted from Hendee *et al.* (1978, in Vaske *et al.* 1995: 41).

pressure is an obvious option, as are measures to separate or limit uses at popular sites.

Paradoxically, one way to approach the problem of excessive tourism pressure in national parks is to concentrate visitor use even more (e.g. see Hammitt and Cole 1991, 1998). Concentration of use can help control general site deterioration by attracting visitors to selected locations able to sustain high levels of use. Alternatively, additional opportunities for tourists can be created by diverting some visitors to underused sites, and by efforts to reduce seasonal or daily peaks in visitation through the use of incentives to extend operations into slack periods.

When essentially voluntary means of bringing about dispersal of use fail to achieve that objective, it becomes necessary to adopt a more direct approach to regulating visitor behaviour. Regulation of use implies some restriction over what tourists are permitted to do. Attempts at 'people control' come down to a choice between 'do' and 'don't' – the 'carrot or the stick'. Most park managers would be aware of the value of allowing the user to retain some sense of freedom of choice, and the role of interpretation in modifying of visitor behaviour is discussed below. However, with certain management problems, such as littering, vandalism or use of vehicles in sensitive areas, enforcement of rules, backed up by strenuous efforts at detection and punishment of offenders, may be the most effective means of control.

With regard to managerial directives, several of the tactics applied in resource management also require visitor regulation as a concomitant of site protection. The admonition to 'Keep off the Grass', for example, is clearly designed to bring about a particular response in use patterns, and thereby help maintain the condition

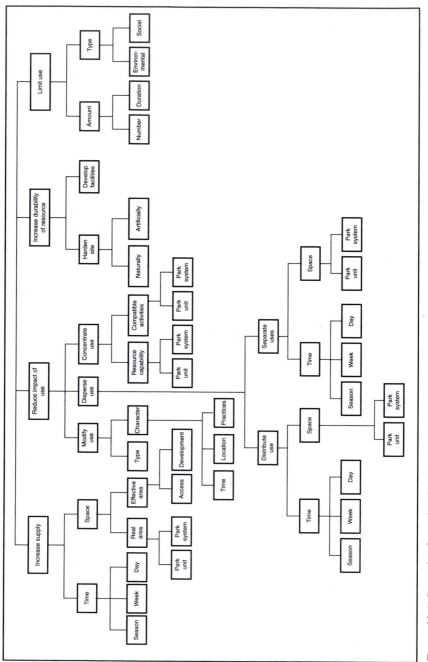

Figure 10.4 Strategies for national park management

Source: Manning (1979: 14, in Pigram 1983: 175)

of the site. Regimentation for its own sake, however, is indefensible and, in most instances, a positive approach is possible and preferable to an endless array of signs informing users of what they cannot do (Figure 10.5). Indeed, clever interpretation strategies may promote what cannot or should be done in a positive manner, especially by highlighting possible impacts.

On-site control involves, among other things, site-hardening by way of trail establishment and maintenance, and directions as to location, time and duration of park visits and activities, in order to attain the desired intensity of use for the area (Table 10.5). Among the most common procedures for visitor control are zoning or scheduling, and rationing. These have the advantage, not only of limiting use, but of promoting dispersal of use, and reducing conflicts by separation of incompatible types of tourism activity, such as fishing and water-skiing.

Zoning involves the clustering of compatible uses in selected parts of a site. National parks are themselves a form of zoning, and certain areas within parks are designated for special purposes (for example, see Chapter 9). Different stretches of a river or lake can also be zoned for different uses, sometimes on the grounds of safety, or because the resource attributes do not lend themselves to all types of recreation, or simply to avoid mutual interference and maximise satisfaction between users. Spatial zoning is likely to be more successful where there is a logical and accepted basis for partitioning of the site. It is useful, too, if zone boundaries can be aligned to some natural or recognisable feature (e.g. different activities allocated to the opposite banks of a river).

Scheduling, or zoning by way of time limitations, is another useful procedure for visitor control in national parks. Recreation activities using the same site are allocated to specific time periods on an hourly, daily, weekly or seasonal basis in order to reduce conflicts and to ensure adequate rotation of use. The timeframe chosen depends upon the degree of conflict and the level of competing uses. A variation of time zoning, especially with linear resources, is the staggered scheduling of departure times of such activities as river tours. Ideally, schedules

Table 10.5 Strategies and actions for reducing impact on campsites

Strategy	Possible actions
Reduce amount of use	Limit number of parties entering the area
Reduce per capita impact	
• Use dispersal	Persuade parties to avoid camping on highly impacted campsites
• Use concentration	Prohibit camping anywhere except on designated sites
• Type of use	Teach low impact camping techniques
• Site location	Teach parties to use resistant sites for camping
• Site hardening/shielding	Build wooden tent pads on campsites
Rehabilitation	Close and re-vegetate damaged campsites

Source: Hammitt (1990: 21).

"...but, Harold, there must be something we can do here...it's such a lovely spot..."

Figure 10.5 A question of interpretation?

should be drawn up after consultation with user groups, and, if possible, tailored to fit normal recreational patterns.

Rationing refers to the mechanism through which opportunities to use designated recreation resources are distributed to users. Implementation of rationing assumes: that reasonable estimates of tolerance to use, or carrying capacities, can be established; that in the absence of rationing, use would exceed capacity at some sites; and that a reduction in use through rationing is the preferred management option (Grandage and Rodd 1981).

Recreational use can be rationed by various means. Chubb and Chubb (1981) suggest three broad approaches:

1   Sharing the limited supply of recreational opportunities among potential participants by measures such as the issue of permits and licences; use of a reservation system; queuing and participation on a 'first-come, first-served' basis; or allocation of opportunities by lot;
2   Providing a limited experience to as many participants as possible, even though this may affect the quality of the experience (e.g. placing limits on the size of parties, or rotation or regulation of use to encourage high turnover).
3   Making it more difficult for people to participate. Use can be discouraged by restricting access; making reservations harder to obtain; insistence on merit standards (certain skills or knowledge) as the basis for entry; and the imposition of fees. Differential fee structures can also be used to redistribute recreational use patterns.

The Milford Track in New Zealand is one of New Zealand's most famous walking tracks (see Figure 10.6) and has highly regulated use from October to April. On the track at any one time are 48 'freedom walkers', who walk the entire track and spend three nights staying in accommodation provided by the Department

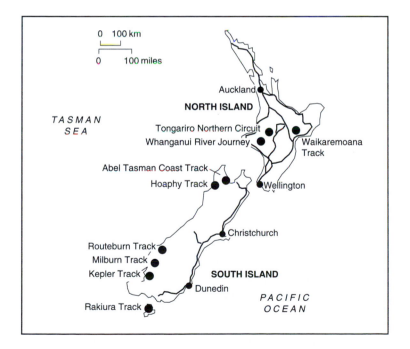

Figure 10.6 Great walks in New Zealand

Source: Adapted from New Zealand Department of Conservation (http://www.doc.govt.nz/ Explore/002~Tracks-and-Walks/Great-Walks/index.asp, accessed 20 March 2005).

of Conservation (DOC); 48 people participating in 'guided walks', staying over-night in private huts; and 48 day-walkers. All walkers are required to hike from Glad Wharf in the east to Sandfly Point, Milford Sound. Walkers are required to stay at their designated huts at their designated times so walkers from different groups do not meet up. A boat transports people to the walk and from its end point, although the walk can be accessed via a lengthy and strenuous walk estimated to take eight or more hours, depending on conditions at the time. Less than 50 people made the walk in the late 1880s. In 2003, approximately 11,000 walkers completed the walk. Commercial pressures on the Department of Conservation to relax their restrictions have been mounted periodically by tourism operators, but DOC has not yielded to their requests.

De-marketing is a concept applied by Kotler (1971) which, as the term suggests, leads to a decrease in demand for a site. Groff (1998) identified three circum-stances governing the application of de-marketing strategies:

- *Temporary shortages* – due to either lack of supply or underestimation by management of demand for particular settings or programmes;
- *Chronic overpopularity* – can seriously threaten the quality of the visitor experience and also damage the natural resource that attracts the visitors;
- *Conflicting use* – encompasses issues of visitor safety, compatibility of use with the available resources, and the different uses and programmes demanded by the public.

(Groff 1998, in Jenkins and Wearing 2003: 333)

Park managers may market and de-market a park, specific site or particular activities simultaneously (Crompton and Lamb 1986, in Jenkins and Wearing 2003). De-marketing involves the adoption of one or more strategies:

- increasing prices in such a way that they increase disproportionately as time spent in the park or at a specific site increases;
- creating a queuing system;
- directing marketing and promotional activities to particular markets using relevant media channels;
- promoting the importance of the park or site by educating users and other members of the public about the need to conserve the resource and to minimise impacts;
- publicising a spectrum of recreational opportunities in the region;
- highlighting the environmental degradation that could occur if too many people frequent the area; and
- publicising restrictions and difficulties associated with accessibility.

(Groff 1998, in Jenkins and Wearing 2003: 228)

Any form of rationing discriminates to a certain extent against some participants and some forms of recreation. The use of fees or eligibility standards to apportion recreational opportunities has been criticised on the grounds of equity and effectiveness. However, fees are becoming a greater source of necessary funding for park management activities. Many protected areas are confronted with a dilemma – more use (or tourists) might assist parks achieve their dual goals, enhance environmental awareness and understanding, and encourage support for national parks and conservation objectives. Yet, user fees collected from visitors rarely go even a small way to site maintenance, development and management. Governments in Australia, for example, have established hundreds of parks in the last decade, but government funding and revenues from user fees have not matched this growth. Dickie (1995) argued there was little infrastructure in parks, and much of what was there was in a poor state. It is not unusual to find tracks or infrastructure closed for lengthy periods of time because of neglect and the subsequent need for substantial works programmes amid concerns arising over safety and risk.

Fees and charges employed by parks agencies include:

- *User fees*: charges on 'users' of an area or facility such as park entrance or tramping fees;
- *Concessions*: a fee for the permission to operate within a location for groups or individuals that provide certain services to visitors (e.g. food, accommodation and retail stores);
- *Sales and royalties*: fees levied on a percentage of earnings that have been derived from activities or products at a site such as photographs or postcards;
- *Taxation*: an additional cost imposed upon goods and services used by visitors (e.g. airport taxes);
- *Donations*: tourists are often encouraged or requested to contribute a donation to help maintain a facility (see Hedsrom 1992; Marriott 1993; Lindberg and McKercher 1997, in Jenkins and Wearing 2003: 225).

Buckley (2000) suggested a model involving a range of funding sources, in which:

> funds for conservation management and the provision of basic recreation facilities are provided centrally through the government budget process, and funds required for marginal expenditure associated with increasing levels of tourism may be raised from tourists and tour operators, and retained locally for immediate management expenditure.
>
> (Buckley 2000: 37)

Unfortunately, many tour operators do not adequately reimburse parks agencies for their use of parks for commercial enterprises. In summary, all rationing systems have shortcomings, so that, once again, monitoring of visitor behaviour and user preferences plays an important part in visitor management. In addition, Figgis believes that the increasing dependence of parks authorities on tourism charges to meet budget shortfalls will lead to a shift to too great an emphasis on tourism-centred management (Figgis 2000).

Only through knowledge of what visitors do and want to do can appropriate management tactics be devised. Monitoring may be formal or informal, overt or covert, and it should be a two-way exercise concerned as much with the impact of the site on the visitor as how the visitor affects the site. The managerial process is complicated because it is not merely a question of ecological carrying capacity and overuse; social carrying capacity can also be involved, along with the various sources of conflict identified earlier.

To complement procedures for visitor regulation, management can seek to modify recreation behaviour by persuasion and the provision of information through interpretative services.

Information and interpretation

> Interpretation communicates information and perception about a park, forest, structure, battlefield, or an entire region's distinctive features and influences. It helps visitors understand and appreciate the special natural and cultural resources, how these have influenced the way the region has evolved, and how its protection or disappearance will affect the area's future. The individual should thus enjoy a richer life through enhanced perception of landscapes and historical artefacts [or heritage].
>
> (Knudson *et al.* 1995: xix)

Interpretation has a lengthy history (Dewar 2000). As a profession, it dates back to 1896,

> when the Boston Museum of Fine Art and the Louvre in Paris began offering interpretive tours to the public ... In 1907, the American Museum of Natural History appointed Agnes L. Roseler as the first full-time professional interpreter ... C.M. Groether is credited by Weaver for introducing nature interpretation into North America from Europe (Weaver 1952: 18).
>
> (Dewar 2003: 266)

Wilcox (1969) wrote, in reference to America, that interpretation was a most challenging and provocative area of growth, with potential good for society in how it sparks imagination, and in how it can act as a tool to build a more meaningful life. Not much has changed.

The provision of an appropriate interpretation programme is an important supportive aspect of park planning and management, be it strategy or tactic. The main aim is to communicate to park users the objectives of management and the rationale for the various measures undertaken. In the long-run, a sound interpretation policy may provide the key to resolving the dilemma between park preservation and use by developing in park users a deeper regard for national parks and a desire for a meaningful role in their care and management.

Interpretation has been described by Tilden (1977) as more than just instruction or communication of information. Tilden sees the chief function of interpretation as provoking and stimulating interest and awareness among visitors to a recreation site. This is to be achieved by revealing meanings and relationships in nature by reference to original objects, and by first-hand experience with common, easily understood examples and materials (for practical examples of interpretation plans and designs for trails, parks, historic and other sites, see Trapp *et al.* 1994; Veverka 1994; Knudson *et al.* 1995).

A second function of interpretation is to assist in accomplishing management objectives. It can do this by encouraging appreciation of the recreation environment and promoting public cooperation and responsibility in conserving recreational values. Much destructive behaviour results from ignorance rather than malicious intent, so that increasing the flow of information to the public is a preferable and probably cheaper means of reducing depreciative acts than prohibitions and censure (Lime and Stankey 1971). As Clark (1976) and Harrison (1977) explain, the key is often in pointing out 'why', when certain norms of behaviour are required:

- Why can't cars be driven off parking pads?
- Why can't tables be moved ...?
- Why can't a tree be chopped down for firewood?
- Why can't initials be carved on benches or tables or trees?

(Clark 1976: 66; also see Beckman and Russell 1995)

Many people who visit museums, parks, forests and similar areas, welcome interpretation (Walsh 1991, in Knudson *et al.* 1995). In the dissemination of information about recreation opportunities, Jubenville (1978) makes the distinction between advice reaching the potential visitor before arrival at the site (Regional Information System) and information provided at the site (Area Information System). Prior information should reach the individual when choices concerning recreation participation are being considered. It is more often the task of government or regional organisations than the specific site manager, and can even involve zero, minimal or negative information, aimed at diverting attention or making heavily patronised sites less attractive (i.e. de-marketing). Thus, certain sites or facilities may be omitted from a map, or reference may be made to popularity and associated crowding; conditions which some will try to avoid.

Much more effort is directed positively into inducing desired patterns of behaviour on-site by increasing public awareness through publicity, education, interpretation and other less obtrusive methods of persuasion. Freedom of choice is seemingly not directly involved, yet the behavioural response sought is produced. Various means are available for transmitting information and communicating information between management and visitors. Typical approaches involve the use of maps and signposting, publications and brochures, electronic media and on-site contact by way of visitor information centres and guide services. These last methods for getting the message across are more often in the nature of interpretation than mere passive provision of information (see Trapp *et al.* 1994; Alderson and Low 1996).

On the basis that 'an informed public is a caring public', the former Countryside Commission in Britain designed a number of self-guided trails around forests, farms, urban centres, ancient monuments and natural areas. The aim was to increase understanding and appreciation of these features, and thereby engender improved standards of behaviour and greater respect for the environment. Innovative interpretation schemes have been implemented in such countries as Australia (e.g. Beckman and Russell 1995; Hall and McArthur 1996; Black and Weiler 2003; *Annals of Leisure Research* 2004), Canada (e.g. Graham and Lawrence 1990), and the US (e.g. Trapp *et al.* 1994).

The Leave No Trace Program involves a partnership of non-profit organisations, corporations, retailers, universities, youth programmes and other groups. Leave No Trace is 'dedicated to promoting and inspiring responsible outdoor recreation through education, research and partnerships. Leave No Trace builds awareness, appreciation and respect for our wildlands' (Leave No Trace, http://www.lnt.org/about/index.html). The Leave No Trace Program has seven principles (see Table 10.6) and offers an extensive array of educational information disseminated through

Table 10.6 Principles of the Leave No Trace Program

Plan Ahead and Prepare

- Know the regulations and special concerns for the area you'll visit.
- Prepare for extreme weather, hazards and emergencies.
- Schedule your trip to avoid times of high use.
- Visit in small groups. Split larger parties into groups of 4–6.
- Repackage food to minimise waste.
- Use a map and compass to eliminate the use of marking paint, rock cairns or flagging.

Travel and Camp on Durable Surfaces

- Durable surfaces include established trails and campsites, rock, gravel, dry grasses or snow.
- Protect riparian areas by camping at least 200 feet from lakes and streams.
- Good campsites are found, not made. Altering a site is not necessary.
- In popular areas:
  - Concentrate use on existing trails and campsites.
  - Walk single file in the middle of the trail, even when wet or muddy.
  - Keep campsites small. Focus activity in areas where vegetation is absent.
- In pristine areas:
  - Disperse use to prevent the creation of campsites and trails.
  - Avoid places where impacts are just beginning.

Dispose of Waste Properly

- Pack it in, pack it out.
- Inspect your campsite and rest areas for trash or spilled foods. Pack out all trash, leftover food, and litter.
- Deposit solid human waste in catholes dug 6 to 8 inches deep at least 200 feet from water, camp, and trails. Cover and disguise the cathole when finished.
- Pack out toilet paper and hygiene products.
- To wash yourself or your dishes, carry water 200 feet away from streams or lakes and use small amounts of biodegradable soap. Scatter strained dishwater.

many partners. The need to tailor recommendations to each site is noted and advice is clear and concise. Those looking for more detailed scientific information can readily access it. Moreover, there are accredited educational courses and awareness workshops conducted in association with the programmes.

A basic objective of recreation management is to provide a sustained flow of benefits for users. Those benefits are clearly recreation-related, but these can be enhanced through thoughtful educational and other programmes. Concern for the quality of the visitor experience, then, is another justification for effective interpretation programmes.

> Increasing our contact with visitors can help them find out what the range of recreation opportunities and attractions is … recreational experiences may also be enhanced if visitors can be taught an understanding of the basic concepts of ecology and other outdoor values … By deepening their sense of appreciation and awareness of the natural environment, more recreationists could take better advantage of an area's recreation potential.
>
> (Lime and Stankey 1971: 181)

Leave What You Find

- Preserve the past: examine, but do not touch, cultural or historic structures and artifacts.
- Leave rocks, plants and other natural objects as you find them.
- Avoid introducing or transporting non-native species.
- Do not build structures, furniture, or dig trenches.

Minimize Campfire Impacts

- Campfires can cause lasting impacts to the backcountry. Use a lightweight stove for cooking and enjoy a candle lantern for light.
- Where fires are permitted, use established fire rings, fire pans, or mound fires.
- Keep fires small. Only use sticks from the ground that can be broken by hand.
- Burn all wood and coals to ash, put out campfires completely, then scatter cool ashes.

Respect Wildlife

- Observe wildlife from a distance. Do not follow or approach them.
- Never feed animals. Feeding wildlife damages their health, alters natural behaviors, and exposes them to predators and other dangers.
- Protect wildlife and your food by storing rations and trash securely.
- Control pets at all times, or leave them at home.
- Avoid wildlife during sensitive times: mating, nesting, raising young, or winter.

Be Considerate of Other Visitors

- Respect other visitors and protect the quality of their experience.
- Be courteous. Yield to other users on the trail.
- Step to the downhill side of the trail when encountering pack stock.
- Take breaks and camp away from trails and other visitors.
- Let nature's sounds prevail. Avoid loud voices and noises.

Source: Leave No Trace: Centre for Outdoor Ethics, http://www.lnt.org/about/index.html, accessed 24 February 2005.

Care is needed in the selection and implementation of on-site interpretative methods. Mention has already been made of some of the means of communication available. Basically, the choice is between personal services (e.g. talks, demonstrations, guided tours and information services at a visitor centre) and self-directed services (e.g. printed materials, sign-posting, audio-visual media and the internet) (see Tilden 1977; Knudson *et al.* 1995). The advantages and disadvantages of such methods have been evaluated by several authors (e.g. Alderson and Low 1996). The choice of media will be influenced by philosophical and practical considerations pertinent to a particular site. The method selected should be capable of interpreting the recreation environment to the anticipated audience in an appropriate (exciting, interesting, comprehensible) fashion, as well as being reliable, flexible, compatible with other media and reasonably vandal-proof (New Zealand National Parks Authority 1978).

Interpretation is more than mere mechanics, and a certain amount of caution is called for in the implementation of an interpretation programme. Too much interpretation can be counter-productive and destroy the sense of spontaneity and discovery in recreational activities. Participation can become 'over-programmed', and people may resent what they perceive as attempts at 'brain-washing' and efforts to force them into designated modes of use and enjoyment. Managerial attitudes can also intrude: elitist overtones and preconceived obsolete notions of what constitutes acceptable patterns of recreation behaviour (or of deviance), can distort the orientation of interpretation initiatives. Despite the benefits of interpretation and the problems that can arise in practice, rarely are the programmes or the interpreters themselves evaluated. A key question is: 'Precisely how and why does a particular interpretation programme change visitor behaviour before, during and after the on-site visit?'

All procedures aimed at recreation visitor management, whether direct regulation or indirect modification of user behaviour, involve some loss of freedom. Some trade-off is required between freedom of choice and the adequacy of the resource base to meet the requirements of users and the objectives of management. However, positive manipulation of the physical and social environment to create and enhance opportunities for tourism and recreation, is surely preferable to reliance on negative forces of congestion, frustration, dissatisfaction and ultimately self-regulation, to produce their own solution.

Management guidelines

The International Union for the Conservation of Nature proposed guidelines for the management of protected areas (Harrison 1992) (see Table 10.7). These guidelines arose out of the 4th World Parks Congress in Caracas and formed the basis for a series of documents directed towards protected area managers (see Thomas and Middleton 2003). At least two of the guidelines called for further consideration:

- involving local communities in park management (see below);
- managing park boundaries and transboundary zones (discussed earlier in this chapter and in Chapter 9).

Table 10.7 Proposed protected area management guidelines

1   Involving local people in protected areas management
2   The preparation of protected area systems plans
3   Data management guidelines for protected area managers
4   Research tools for enhancing management of protected areas
5   Extending the benefits of protected areas to surrounding lands
6   Using protected areas to monitor global environmental change
7   Management of protected areas by private organisations
8   The application of an international review system for protected areas
9   Monitoring management effectiveness and threats in protected areas
10 Establishing and managing genetic resources
11 Integrating demographic variables in the planning and management of protected areas
12 Applying science to the establishment and management of protected areas
13 Effective management of marine protected areas
14 Expanding the world's network of protected areas
15 Effective management of transfrontier protected areas

Source: Harrison (1992: 23).

Community-related management of parks

The notion that local communities should be involved in park management follows logically from the view that parks should not exist in isolation from their surroundings, but should be considered and managed in relation to them. 'Parks and neighbouring communities are bound together by their individual actions into a collective future' (Field 1997: 425). Unfortunately, parks and their neighbours are often in an adversary situation, as evidenced by the potential for conflict noted earlier in this chapter. It is far preferable for the relationship with parks and protected areas to be viewed as an asset, offering positive and productive benefits (Renard and Hudson 1992). The benefits of community-based approaches to park management include:

- building popular support for parks;
- addressing concerns of communities affected by parks;
- ensuring that benefits from parks reach local communities;
- supplementing public funds and personnel needs in park management;
- integrating community knowledge of park resources into management;
- being responsive to variations in social and environmental conditions;
- providing training and opportunities for skill development for communities to participate in park management.

(Renard and Hudson 1992: 4)

For these benefits to be realised, meaningful partnerships between the park and the local community must be forged, reflecting the rights, aspirations, knowledge, skills and resources of that community. Co-management is based on:

- participatory planning, collaboration and shared responsibility;
- access to information;

- appropriate institutional arrangements; and
- sound legal, technological and financial support.

Advances along these lines are being made, with significant input to park management from local indigenous people. In Australia's Uluru–Kata Tjuta and Kakadu National Parks, for example, local Aboriginal communities form a growing component of park management personnel, contributing important elements of their knowledge and culture to interpretation programmes, especially those of spiritual significance. Tours, trails and brochures reflect Aboriginal themes, and offer authentic aspects of the Aboriginal lifestyle by way of handicrafts and 'bush tucker' (native foods). Aboriginal participation in park management, in turn, gives the local communities a much greater sense of control and self-determination over environmental and conservation issues (Absher and Brake 1996; Wearing and Huyskens 2001).

In sub-Saharan Africa there is now a widely held view that local people must be involved in protected area management (Ite and Adams 2000). Integrated-conservation-development-projects (ICDPs) or people and park projects are evidence of the need to avoid confrontations between park managers and local communities and to consider the needs of local communities. According to Ite and Adams (2000),

> It is self-evidently true that in areas such as southeast Nigeria with a long history of local dependence on forest land, protected areas such as the CRNP (Cross River National Park) will be successful in conserving forest environments only if they are able to meet the legitimate socio-economic development aspirations of the people living in and around such areas.
>
> (Ite and Adams 2000: 340)

Unfortunately, the ICDP has failed in its implementation at CRNP. The principles and processes upon which the Project was based were untested and based on a top-down approach (Ite and Adams 2000).

Closer to the cities, partnerships can be built up between urban populations and neighbouring national parks. Once again, the inevitable interaction between residents, tourists and protected areas adjoining cities, is the rationale for collaboration. In Britain, the Countryside Agency is active in promoting links between city communities and neighbouring parks. An experimental project run in conjunction with the former Countryside Commission and the Birmingham City Council demonstrated the reciprocal benefits of the scheme – for the national parks in increasing awareness of what they had to offer, and for the city in opening up opportunities to experience the rural environment.

Natural-area protection has long been considered the sole responsibility of government and the public sector. Increasingly, financial stringency has raised the possibility of a role for private enterprise in natural-area management, if not in privatised national parks, at least in providing some of the facilities and activities of management required to service park visitors. Such a prospect raises serious questions about the respective roles of the public and private sector, and whether a balance can really be attained between nature protection and profit (Charters *et al.* 1996). Certainly, scepticism surrounded a recent proposal in the US, to allow a private developer to build and operate a new visitors' centre at the historic military park in Gettysburg, Pennsylvania, at no cost to the Park Service. However, the private tourist developments described in the Australian studies suggest that commercial operations can co-exist compatibly with nature conservation, even in environmentally sensitive areas used for ecotourism. At the same time, an enhanced role for the private sector in park management will need a fundamental shift in the attitudes of government agencies and environmental interests before the opportunities for collaboration can be seriously addressed. It will also call for more environmentally sensitive practices by tour operators.

No longer is concern for the natural environment the preserve of public agencies. Increasingly, landholders in the private sector are taking up the challenge of setting aside portions of their land for nature conservation (e.g. Earth Sanctuaries Limited 2000). Future management of parks and protected areas may well proceed in tandem with the private sector and community groups, in a more collaborative effort to ensure sustainability and biological diversity of the natural environment.

Guide to further reading

- Tourism and recreation management in national parks and reserves: Black and Breckwoldt (1977); Boden and Baines (1981); Kelleher and Kenchington (1982); Edington and Edington (1986); Bateson *et al.* (1989); Olokesusi (1990); Eagles (1996); Butler and Boyd (2000); Eagles and McCool (2002); Buckley (2003b, 2004).
- Marine areas: Davis *et al.* (1997); Cater and Cater (2001); Garrod and Wilson (2003).
- Interpretation: three excellent and widely sourced works are Tilden (1977); Knudson *et al.* (1995); Dewar (2000). Also see, Black and Weiler (2003); Dewar (2003); Tubb (2003); *Annals of Leisure Research* 7(1).
- Economic, physical and social impacts in national parks and protected areas: Darling and Eichhorn (1967); Wall and Wright (1977); Jefferies (1982); McDonald and Wilks (1986a, 1986b); Landals (1986); McNeely and Thorsell (1989); Buckley and Parnell (1990); McIntyre and Boag (1995); Liddle (1997); Worboys *et al.* (2001); Weaver (2001b – see Chapters 23–7); Jenkins and Wearing (2003).

Review questions

1   What are the important principles governing the declaration of national parks in a country of your choice? How is a national park defined in that country? Is the management of national parks in that country in line with IUCN definitions?

2   What are the major environmental impacts that threaten national parks? Have these impacts been addressed by parks management? If so, how? If not, why not? In your answers, make reference to case studies of one or more national parks.

3   Should national parks authorities be permitted to charge entrance/user fees? Explain your answer with reference to case studies.

4   Describe some of the main political issues arising out of recreational and tourist use of national parks both generally, and with specific reference to one or more case studies.

5   What are some of the fundamental differences in land ownership, and recreational use and management of national parks in the UK, the USA and Australia?

11 Outdoor recreation, tourism and the environment

> Rather than opposing change, or merely accepting and accommodating change, the tourism industry must manage change to its advantage and that of the environment which nurtures it. Endorsement and application of the concept of sustainability and of best-practice environmental management offer compelling evidence of how change can be harnessed to contribute towards the achievement of environmental excellence. Although tourism flourishes best in conditions of peace, prosperity, freedom and security, disturbance to these conditions is to be expected. The industry response must be sufficiently resilient to generate opportunities for the growth of tourism in keeping with the dynamics of a changing world and increasing concern for ecologically sustainable development.
>
> (Pigram and Wahab 1997: 17)

The environment is the aggregate of resources available to human beings. It comprises the natural environment of earth – water, air, flora, fauna and ecosystems and their associated processes – and the social, cultural and economic environments of human beings. Clearly, the environmental impacts of tourism must be viewed in the broadest sense, taking into account the full range of economic, ecological, social and cultural factors.

Tourism is an important human activity which impacts on physical, socio-cultural and economic environments. As noted in Chapter 1, discussion of tourism in the context of outdoor recreation is logical. Much tourism is recreational, in that tourist activities are engaged in during leisure time, commonly outdoors, for the purpose of pleasure and personal/group satisfaction. Similarly, outdoor recreation overlaps with tourism in the distinctive characteristics and behaviour associated with each.

Recreation, tourism and the environment are interdependent, but their relationship is not constant, varying over space and time. The forces underpinning the growth and changing patterns and processes with respect to leisure, recreation and tourism, coupled with such factors as greater environmental awareness, industrial growth, pollution, limited knowledge about ecological processes, and other factors, mean that this relationship is very complex and dynamic. Recreation and tourism can cause negative and positive environmental change, with debate about acceptable limits of change likely to emerge. Critical concerns, then, are understanding the relationships between recreation, tourism and the environment, and identifying ways of managing resources in harmony with the attractiveness of many recreational and tourist activities.

Concepts and definitions

Definitions of 'tourist' and 'tourism' are many and varied and are discussed widely in textbooks and other publications, but most incorporate the notions of distance travelled, and duration and purpose of travel. Certainly, the term implies more than the French derivations – *tour* (a circular movement) and *tourner* (to go around).

Tourism is a form of human behaviour (Przeclawski 1986), and 'a category of leisure with special significance in individuals' total leisure patterns' (Simmons and Leiper 1993: 204). According to Leiper (1995),

> Tourism can be defined as the theories and practices of travelling and visiting places for leisure-related purposes.
>
> Tourism comprises the ideas and opinions people hold which shape their decisions about going on trips, about where to go (and where not to go) and what to do or not to do, about how to relate to other tourists, locals and service personnel. And it is all the behavioural manifestations of those ideas and opinions.
>
> Leiper (1995: 20)

Leiper's definition is, at least conceptually, somewhat closely aligned with Jafari (1977: 8), who stated that 'tourism is the study of man away from his usual habitat, of the industry which responds to his needs, and of the impacts that both he and the industry have on the host, sociocultural, economic and physical environment'. Pearce (1987: 1), too, presents a robust conceptualisation of tourism, stating '... tourism may be thought of as the relationships and phenomena arising out of journeys and temporary stays of people travelling primarily for leisure or recreational purposes'. Distinctions between recreation and tourism appear founded on the assumption that outdoor recreation appeals to the rugged, self-reliant element in the population, whereas tourism caters more overtly for those seeking diversion without too much discomfort. As noted in Chapter 1, this gulf is an artificial one.

> Apart from emphasis on an alien environment somewhat removed from the place of residence, tourism is carried on within an essentially recreational framework. Differences then become a matter of degree and motivation with tourism calling for a more sophisticated infrastructure and tending towards the opposite end of a time-space-cost continuum to recreational travel.
>
> (Pigram 1983: 184)

Tourism, then, is a specialised manifestation of recreation, and in this book, of outdoor recreation. Recreation and tourism share many features, but the scale of tourism as a concept involving the movement of more than one billion people around the world has perhaps greater potential for lasting repercussions on the landscape and the environment and thus ample scope for resource management and planning.

Global tourism

Tourism Satellite Accounts (TSAs) are now used to measure tourism in national economies and to provide a means of comparing countries (WTO 2001; Australian Bureau of Statistics (ABS) 2005b). These Accounts have been developed to provide a supposedly more accurate measure of the magnitude of tourism than previous measures and yet there are still some concerns about what should be included in these accounts. In other words, governments and industry are still debating definitions of tourism and concepts relevant to tourism (e.g. 'tourist'), and the Australian Government's Productivity Commission Report suggests the contribution of tourism to the Australian economy is far less than the TSAs indicate.

Despite the dangers of comparing data across time, the indications are clear; tourism (however it is defined) has expanded rapidly since the Second World War, and tourism will continue to grow for some time to come. Some countries and regions and local destinations will grow faster than others; some will experience dramatic declines; some will grow without much hard work; some will work hard and struggle to survive because of minor to major incidents.

Tourism has become one of the largest (if not *the* largest) single item in world trade. From 1950 to 1972, annual tourist arrivals in all countries grew from 25 million to almost 200 million, an average growth rate of about 10 per cent per year. In the same period, total foreign exchange earnings from tourism rose from $2.1 billion to $24 billion, an average annual increase of about 11 per cent. By 1976, the number of global visitor arrivals was estimated at 220 million, an increase of more than 90 per cent in a decade, while travellers spent, in all, about $40 billion. Arrivals had grown to 264 million by 1978, and expenditures to around $63 billion in that same year (WTO 1979).

Tourism remains one of the highest industry growth areas, in terms of both expenditure and foreign currency generation. In 2002, global tourist arrivals increased to 703 million, and world tourism receipts, excluding air fares, amounted to $474 billion. In 2003, there was a slight decline as international visitor arrivals fell by about 1.2 per cent to just over 690 million visitor arrivals. Nevertheless, according to several estimates, and depending upon how it is defined, tourism has become the world's largest business enterprise, overtaking the defence, manufacturing, oil and agriculture industries. International tourism will continue to grow. WTO expects international visitor arrivals to exceed more than 1 billion people by 2010 and 1.6 billion by 2020, with roughly two-thirds of these being intra-regional travellers.

Trends in world tourism since 1950 reveal a heavy geographical concentration of both tourist arrivals and tourism receipts. Currently, about three-quarters of world travel is intra-regional travel. Europe dominates international tourist arrivals and world tourism receipts, although that region's share of international visitor arrivals is predicted to decline proportionally by about 2020. Much of the growth in international travel is forecast to occur in the East Asia/Pacific region. On a country-specific basis, 'The United States has been the largest recipient of tourist travel income for several years, and will probably continue to be, receiving more than twice as much international tourism income as its nearest competitor, France' (Lundberg *et al.* 1995: 8) (also see WTO 2000; Weaver and Lawton 2002; for a detailed discussion see Vanhove 2005).

Distance and costs are constraints to travel, especially international travel. However, some countries of the Asia and Pacific regions, notably China, Japan and Korea, recently emerged as potential areas for tourist development, and as sources of international visitors. The rapid growth of tourism in the Asia and Pacific regions, particularly since the early 1980s, has been attributed to the increasing numbers of intra-regional tourists (Mak and White 1992; Forsyth and Dwyer 1996; Vanhove 2005). Western Europe and North America will continue to dominate world visitor arrivals and tourism receipts, and indeed there have been unexpected increases from these markets to such countries as Australia. This increase could be attributed to reductions in the value of the Australian dollar, intensified and carefully targeted international marketing and promotion, and the staging of large-scale events. The East Asia and Pacific regions will be major tourist growth regions in terms of international travel and tourism development. Their recovery from recent disasters and terrorist attacks and alerts is almost certain to be rapid.

Tourism systems

Tourism involves many elements, including tourists themselves, transport systems, the regions of tourist origins, the destination region and the linkages in-between. Some writers have suggested that a systems framework is the most suitable means of drawing these facets together for study.

Leiper (1979, 1981) proposed an open system of five interacting elements, encompassing a dynamic human element – the *tourists*; three geographical elements – the *generating region*, the *transit route* and the *destination region*; and an economic element – the *tourist industry* (see Figure 11.1). Leiper's model recognises that the central element of the system is people – the tourists themselves. They comprise the energising source, and their attributes and behaviour help define the role of other elements in the system. The generating region is the origin of potential tourist demand, linked by transit routes to the destination or focus of tourist activity. Subsumed within these three geographical elements are the industrial component and service infrastructure of tourism, comprising all the firms, organisations and facilities that are intended to serve the specific needs and wants of tourists, before departure, en route and at the destination(s).

Leiper (1995: 26) suggested 'that any of the five elements can be used as a focal topic. Studying tourists involves considering tourists in relation to the other four elements ... Studying places as tourist destinations involves considering that element in relation to the other four, and so on' (see Figure 11.2). In this way,

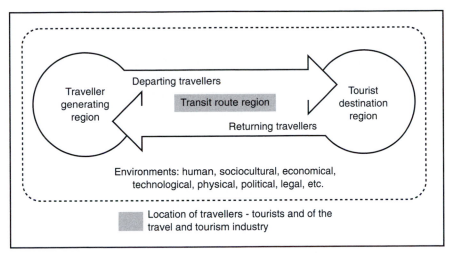

Figure 11.1 A basic whole tourism system

Source: Leiper (1995: 25).

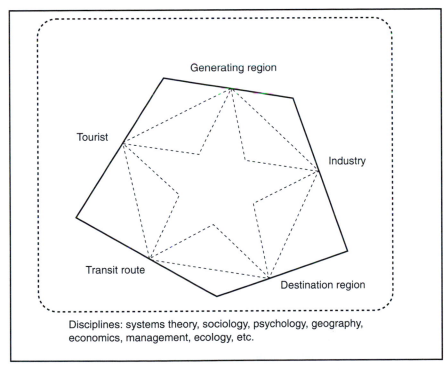

Figure 11.2 A systematic model for interdisciplinary tourism studies

Source: Leiper (1995: 26).

Leiper presented a means for the enhancement of tourism knowledge, and a basis for good scientific theory. He stated:

> Evidence about tourism comes from looking at the elements (tourists, places, organisations), at their interaction with one another and with environments, and from making observations using appropriate techniques from a range of disciplines. This is balanced by the systemic structure explicit in the model for interdisciplinary tourism studies.
>
> (Leiper 1995: 27)

Leiper's model can be criticised, in that it might be argued that the destination region and its distinguishing characteristics should receive more prominence, and the generating region less. Obviously, the latter is the scene for a good deal of advertising and promotional activity, designed to stimulate tourism. Market research has also been directed to discovering what it is about the environment at the origin which helps generate an exodus of tourists. Unfortunately, however, tourists 'at home' are largely indistinguishable from the rest of the population, and even if they were distinguishable, their presence and often humdrum everyday existence, holds no special significance for the generating region. The destination, on the other hand, receives and reflects the full impact of the influx of visitors. This is where most tourism studies have been directed, and rightly so. Leiper (1979) concedes that it is the destination region where the most significant and dramatic aspects of tourism occur. Its attractions and facilities are essential to the tourism process, and it is the location of many of the important functional sectors of the tourist industry. To a significant degree, the environment and landscape of the destination region could be said to be an index of all the positive and negative features of modern tourism. However, we must also consider other matters such as why tourists visit the places they do, and how they behave and why.

Tourist motivation

The subject of tourist motivation involves questions about why people travel. However, identifying clearly the relationships between an individual's motivations and selection of a destination is a difficult task. Krippendorf (1987), for instance, identified a number of tourist motivations, including:

- recuperation and regeneration;
- compensation and social integration;
- escape;
- communication;
- broadening the mind;
- freedom and self-determination;
- self-realisation;
- happiness.

Collectively, these motivations reflect that 'the traveller ... is a mixture of many characteristics that cannot be simply assigned into this category or that one' (Krippendorf 1987: 28).

Tourist motivations have occupied an important place in tourism literature. One of the most widely cited publications on tourist motivation was that of Gray (1970), who presented two basic reasons for pleasure travel – 'wanderlust' (people's desire to leave familiar surroundings and experience things exciting and different or unknown) and 'sunlust' (seeking out places that have better attributes for specific purposes than are available locally, and which may literally mean a 'hunt for the sun') (Pearce 1987: 2). 'Wanderlust may be thought of essentially as a "push" factor whereas sunlust is largely a response to "pull" factors elsewhere' (Pearce 1987: 22) (also see Chapter 2 for a more detailed discussion of recreation motivations and choice).

Initially, the predominant approach to the study of tourist travel motivation was to attempt to characterise 'push' factors as determinants of travel behaviour, such factors typically being conceptualised in terms of needs. For instance, the role of escapism was central to the work of Dann (1976), who argued that fantasy motivators form an important element of travel demand, and demonstrate its individualistic nature. As Leiper (1984) similarly argued:

> all leisure involves a temporary escape of some kind ... tourism is unique in that it involves a real physical escape, reflected in travelling, to one or more destination regions where the leisure experiences transpire ... A holiday trip allows changes that are multi-dimensional: place, pace, faces, lifestyles, behaviour and attitude. It allows a person temporary withdrawal from many of the environments affecting day to day existence.
>
> (Leiper 1984)

Many other theories and means of explaining tourist motivations have been described. These include:

- *Maslow's hierarchy of human needs*, ranging from physiological needs (e.g. air, food, water, sleep), to safety (e.g. security, freedom from danger), to social/ love (e.g. love, affection), to ego/esteem (e.g. self-respect, reputation, prestige), to self-actualisation (e.g. self-fulfilment) (see Ross 1998: 20);
- *Crompton's nine motivations* – escape from a mundane environment; exploration and evaluation of self; relaxation; prestige; regression; enhancement of kinship relations; facilitation of social interaction; and two cultural or pull motives, novelty and education (Hall 1998; Ross 1998);
- *Pearce's travel career ladder* builds on Maslow's model. It suggests that travel behaviour reflects a hierarchy of travel motives, where people start at different levels, change their levels during their lives, and are influenced by such factors as money, health, and other people. People may have more than one motive for visiting a destination so that more than one level of the travel career ladder is involved in any one trip. Motivations are therefore complex and one variable

is unable to explain any individual's travel behaviour at a point in time much less over several years or one's life span. Pearce's model notes that people tend to ascend the travel career ladder towards the need for self-actualisation as they gain travel experience, and that higher level motives necessarily include lower level motives, though at any one time one motive tends to be dominant (Pearce *et al.* 1998: 42–4).

The discussion of tourist motivations, though brief here, highlights the critical role of market segmentation in tourism planning and development (e.g. Novelli 2005). Tourist areas cannot be all things to all people. They require careful planning and management, while providing appropriate resources and facilities for those travellers destination regions wish to attract.

Tourism–environment interaction

Tourism can have beneficial and negative consequences for the environment; tourist development can contribute to substantial upgrading of the recreational resource base, and thus add to visitor and local resident enjoyment. It can also lead, for example, to improved transportation systems (an important component of the tourist experience) through advances in vehicle and routeway design (Gunn 1994). This allows greater opportunity for pleasurable and meaningful participation in travel, and, simultaneously, creates external economies. 'Improvements in transportation networks, water quality and sanitation facilities may have been prompted by the tourist industry, but benefit other sectors of the economy. An international airport … provides improved access to other regions for locally produced goods' (Vanhove 1997: 67) (also see Gunn 1994; Page 1994), but its impacts can be devastating for people living in nearby areas (e.g. dispossession, noise, traffic, congestion, pollution).

Enhanced understanding of the resource base is another positive outcome of pleasure travel, brought about by the application of various management techniques to interpret and articulate the environment to visitors (e.g. see Hall and McArthur 1993, 1996; *Annals of Leisure Research* 2004). Beneficial modifications or adaptations to climate in the form of recreational structures, clothing and equipment, have also been developed in response to the stimulus from tourism. Positive response can also be seen in the broadening of opportunities to view and experience both the physical and cultural world. Ready examples are the opening of national parks, wilderness areas and forests for recreational use, and continued agitation for better access to water-based recreational resources along streams and coastlines.

For many years, improved habitats for fish and wildlife, and control of pests and undesirable species have become possible through the economic support and motivation of increased use. According to McNeely (1988, in Lindberg 1991), African fauna (e.g. African elephants, lions, mountain gorillas and rhinos) are protected and managed as tourist resources. Lions and elephant herds were individually estimated to be worth about $27,000 and $610,000 per annum, respectively. Viewing of flamingos is the most popular visitor activity in Lake

Nakura National Park, Kenya, and is estimated to be worth up to $15 million per annum, but substantial costs are often involved in the management of wildlife and in the impacts that management and the wildlife themselves have on local communities. Unfortunately, villagers living near parks and other protected areas have been killed by buffalo, elephants and lions, while communities may be dispossessed as a result of the declaration of parks (Munyi 1992, in Lillieholm and Romney 2000). In other words,

> Nature conservation can create an allocation imbalance between the costs and benefits of protection, where protected area benefits accrue to broad, global constituencies and future generations, while costs are assumed by rural residents who are least able to afford them … This is particularly true in sub-Saharan Africa, where 29 of the world's 36 poorest nations are located. Amid this backdrop of hunger, preventable poverty-related disease, and high infant mortality rates, it can be difficult to justify setting aside productive lands for wealthy foreigners to observe wildlife.
>
> (Lillieholm and Romney 2000: 143)

The many ramifications of tourism give much scope for interaction with the environment. Some observers (e.g. Eckbo 1967; Relph 1976), while conceding beneficial spin-offs in the economic, political and cultural spheres, remain convinced that, 'in the long run, tourism, like any other industry, contributes to environmental destruction' (Cohen 1978: 220), or conserves only the things that are of potential and actual tourist interest (i.e. flora, fauna, cultures and landscapes that tourists want to see). Gartner (1996) points to critics who oppose tourist development on the grounds of declining water quality, stemming from visitor overuse in the Mediterranean, the Adriatic and other popular destinations (for a detailed discussion on seacoasts and tourism, see German Federal Agency for Nature Conservation 1997). Gartner, and others, like the authors of this book, believe that with proper planning and management, incorporating sustainable environmental goals, tourism can help maintain or even enhance the environment, and be a positive influence in the process of cultural dynamics, while simultaneously contributing to a region's economic development. Despite the protracted debates about tourism and its economic significance in developed and developing countries, there is considerable disagreement as to whether the incidence and magnitude of the effects of tourism can be accurately measured (e.g. the problems in determining carrying capacities that are noted in earlier chapters).

The tourism–environment relationship is often expressed in terms of opposing alternatives – either protecting the environment for tourism, or protecting the environment from tourism. However, these objectives need not be mutually exclusive. There appear to be several modes of expressing the impact of tourism: the net effect may well be tourism-related *and* environmental enhancement. A balanced appraisal of the growing links between heritage and tourism in European cities (Ashworth and Tunbridge 1990) identified both the benefits and problems in such links, and the importance of delicate planning and management if the two were to exist in a mutually beneficial manner. The development of ecotourism (see below) has been influential in this regard. Increasing cultural awareness and consciousness have stimulated restoration of historic sites and antiquities. Referring particularly to Europe, Haulot (1978) raises questions about how this heritage – the landmarks, castles and artefacts of past eras – could be kept intact, if its existence and preservation had not become the ongoing concern of a great audience of tourists, resulting both in substantial financial contributions from the visitors themselves and generous state support. The Welsh castles and town walls of Edward I in Gwynedd, The Tower of London, Hadrian's Wall, Stonehenge, Avebury, and the megalithic sites of Salisbury Plain, Wiltshire, are World Heritage Sites in Britain which attract millions of international and domestic visitors.

Design of contemporary tourist complexes appears to benefit from the demands of a more discerning tourist population. While there remain many examples of unfortunate additions to the tourist landscape, modification of the built environment for today's tourist is marked increasingly by quality architecture, design and engineering. Higher standards of safety, sanitation and maintenance also help to reduce the potential for pollution, spread of disease and environmental damage. These advances demonstrate that tourism need not destroy natural and cultural values, and, in fact, can contribute to an aesthetically pleasing landscape.

The potential for enlightened development is exemplified in the many sophisticated tourist developments in various countries of the world. However, there are environmental costs in tourism development: the distribution of the economic benefits of tourism is uneven, and ecological damage and destruction are widespread.

The environmental benefits and costs of tourism raise the question of externalities – 'those costs or benefits arising from production or consumption of goods and services which are not reflected in market prices. Because of this there is little incentive for firms to curb external costs, since they do not have to pay for them' (Tribe 1995: 244) (see Table 11.1). Examples include aircraft noise disturbing residents around airports serving tourist routes; the loss of a mangrove swamp when a tropical island resort is built; and inadequate fauna preservation through establishment of African safari parks (e.g. see Bull 1995: 163). Tribe (1995: 245) illustrates the point with reference to sewage discharges into the sea:

Table 11.1 Positive and negative externalities

Negative externalities	Positive externalities
Water pollution	Image effect
Air pollution	Promotion effect
Noise pollution	Improvement of international liquidity
Traffic congestion	position
Security costs of events	Increase of property value
Destruction of landscape	Free fees
Extension	
Sight pollution of windmills	

Source: Vanhove (2005: 214–15).

Whilst there is little marginal private cost to the water companies for pumping sewage into the sea, it represents a loss of well-being to people who want to use the sea. There is a considerable marginal external cost, which takes the form of cleaning costs to surf equipment, medical costs to treat infections and loss of earning caused by sickness. These are readily identifiable costs to which must be added the general unpleasantness of contact with sewage.

(Tribe 1995: 245)

'By analysing externalities, the public or social benefits and costs of tourism may be added to, and subtracted from, its commercial market value to an economy' (Bull 1995: 163). Clearly, then, 'Recognition of the importance of unpriced values and externalities in tourism at least warns us to treat carefully any statistics on the commercial importance of this sector to an economy' (Bull 1995: 176). Rarely, however, are the external benefits and costs of tourism accounted for in financial terms.

Although it can be conceded that tourism has much (perhaps unrealised) potential for environment enhancement, negative impacts do occur in a number of areas, particularly from the predatory effects of seasonal migrations of visitors, and resulting disturbance to, or destruction of, flora and fauna. The most obvious repercussions are likely to be in natural areas, but the built environment and urban areas may also be impaired, and the social fabric of communities can be widely disrupted.

Pollution, both direct and indirect, and in all its forms (from aircraft emissions, to architectural insensitivity, to destruction of ecosystems), is a conspicuous manifestation of the detrimental effect of tourism (Young 1973; Gunn 1994; Newsome *et al.* 2002) (see Box 11.1). Erosion of the resource base is a particularly serious environmental aspect (Wall and Wright 1977). This can range from incidental wear and tear of flora and structures, through soil erosion, to vandalism

---

Box 11.1 **Pollution**

Pollution is deterioration of part of the environment because of the occurrence of substances or processes of such types and in such quantities that the environment cannot assimilate them before they cause damage. Some observers suggest that this definition does not go far enough, and that any discharge of effluents or emissions pollutes the receiving environment in that it changes the state and probably the quality of that environment. In the real world, complete elimination of pollution cannot be achieved. Realistically, all that can be done is to reduce pollution to the minimum in socio-economic terms, control the types and levels of pollutants acceptable and determine selectively the location of sites where certain pollutants are to be released. Pollution management is expensive, both in terms of technology and procedures called for, and trade-offs may be required in production and capacity levels.

continued...

***Box 11.1* Pollution (continued)**

Pollution can occur most commonly in air, soil and water. The major sources of contaminants are plant nutrients (nitrates, phosphates, ammonia), toxic chemicals (heavy metals, pesticides, petrochemicals), organic wastes (e.g. from sewage, food processing industries and intensive livestock enterprises), heated effluence, polychlorinated biphenyls (PCBs) and salinity. Although many of these are naturally present, loadings can be significantly increased by human activities such as industry and mining, waste disposal, and urban runoff. Wet-weather pollution of urban waterways is a continuing problem following stormwater runoff and may be aggravated by overflows of sewage from surcharging sewer mains.

Although the scale of most recreation activities would not always lead to a significant, identifiable pollution problem, they can be the source of liquid or gaseous substances which potentially are a hazard to health and the environment. Among such sources are the discharge of sewage into water-bodies and the ocean; emissions from heating and refrigeration units; discharge of hazardous substances through the sewerage or drainage system; emissions, odours and spills from land or water-borne vehicles; and noise and light pollution. In many situations, practices which are potentially polluting are controlled by regulation. Even where this may not apply, good relations with visitors and host communities call for a mode of operation at a recreation site which will minimise the release of harmful or undesirable substances into the environment.

Effluents and emissions can be reduced to an acceptable level by phasing out the use of hazardous substances such as chlorine additives in water-based recreation facilities; eliminating the use of leaded petrol and toxic detergents; introducing cleaner technologies; installing treatment and filtration equipment; and adopting proper procedures for storage, use and disposal of waste products. Relatively simple amendments to operating practices can reduce or eliminate nuisance to susceptible environments.

It is important that measures to manage pollution at a recreation site be subject to an environmental auditing system. Monitoring and follow-up are essential to ensure that the measures are effective. Self-regulation is an important addition to mandatory inspections by a regulatory agency. Pollution management is an integral part of best practice environmental management in the leisure and recreation sector, and will be cost-effective in long-term operational savings and community relations.

Source: Pigram (2003b: 379–80).

and deliberate destruction or removal of features which constitute the appeal of a setting. Erosive processes are accelerated at times by use of incongruous technological innovations and by inferior design and inappropriate style in the construction of tourist facilities. Tangi (1977), for example, described some of the tourist resorts of the Mediterranean as architectural insults to the natural or historic sites where they are located. It is as well to remember, of course, that the strange architecture of today, which may be challenged by so many, may become the heritage of tomorrow and challenged by few (as is the case with the Sydney Opera House, Australia).

In many destination areas, the environment must serve not only conflicting tourist uses, but also the resident community, many of whom take a proprietorial attitude towards their surroundings. Congestion and overtaxing of infrastructure and basic services, which are particularly prevalent in high seasons, can generate dissension between visitors and the local population, the latter coming to resent the intrusion of tourism (e.g. see Doxey 1974; Pearce 1978, 1979). In a related study, Rothman (1978) listed municipal services and facilities, access to recreational sites, and personal and social life, as features of the socio-cultural environment seasonally curtailed by vacationers. Social interaction between residents and tourists was reported as minimal, with large numbers of visitors opting for an exclusive environment requiring the least cultural adjustment on their part. In developing countries, too, aspects of tourism may have long-term disruptive effects on the lifestyles and employment patterns of host communities, especially where the actual dispersal of visitors is minimal and they stay largely within the confines of 'the resort'.

Tourists help spread, among other things, AIDS, sexually transmitted diseases, and flu. Tourists themselves face health risks, and may transmit diseases and illness on their return home. Prostitution has strong links with tourism. There are many accounts of sex tourism in Thailand and other Southeast and East Asian countries in particular (e.g. see Ryan and Hall 2001), and most recently in Africa (Kibicho 2004a, 2004b). Child sex tourism is an ugly turn on this theme. Child sex tourism has been around a long, long time, but only recently has it become the centre of academic, media and government interest. At a conference on the child sex trade in June 1994, Chuan Leekpai, Prime Minister of Thailand, stated:

> ... this problem has not arisen just in the last year or two. It started long ago, but in the past it was not taken as a serious matter. The world didn't pay much attention to it; there was no organization working on this problem; there was no governmental policy, either written or spoken, regarding this problem and there was no international traffic of prostitutes from one country to another. However, all these things have now occurred and Thailand (like other countries in the region) must face the problem
>
> (Szadkowski 1995)

ECPAT, the organisation to End Child Prostitution in Asian Tourism was created in 1991 and became a central institution in this task. It took only a couple of years for ECPAT to spread its wings to almost thirty countries, but its energies have been directed mainly to Sri Lanka, the Philippines, Taiwan and Thailand.

The potential for tourist activity to disrupt host communities often varies seasonally. The subject of seasonality with respect to tourism is complex, but its causes and effects have received insufficient critical attention. As Butler and Mao (1997) point out,

> The nature of the relationship between seasonality and the motivation of visitors is not known, and issues such as whether dissatisfaction with conditions in the origin region or desire for the attractions of the destination play a greater role in shaping the seasonal patterns of tourism is also a mystery. It is also not known why tourists travel in peak seasons, because a number of forces are likely to be acting on tourists' motivations and choice at any one time.
>
> (Butler and Mao 1997: 21–2)

With so many variations on the theme, it is difficult to generalise on the relationship between tourism and the environment. The relative importance of each influential factor varies with the location and situation, and negative effects need to be balanced against positive impacts. Certainly, the ugly face of tourism receives wide exposure, and the relationship depicted in Figure 11.3a could well apply, with an increase in tourism bringing about a decrease in environmental quality. However, change does not necessarily equate with degradation, and tourism and environmental quality are not mutually exclusive goals. The net effect may be marginally negative (Figure 11.3b), or the two may be organised in such a way that both benefit and give each other support (Figure 11.3c).

Clearly, tourism and protection of the resource base are more alike than contradictory; the demands of tourism, instead of conflicting with conservation, actually require it (Gunn 1972). If this is not the case, the very appeal which lures the visitor to a site will be eroded, and with reduced satisfaction, any chance of sustained viability for the destination will disappear.

Environmental influences

The nature and extent of tourism's impact on the physical environment are determined by many factors, including:

- *the length of time since tourist development was initiated and the aspirations of developers* – short-term goals characterise much tourist development, which is largely speculative in nature, and which is facilitated by entrepreneurs who are often either ignorant of, or blatantly ignore, the consequences of their actions and their cumulative effects;
- *the number of tourists and the intensity of on-site use* – all things being equal, as visitor numbers increase, it is likely that there will be greater transformation of the environment. Of course, the resiliency of the ecosystem (see below) and management regimes will affect this relationship so that it is non-linear;
- *the nature and resilience of the destination's ecosystem* – certain types of vegetation and soils can withstand greater visitor numbers, while climate and seasonal variations in temperature, rainfall and humidity influence plant growth rates. Similarly, Western cultures are perhaps less likely to be influenced by

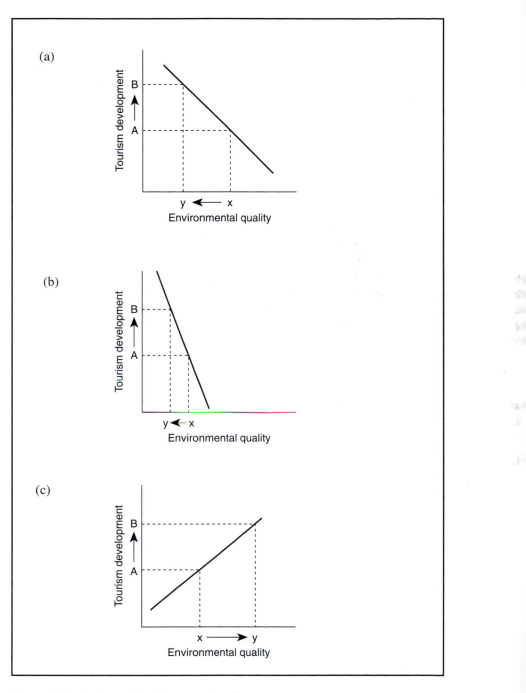

Figure 11.3 Tourism and environmental quality

Source: Pigram (1983: 210).

Western visitors' attitudes and behaviour than less developed or Asian countries and vice versa. All in all, there is great variation among ecosystems and cultural systems even at the local level;

- *the dynamic nature of tourist demand and the dynamism of tourist development* – tourist motivations and choices change over time, so that destinations rise and fall in popularity, while the nature of attractions, services and facilities at destinations change.

Aside from the above factors, a broad range of environmental factors and their relationship to tourism warrant discussion.

Weather, climate and environmental uncertainty

The relationships between tourist and recreation activity and the environment (e.g. aesthetics, temperatures, topography and wildlife) have received insufficient attention. Weather and climate may be regarded as inputs to the amenity index of a place or region (Pigram and Hobbs 1975). In this context, they are important influences on tourist choices and behaviour. 'Environmental changes, including ... shifts in climatic patterns, have the ability to affect destinations, by directly changing some attributes of the tourism product itself and by altering locational advantages of different types of destination' (Craig-Smith and Bull 1990, in Bull 1995: 238). Acid rain, global warming and ozone depletion are indications that the atmosphere may be changing at an unprecedented rate, with implications for flora, fauna, and human activities, including tourist development. Extreme events such as floods, droughts, tsunamis and hurricanes may increase or decrease, and the insurance industry is playing a growing role in influencing what can be built where. Many recreation activities take place on coasts and in mountains, both particularly vulnerable locations in an era of change.

> Global climate change, including ozone depletion, may modify tourism and recreation resources and how the potential clientele perceives them and thus require businesses and destination areas to adjust to changing circumstances ... with respect to the climate, the past may no longer be an adequate guide to the future ... climate change may be one factor among many worthy of inclusion in tourism planning and investment decisions.
>
> (Carmichael *et al.* 1995: 513)

Despite recent indications that many tourist areas may be affected by the warmer temperatures stemming from global warming, research concerning (1) the relationships between tourist and outdoor recreation activity and environmental (and, more specifically, climate) change, and (2) industry response strategies, is lacking (e.g. König 1998). Outdoor recreation is an activity that will probably adapt to different conditions as tourists substitute one activity for another, or one resort for another, as conditions and seasons change (see below). Downhill and cross-country skiing are two activities likely to be most affected by global warming, not to mention coastal areas and associated activities and resources, if sea levels rise (see Hall and Higham 2005). Other activities, too, could be significantly affected. As the relationship between tourism and the environment becomes more widely recognised, so, too, does the relationship between tourism and nature conservation, and the need for environmentally sustainable forms of tourist development.

Numerous factors have influenced the patterns and processes of international and domestic tourism growth, and, in some areas, decline, as well as more specific world and regional trends. These factors include industrialisation; freer trade (as policies of high and extensive protectionism are abandoned); widespread growth in wealth and leisure; increased environmental awareness; growing conflict among competing resource users; ageing populations; the ease and increased speed with which people can travel further; and changes in employment structures (e.g. decline in agricultural employment). These have significant implications for the resource base for tourism and outdoor recreation.

The resource base for tourism

The complex pattern of tourism across the globe reflects the diversity of environments which constitute tourist resources, and the varied experiences which travellers seek. A common element is the contrast between the home region and the destination. If there were no perceived difference from place to place (natural or fabricated), tourism would not exist. Contrasts may be sought and discovered in the physical environment, the cultural and historic landscape, the people, artificially created attractions, and festivals and events.

Although socioeconomic, political and technological forces, perhaps of all the factors affecting the development of tourism, the most important are physical. Some of the strongest flows of tourists are from cool, cloudy regions, to places highly regarded for their warm, sunny climate. For many tourists, 'wanderlust' appears to take second place to 'sunlust'. By contrast, the popularity of winter tourist resorts rests in great part on cool (though hopefully sunny) weather and the assurance of adequate and long-lasting snow cover.

Yet, one of the factors little considered in tourism literature until very recently, is the potential impact of climate change. For instance, studies on the impacts of climate change, due to an enhanced greenhouse effect, on the snow pack in Australia, suggest that climate change would increase the frequency of winters with little natural snow (Haylock *et al.* 1994 and Whetton *et al.* 1996, in König 1998). Climate change due to an enhanced greenhouse effect is predicted to have the biggest impact on the Australian ski industry, and the highest resorts with the best natural snowfalls and the best conditions for snowmaking. This would create 'two classes' of resorts: (1) smaller resorts at lower altitude, which will lose their downhill ski operation first; and (2) larger resorts, at high altitude, where downhill skiing remains possible. However, in the long-run (assuming a worst-case climate scenario for 2070), none of Australia's resorts will be snow-reliable (König 1998).

Another physical factor with obvious implications for tourist development is the appeal of the coast. A huge proportion of recreational and tourist travel is to coastal environments. The increasing availability of all manner of products that enhance the attraction of the coast, from scuba diving to wind-surfing equipment, mean that people have many more different means of enjoying the water. Mercer (1972) explains the coastal location of many resorts in terms of the attraction of edges or junctions in the landscape – the coastline representing the interface between land and sea. The success of coastal resorts reflects the attraction of the beautiful setting. Unfortunately, however, the impacts on coastal zones are increasing in their variety and intensity – wildlife is disturbed; erosion of shorelines is exacerbated through use of craft that generate waves; pollution, litter and sedimentation are prominent in some destination areas. The impacts are, however, sometimes positive.

> A number of coastal environments have become marine protected areas (e.g. marine parks) as a result of their value as recreation and conservation resources. The rehabilitation of dunes, wetlands, beaches and islands for recreational and tourism purposes has shown that recreational use can have positive environmental outcomes if suitable management regimes are adopted and implemented.
>
> (Orams 2002: 56)

One component of the physical environment which has more limited significance for tourism is the presence of mineral springs or spas. In historical times, conviction in the medicinal properties of mineral waters for drinking or bathing, stimulated the earliest visitors to places like Bath and Tunbridge Wells in Britain, and Spa, itself, in Belgium. Despite advances in modern medicine, 'taking the waters' at spas and similar health resorts continued to attract a considerable clientele. Increasingly, however, with the development of additional facilities close by for amusement and diversion, the function of spas became as much, if not more, social than therapeutic. One health resort in the United States, French Lick, in southern Indiana, even became the focus for thriving illegal gambling and liquor activities in the 'prohibition' era. Nonetheless, tourism for health purposes remains important for many people, so that clinics and sanatoria continue to attract significant numbers of patrons.

So medical travel is not new, and travelling abroad to access cheaper health procedures is becoming big business. In 2003, Singapore had 230,000 health-related visitors and Bangkok Hospital in Thailand treated more than 52,000 foreigners (Taffel 2004). In Australia, medical tourism, much of it for cosmetic surgery, now rivals the tertiary education market, the country's third biggest export industry. A related phenomenon with implications for tourism is the drawing power of religious shrines, like Lourdes in France and Knock in Ireland, based in part on beliefs in the miraculous powers of water from local springs, which had their origins in visions in the last century. Spiritual reasons have always been a powerful stimulus to travel, and large numbers of pilgrims continue to visit Mecca and other Moslem holy places annually. Religious centres such as the Vatican, Jerusalem and Benares also attract pilgrims in large numbers, as demonstrated by the millions attending the funeral of Pope John Paul II, in Rome, Italy, in 2005.

Many tourists are genuinely interested in foreign places and people, so that aside from the physical environment, the opportunity to make contact with other people's culture and way of life is a strong influence on tourism. The appeal of traditional architecture, folklore, unusual customs, crafts and foods, is well-documented. Not all of these are authentic, and there is considerable potential for tourism to distort the cultural tradition of host communities.

Interest in past cultures is also the basis for historical tourism, whereby the primary focus is on inspecting the legacy of a bygone age. Features of historic interest have a proven fascination for tourists, whether these be the magnificent homes and castles of Britain and Europe, artefacts and ruins of the ancient world, sites of military battles, picturesque villages mirroring a past lifestyle, restored railways and steamships, or the collections of miscellaneous junk which pass for museums in some small, isolated settlements in outback Australia. Countries with a relatively short history (e.g. Australia and New Zealand), often find it more practical and rewarding from a tourism perspective to re-create features and settlements of the past, and present these in something of an outdoor museum setting. Thus, Sovereign Hill promotes itself as a re-creation of one of the early goldmining towns in the State of Victoria. Historical theme parks also flourish in the United States, where attractions like Knott's Berry Farm and Disneyland in Los Angeles rely to a great extent on revivals of the past.

Clearly, tourism, nostalgia and culture can have a mutually beneficial relationship; interest in history stimulates tourism, which, in turn, makes historical (heritage) preservation possible. Handled correctly, preservation certainly pays in terms of tourism. As Newcomb (1979: 232) puts it: 'Our visible past is like a fire which ... if we tend it carefully ... will illuminate our pleasure and ... touch our imagination and our hearts'.

Tourism landscapes

Reference was made earlier to the landscape of tourism, not so much in the sense of attractive scenery, but in the association of distinctive physical and cultural features characteristic of tourist development. Used in this way, the term is analogous to agricultural or residential landscapes. The landscape of tourism reflects the imprint, both good and bad, of mass travel on the environment, and the relationship is inescapable. The landscape makes tourism and, in turn, tourism makes the landscape.

Given the diverse nature of resources and experiences which appeal to travellers, the range of recipient landscapes created for, and emanating from, tourism is wide. The natural beauty to be found in the west of Ireland, the glittering facade of Las Vegas, or the simulated atmosphere of the South Seas re-created in Hawaii, all represent particular landscape types orientated to tourism. Whereas it is easy to deplore the 'look-alike' landscapes spawned by mass tourism across the globe

(Eckbo 1967), it is another matter to attempt to interpret and explain their evolution from a generic point of view (Price 1980, 1981). Some interesting work has been carried out on the townscapes of tourist destinations.

Lavery (1974) outlined the historical background to the development of holiday resorts in Western Europe (also see King 1997), and, in particular, alpine resorts, spas and seaside resorts. He proposed a typological classification of resorts, based on their function and the extent of their visitor hinterland. A hierarchy of eight categories was identified, encompassing: capital cities; select resorts; popular resorts; minor resorts; cultural/historic centres; winter resorts; spas/watering places; and day-trip resorts. Lavery concedes that the classification is subjective and that obvious omissions are, specifically, seaside resorts, religious/spiritual centres, and 'created' resorts such as Disneyland in Florida. Some resorts would also fit several categories, while others have progressed from one orientation to another.

Undoubtedly, tourist destinations, like resources in general, pass through cycles linked to fashion and tourist behaviour. The popular appeal of established destinations fluctuates as changed circumstances trigger new sets of interests and different clients. Innovative forms of tourism may emerge and lead to the eclipse of redundant tourist outlets and the discovery of fresh attractions and venues. Explanation of such cycles has been linked to the behavioural characteristics of travellers. Two major human polarities have been identified:

- *Allocentric persons* (more recently labelled venturers) – self-confident, successful, high earners and frequent travellers, who prefer uncrowded destinations and exploring strange cultures;
- *Psychocentric persons* (more recently labelled dependables) – unsure of themselves, low earners and infrequent travellers who seek the security of tours and familiar destinations (Plog 1972, 1998).

The great majority of people are mid-centric (or centric): they fall between these two extremes and favour budget tours, heavily-used destinations, familiar food and chain-type accommodation. According to this hypothesis, resorts tend to rise and fall in cycles which match their appeal to particular categories of tourists (see Figure 11.4).

> They move through a continuum ... appealing first to allocentrics and last to psychocentrics ... As the destination becomes more popular, the mid-centric audience begins to pick up ... [which] leads to further development of the resort, in terms of hotels, tourist shops, scheduled activities for tourists and the usual services that are provided in a 'nature' resort area ... continued development ... carries with it the threat of the destruction of the area as a viable tourist resort ... Destination areas carry with them potential seeds of their own destruction, as they allow themselves to become more commercialized and lose the qualities which originally attracted tourists.
>
> (Plog 1972: 4)

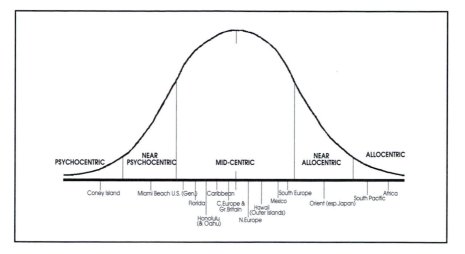

Figure 11.4 The cycle of tourist resort development

Source: Kaiser and Helber (1978: 8, after Plog 1972).

It is important to note that decline of a resort is not inevitable. With appropriate planning and sound management, it is possible for success to be predicted, achieved and sustained. The possibility of rejuvenation is also stressed by Butler (1980), who cites the introduction of gambling casinos into Atlantic City, New Jersey, as an attempt to tap a new resort market. Other studies (e.g. Christaller 1963; Hovinen 1981, 1982) have also examined patterns in the development of tourist destinations.

From the point of view of landscape, it could be expected that each category of resort would develop its own recognisable blend of structures, activities and functions making up a tourist environment responsive to the requirements of the predominant type of visitors. The distinctiveness of tourist centres as special-purpose settlements is perhaps best seen in the morphology and townscape of seaside resorts, especially those of Britain and Western Europe.

In a study of English and Welsh seaside resorts, Barrett (1958) identified several common morphological features or characteristics. In particular, he noted the significance of the seafront in the structure and location of the commercial core, and a marked zonation of vacation accommodation and residential areas. Moreover, because growth along one axis was precluded or restricted, elongation of settlement occurred parallel to the coast. In Barrett's study, the core shopping and business district was offset symmetrically to a frontal retail and accommodation strip, which was the focus of resort activities, and which was functionally and socio-economically distinct from the rest of the town (Figure 11.5). All these features were subject to modification, because of terrain and pre-resort transport and land-use patterns.

Studies of New Jersey seashore resort towns also identified linearity in the various functional zones in response to location of principal routeways and proximity to the beach, and recognised a specialised frontal trading zone, termed

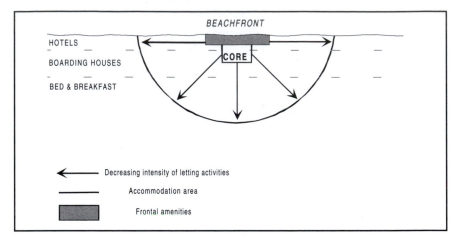

Figure 11.5 Schematic model of an English seaside resort

Source: Barrett (1958: 36).

the Recreational Business District (RBD) (Stanfield 1969; Stanfield and Rickert 1970). This zone was spatially and functionally distinct from the Central Business District (CBD), and comprised an aggregation of seasonal retail establishments catering exclusively for leisure-time shopping. Stanfield nominated the boardwalk as a uniquely American phenomenon, and grouped it with the British pier and promenade as a major contribution to the morphology of resort settlements.

Lavery (1974) put forward a schematic representation of a 'typical' seaside resort, with prime frontal locations occupied by the larger accommodation facilities, and a gradation in land values and tourist-oriented functions away from the seafront, the main focus of visitor attraction. Lavery also noted the spatial and functional separation of the CBD from the RBD, and associated the latter with the main route from the public transport terminal (e.g. railway station), in contrast to the emphasis given by Stanfield to vehicular access.

In Australia, an attempt was made to establish the extent to which the 'model' features found in British and North American seaside settlements were present in Australian beach resorts (Pigram 1977). The study was carried out at the Gold Coast on the Queensland/New South Wales border, which has become the focus of intensive tourist development catering to over two million visitors annually (Figure 11.6). Several interesting parallels can be drawn between the urban structure of Gold Coast settlements and that outlined above. The attraction of the coast and beaches, the role of routeways and termini, the importance of topographical features, and the influence of pre-resort form and function are readily discernible.

An interesting aspect in the Australian study was the development of paired resort nodes at either end of the Gold Coast tourist complex. At the northern extremity, Surfers Paradise dominated the amusement and entertainment scene (RBD), whereas Southport is the regional and commercial centre (CBD). In the south, Coolangatta specialises in recreational business, while the CBD is across

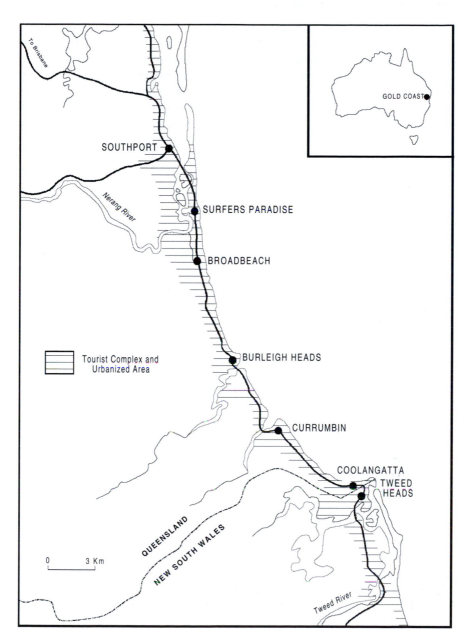

Figure 11.6 The Gold Coast tourist complex, east coast of Australia

Source: Pigram (1983: 203).

the State border in Tweed Heads. The end result is paired beach resorts which reflect, in part, the antecedents of European settlement in Australia, yet show clearly the effects of modern forces in shaping the tourist landscape.

More recently, in extending the work of those authors cited above, Meyer-Arendt (1990) developed a model of the morphology and evolution of seaside resorts in the Gulf of Mexico. In a departure from previous studies, Meyer-Arendt suggests that the development of seaside resorts actually resembles a T-shape pattern, with the initial beach access point becoming the main point (or locus) of tourist activities, eventually evolving into the recreational business district (RBD). This pattern also forms the basis of the beach resort model suggested by Smith (1992), who, in recognising the importance of second homes and low-budget accommodation, proposed eight stages through which a beachside resort develops.

Resorts are receiving increasing research attention (King 1997).

> Since resorts often aspire to being self-contained tourism destinations in their own right, the literature has studied the phenomenon from a diversity of angles, not confined to the study of resort facilities *per se*. These approaches include resort development (Dean and Judd 1985; Stiles and See-Tho 1991), planning (Smith 1992), the assessment of local attitudes (Witter 1985), marketing (King and Whitelaw 1992), the resort life cycle (Butler 1980), resorts as communities (Stettner 1993), architecture (England 1980), landscaping (Ayala 1991a, b) and key success factors (Wober and Zins 1995). A significant dimension of this literature is the role of resorts as 'enclave' developments, separated from the reality of daily life in adjacent areas or regions.
>
> (Freitag 1994, in King 1997: 11)

Integrated resorts are properties which incorporate a wide range of recreational facilities and accommodation types. 'Historically the evolution of tourism has been closely identified with the beginnings and subsequent development of resorts' (Medlik 1993: 126). 'The resort concept is based on providing leisure and recreation opportunities. Many resorts are self-contained destinations providing accommodation, food service, shopping and developed recreation opportunities. Some resorts rely on the natural resource base of the area for access to recreational opportunities' (Gartner 1997: 135). Resorts can be classified in many ways according to their specific location, their season of use, and/or the recreational opportunities they offer – island resorts (see King 1997), seaside or beach resorts, mountain or ski resorts, or health resorts. The term 'resort' has been used to describe tourist destinations at different scales, each locale combining specific locational, seasonal and recreational characteristics.

Butler (1980) has modelled the evolution of tourist resorts/destinations, using product life-cycle analysis, and identifying the links between the development or otherwise of a tourist resort/destination area and the nature of the travel market (see Figure 11.7). In the early stages of the life-cycle, few people visit the area, and most services are locally provided. As the area increases in popularity, the extent of tourist development increases, and the nature of that development changes

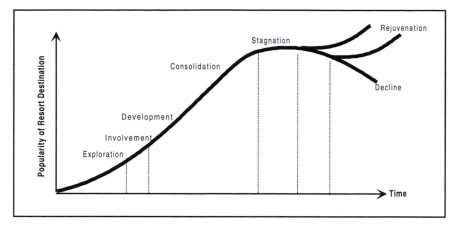

Figure 11.7 The hypothetical model of resort life-cycle

Source: Butler (1980).

as there is a shift from natural physical and cultural attractions to ones which are more contrived and less authentic. In the development phase:

> ... local involvement and control of development will decline rapidly. Some locally provided facilities will have disappeared, being superseded by larger, more elaborate, and more up-to-date facilities provided by external organisations, particularly for visitor accommodation. Natural and cultural attractions will be developed and marketed specifically, and these original attractions will be supplemented by manmade imported facilities. Changes in the physical environment of the area will be noticeable, and it can be expected that not all of them will be welcomed or approved by all of the local population.
>
> (Butler 1980: 8)

Eventually, resort areas reach maturity, where options for planning and development arise and become more pressing. The resort may enter a stage of decline or stagnation, and may then be rejuvenated. As development increases, infrastructure keeps pace with the rising level of visitors. More activity options are also added. Environmental and social impacts result from facility, attraction and infrastructure development. Roads may be built into scenic areas to offer more sightseeing opportunities; seaside resorts may experience a spread of development along the coastline; ski resorts may expand the number of their trails, and so on.

Mathieson and Wall (1982: 121) identified four types of transformation that occur during the development stage: architectural pollution; ribbon development and sprawl; infrastructure overload; and traffic congestion. However, this analysis tends to focus on the negative impacts of tourism on the environmental quality of the area, and does not give due regard to the complex interrelationship between the environment and tourism, which are inextricably linked, and which can produce benefits for each other (see below). Nonetheless, mass tourism, in particular, may have negative impacts: urban sprawl, diminished aesthetic values of the natural landscape, and decline in the levels of local ownership (Kariel 1989). Unrestricted tourism development presents problems. Therefore, tourist development must be well planned, professionally managed and set in a broader context of development and environmental conservation (see Pearce 1989).

Tourism and nature conservation

Conservation is a philosophy which is directed at the manner and timing of resource use (O'Riordan 1971: 8), and may be defined as managing the resources of the environment – air, water, soil, mineral resources and living species, including humans – so as to achieve the highest sustainable quality of life.

Nature conservation is a dynamic concept which is subject to diverse understandings and interpretations, spatially and temporally, and which is supported for many different reasons (e.g. ethical reasons; encouraging environmental sustainability; maintaining genetic diversity; recreation; scientific research; future choices and utility; education; and political reasons).

Recognition of the importance of nature conservation can be seen in the relatively recent rise of the environmental movement, and, simultaneously, the development of a conservation ethic in modern society. That recognition is tangibly evident in (1) the creation and resourcing of public and private sector agencies and interest groups, (2) related legislation and public policy, and (3) the establishment of resource management units such as national parks and wilderness areas, which often serve as important tourist attractions.

Tourism and nature conservation are interdependent, and their relationship has been a lengthy one. Tourism often stimulates measures to protect or conserve nature, but, at the same time (and somewhat paradoxically), presents a significant environmental risk, especially because of its demands on the natural environment, and therefore on responsible agencies. These risks are intensifying as domestic and international tourist demand for natural areas grows in many developed and less developed countries. Furthermore, the nature of that tourist demand is such, that tourists are seeking more spontaneity, independence and participation in their travel experiences (e.g. the growth of nature-based tourism – see below).

Budowski (1976) noted three different relationships with respect to conservation and tourism – conflict, coexistence or symbiosis – which can exist between those promoting tourism and those advocating conservation of nature. Conflict occurs when conservationists see that tourism can have only detrimental effects on the environment. Coexistence is noted when some, though possibly little, positive contact occurs between the two groups (conservation and tourism). Symbiosis is reached when the relationship between tourism and conservation is organised in such a way that both derive benefit from the relationship. Conflict and coexistence are common. Symbiosis is perhaps the least represented relationship in the national and international perspective.

As noted above, tourism provides an economic impetus for the conservation of the environment, due to the fact that protected and/or scenic areas are major attractions for domestic and international tourists. Tourism can also contribute to a wider appreciation of nature conservation by promoting and making more accessible specific sites and aspects of nature.

The role of tourism as a consistent contributor to nature conservation is often debated, because, among other things, tourists trample vegetation, disturb wildlife, carry pathogens and weeds, and do not always behave in ways which promote the symbiotic relationship desired between the industry and conservation (e.g. when engaged in vandalism or littering). Tourism, too, has fostered the intensive viewing (with resulting disturbance or damage) and export of protected and/or endangered species.

The interrelationships between tourism and nature conservation are thus extremely complex and dynamic, with conflict being most acute where tourist development occurs rapidly and without strategic planning. Unfortunately, tourism and recreation research has developed few strong concepts or theories to guide the role and management of tourism in nature conservation. Many studies focus narrowly on the physical impacts of developments at a particular site and neglect the human element, few have a longitudinal basis, and most are reactionary. Even when impacts are identified or predicted, the research focus is largely limited to the effects of tourism on vegetation and, to a lesser extent, on wildlife, with impacts on air and water quality, soils and ecosystems relatively neglected. Therefore, a number of methodological problems concerning research on tourism and the natural environment (and thus nature conservation) can be identified:

- the difficulty of distinguishing between changes induced by tourism and those induced by other activities;
- the lack of information concerning conditions prior to the advent of tourism and, hence, the lack of a baseline against which change can be measured;
- the paucity of information on the numbers, types and tolerance levels of different species of flora and fauna;
- the concentration of researchers upon particular primary resources, such as beaches and mountains, which are ecologically sensitive.

(Mathieson and Wall 1982: 94)

Tourism development must be environmentally sensitive and consistent with long-term nature conservation, otherwise it presents risks to the sustainability of the industry itself, and more generally the natural environment. Tourist pressures on nature conservation will continue to grow, and a clearly established and wide-spread balance between tourism and nature conservation will never be universally accepted, this being perhaps most problematic in wilderness and very sensitive areas. The relationship between tourism and nature conservation is thus a highly political issue, which is in need of much greater research attention if the natural resources upon which tourism so heavily relies, are not to be degraded or destroyed. Successful integration of tourism and nature conservation objectives is of increasing importance, because it can enhance the choices of people and help maintain or even enhance the quality of the environment.

Tourism and environmental compatibility

Environmentally compatible tourism describes a situation whereby tourism and the environment are able to exist in harmony, so that tourism does not detract from, or harm the environment nor vice versa. Increasing environmental awareness and conservation activities around the globe have contributed to efforts to establish environmentally compatible tourism. In some instances, the tourism industry has entered into partnerships with environmental and other groups, has consulted effectively with host communities, resource management agencies and governments, and has directly supported conservation or preservation objectives. However, these situations are all too rare, and with the recent rise of nature-based travel, industry support for environmental conservation or preservation is often merely a means of marketing and promoting business operations. The recent development and promotion of ecotours by many tourist operators are examples where industry regulation with respect to restricting undesirable business and tourist behaviour, and environmental impacts, is lacking, and where suspicions have been cast on the environmental compatibility of tourist operations with the host environment.

In order to reduce the conflicts, and enhance the relationship, between tourism and the environment, environmental impact statements are now required in many countries as part of the approval, and monitoring and evaluation processes for tourism projects. This is particularly the case if projects are large, or located in, or adjacent to, environmentally sensitive areas such as protected areas, rainforests, coastlines or estuaries. Further developments in environmentally compatible tourism approaches can be seen in the implementation of environmental auditing processes in the public and private sectors.

The key to achieving environmentally compatible tourism and, ultimately, a sustainable tourist industry, is recognition of the need for environmentally sensitive policy-making, planning and development. The integration of tourism and the environment is being carried out at different levels in a number of places and for a variety of reasons, with various mechanisms being utilised. Strategies and related activities range in size from small-scale to large-scale projects, and include various economic, nature conservation, cultural, social, heritage, spatial/regional and political benefits and costs. On a broader national and global scale, approaches to integrating tourism and environmental objectives are being developed and promoted by international and national tourism agencies, and to a limited extent by multi-national corporations. The development and promotion of nature-based tourism and ecotourism are notable responses.

Nature-based tourism

Nature-based tourism may be defined as 'domestic or foreign travel activities that are associated with viewing or enjoying natural ecosystems and wildlife, for educational or recreational purposes' (HaySmith and Hunt 1995: 203). Tourism industry leaders and natural resource managers face significant challenges in promoting sustainable development of tourism in protected areas, and in managing impacts on flora and fauna (HaySmith and Hunt 1995). Nature-based tourism, encompassing ecotourism, adventure tourism, outdoor-oriented educational tourism, as well as a whole host of other outdoor-oriented, non-mass tourism experiences, is arguably the fastest-growing segment of the tourist industry in many countries (McKercher 1998: ix), and one which holds much promise for environmentally compatible tourism objectives.

> A review of the principles of ESD [Ecologically Sustainable Development] offers valuable insights into how the tourism industry must act in relatively undisturbed areas. Underlying the entire ESD philosophy is a commitment to operate within the social and biophysical limits of the natural environment. To abide by this tenet, tour operators may have to trade off economic gain for ecological sustainability and, indeed, will have to accept that there are some places where tourism should be excluded.
>
> (McKercher 1998: 191)

Nature-based tourism can only survive when the resources on which it depends are protected. Ecotourism was first described by Hector Ceballos-Lascurain (1987, in Boo 1990: xiv) as 'Travelling to relatively undisturbed or uncontaminated natural areas with the specific objectives of studying, admiring, and enjoying the scenery and its wild plants and animals, as well as any manifestations (both past and present) found in these areas'. According to Whelan (1991: 4), 'ecotourism, done well, can be sustainable and a relatively simple alternative. It promises employment and income to local communities and needed foreign exchange to national governments, while allowing the continued existence of the natural resource base'. This last point gives implicit recognition to the need for adequate and appropriate management regimes (also see Valentine 1991: 5), which foster environmental and cultural understanding, appreciation and conservation (e.g. see Richins *et al.* 1996). Much attention has been given to ecotourism (e.g. see Weaver 2001a, 2001b; Fennell and Dowling 2003) and it has 'emerged as a major component of the global tourism industry and an important focus for academics within the field of tourism studies' (Weaver 2001: 2).

An important element in the development of any management regime or programme, is appropriate research. Yet, much tourist activity in natural areas is permitted without a great deal of understanding of tourism's impacts on the ecosystem. With respect to both flora and fauna, and the landscape itself, this is a critical point. For instance, the impacts of tourism on wildlife are well-documented but largely site-specific, and related findings and management strategies are difficult to apply universally. As HaySmith and Hunt (1995) point out:

> Impacts on wildlife from nature tourism are varied, and are often difficult to observe and interpret. Reactions of animals to visitors are complex. Initially, some species or individuals of a species retreat from visual or auditory stimuli caused by humans but become habituated over time. Other species or

individuals that are more sensitive may alter their behaviour and activities to completely avoid contact with visitors, with potentially long term effects. Other animals cannot escape the disturbance and may be negatively affected, directly injured ... or killed.

(HaySmith and Hunt 1995: 206)

Nature-based tourism can be blatantly invasive toward wildlife when hundreds of observers congregate to view one rare animal or group of animals, when artificial feeding is used to draw animals for tourist viewing and entertainment, and when relationships between species are disturbed. Dolphin and whale watching are two popular forms of nature-based tourism in which there is a heavy reliance on tour operators positioning themselves in time and space so as to provide opportunities for people to catch a glimpse of these beautiful creatures or in some instances to feed them. Tangalooma Wild Dolphin Resort and Tin Can Bay (both in Queensland, Australia) are two locations where people have direct contact with dolphins. At the former, bottlenose dolphins have been fed since 1992, while at the latter, Indo-Pacific Humpback dolphins have been fed for more than 20 years. Mayes *et al.* (2004) questioned the impacts and integrity of the opportunities provided at these two destinations. They noted that at Tin Can Bay there were many shortcomings in the management of the site. For example, there were minimal hygiene levels

Plate 11.1 Dolphin and whale watch tours are now very popular in Port Stephens, Australia. However, serious concerns are being raised about their impacts on dolphin behaviour and very little is known about these impacts. Dolphins are now known to be attracted to boats and other marine craft, and a central issue is whether craft get 'too close for comfort' to these creatures.

and practices, feeding of dolphins varied and was not adequately monitored and controlled, people could enter the water without any staff present, flash photography was not regulated, and oral interpretation material was not evident. This was in sharp contrast to Tangalooma Resort (see Table 11.2).

The findings of Mayes *et al.* are hardly surprising. Some ecotourism operators (accredited and non-accredited) do not follow appropriate standards of practice. They violate agreed practices by approaching animals, cornering them and feeding them in order to give their clients a good glimpse of the creatures with no regard for the animal's welfare and behaviour (breeding, feeding, social). Some are content to let clients smoke and dispose of cigarettes into water or on the ground. Interpretive information is scant or even misleading. On the other hand, there are indeed some very conscientious and ethical organisations operating in nature-based areas. Tread Lightly Eco Tours, based in the Blue Mountains, Australia, runs a variety of ecotours (http://www.treadlightly.com.au/default.htm). The inter-pretive information provided during tours ranges from that relating to the geology of the region, to wildlife, to the risks that adventurers can face, to 'bush tucker', to local tourism and recreation developments. The marketing materials give an accurate indication of the care which the tour leader takes in protecting the environment during any tour. The walks, as they are advertised, truly 'are an educational and sensory experience with full emphasis on minimal impact practices'. Ecotourism is a term that is applied to a wide range of activities. The philosophies, principles and practices which are meant to underpin ecotourism operations are admirable, but the application of ecotourism, even in terms of self-regulation and accreditation, relies on some tenuous propositions, programmes and agreements (see Box 11.2).

Table 11.2 Comparison of interaction management and interpretation programmes at Tangalooma Resort and Tin Can Bay, Queensland, Australia

	Tangalooma	*Tin Can Bay*
Management practices		
Pre-handling hygiene	yes	no
Strict no-touching rule	yes	no
Limited number in the water	yes	no
Strict supervision	yes	no
Allocation of amounts of fish	yes	limited
Designated 'feeding only' area	yes	no
Pre-feeding briefing	yes	no
Interpretation programme content		
Call to action	yes	no
Opportunities to join organisations	yes	no
Take home message to participants	yes	no
Conservation ethics discussed	yes	no
Human impacts upon dolphins	yes	no
Strandings and rescues	yes	no
Dolphin biology and ecology	yes	limited

Source: Mayes *et al.* (2004).

***Box 11.2* Ecotourism self-regulation and accreditation**

Three basic principles underpin most definitions of ecotourism: (1) eco-tourism is nature-based; (2) ecotourism is associated with sustainability and seeks to minimise tourism's negative impacts; and (3) ecotourism has an educational component designed to motivate positive changes in people's attitudes and behaviours regarding environmental conservation (Buckley 2001; Weaver 2001a). Ecotourism has been defined in Australia as 'Ecologically sustainable tourism with a primary focus on experiencing natural areas, that foster environmental and cultural understanding, appreciation and conservation' (EA 2005), but there is no consensus on its precise meaning (Weaver 2001b).

As many protected areas experience increasing tourism demand pressures, and use becomes more intense, the potential for conflict between maintaining environmental quality, maximising recreational accessibility and satisfaction, and promoting economic development is enhanced. Governments and tourism industry operators have been seeking to create circumstances in which market forces, or non-governmental institutions such as industry bodies, perform regulatory functions. Self-regulation and accreditation in the tourism industry is one such example which is receiving international coverage (e.g. Issaverdis 2001; Weaver 2001a, 2001b; Buckley 2002).

Ecotourism Australia (EA) is the Australian ecotourism industry's peak representative body. EA's vision is 'to be leaders in assisting ecotourism and other committed tourism operations to become environmentally sustainable, economically viable, and socially and culturally responsible' (EA 2005). One of EA's largest and most significant projects is the National Ecotourism Accreditation Program (NEAP). NEAP is an industry-driven eco-certification scheme or programme, which enables industry, protected area managers and consumers to identify 'genuine' ecotourism products.

The concepts of accreditation and certification are gaining some reluctant acceptance in the tourism industry, but is considered a means of enhancing tourism product standards and contributing to environmental sustainability and organisational marketing profiles (McKercher 1998; Harris and Jago 2001; Issaverdis 2001). Recent figures suggest there are approximately 3,000 nature and ecotourism operators in Australia (ABS 2003), of which, more than 1,000 are licensed to operate in national parks (Buckley 2003d). Yet, the most recent list of accredited products reveals only 240 companies and agencies with at least one product certified through NEAP. A number of factors are, however, providing further impetus to accreditation:

> the increasing expectation of standards and awareness of service quality by consumers, the increasing expectation by travel intermediaries that the product will be safe and environmentally responsible, an increased industry awareness of sustainable business practice, and a growing interest in tourism research.
>
> (Issaverdis 2001: 582; also see Fennell 1999; Weaver 2001a, 2001b)

Assessments of ecotourism accreditation in Australia and overseas are far from conclusive about its merits and about accredited operators' attitudes and perceptions of related schemes. According to Fennell (2002: 214), 'the jury is still out on the success of Australia's accreditation process'.

Buckley (2001, 2002: 183, 185) argued that 'ecolabels and environmental accreditation are controversial topics in tourism', because they may be used by regulatory agencies to grant permits, and because their main function is to act as a 'market mechanism'. He proposed that 'the most basic test of a tourism ecolabel (e.g. Advanced Accredited Ecotourism operator) is whether it is accepted by tourists as meaningful, reliable, and useful in choosing individual products' (2002: 185). Essential components of an ecolabel programme are, therefore, 'transparent criteria and procedures with detailed information readily available to the public as a backup for the labels themselves' (2002: 186). To be effective and reliable, ecolabel schemes need adequate accreditation criteria, clearly defined procedures for assessment and auditing, and penalties for non-compliance. Furthermore, information on each of these aspects of the schemes needs to be available to prospective tourists and members, and the ecolabels must be recognised by the target market.

Weaver (2001a: 158) argued that the self-funded NEAP scheme has led to Australia being regarded as 'the world leader in ecotourism accreditation'. However, Weaver also noted that the NEAP scheme has a number of limitations: (1) the initial application's reliance on self-assessment by the operator could result in misleading submissions being accredited, and not exposed until an on-site audit is undertaken; (2) the integrity of the auditing process is questionable because the paper audit is based on a self-assessment by the operator; and (3) operators are notified 60 days prior to on-site auditing, and as auditors must be allowed free entry, the audited operator is aware of precisely when the audit takes place. Some changes have been made to certification processes, but operators are still notified prior to on-site audits (see NEAP 2003). In short, in Australia, there is no independent on-site auditing of accredited and non-accredited ecotourism products. Yet, according to Jean-Pierre Issaverdis (2001: 580), who at the time of his writing was a board member of EA, auditing is the necessary component to ensure applied concepts of benchmarking, accreditation and best practice 'are valid and that reliable measures of performance are defined'.

Nature-based tourism is growing – on and under land and water surfaces. Impacts are also growing rather than diminishing and this spells a profound threat to the environment. Biodiversity is declining, many endangered species are under threat from recreational developments, public and private sector initiatives in environmental protection are often 'western' (or non-indigenous) in their orientation, while boundaries and related questions arising over cooperation at international to local scales will continue to present great challenges to society (Mercer 2004). While there may well be more organisations implementing sound ecological practices, as the number of operations, tourists and activities grow, the cumulative effects must be considered.

Nature-based and related forms of tourism will only be successful if comprehensive planning strategies include appropriate and extensive research programmes. Any arguments that nature-based, or any other form of tourist activity, has a particular beneficial or negative relationship with the environment, cannot be sustained without related research. Those who choose to argue one way or another could be easily challenged by questions about the precise nature of the tourism–environment relationship.

Plate 11.2 Viewing marine life is of course not restricted to viewing from vessels. Snorkelling is popular at Shark's Cove, Oahu, Hawaii, where turtles and many varieties of fish can be viewed. Unfortunately, people were seen chasing and handling turtles. Unless people are adequately informed about their potential impacts on marine life, such behaviour is likely to arise. No interpretation was available on-site, and the area is not monitored. Perhaps the industries and businesses that profit from the promotion of such activities should be levied to protect sensitive sites and prevent their exploitation and degradation in the name of human pleasure and profit.

Tourism and sustainability

In a more environmentally conscious world, the tourism industry faces increasingly stringent conditions on development; this reflects the concern for sustainability, and the long-term viability of the resources on which tourism depends. The challenge for the industry is to justify its claims on those resources with a commitment to their sustainable management.

Environmental auditing, whether by regulation or legislation, or when undertaken as a self-regulatory initiative, can be a useful management tool to help achieve sustainable development. As global demands on space and resources grow, with increased population, technological change, and greater mobility and awareness, pressure will increase on the tourism industry to implement appropriate steps for monitoring and evaluating its environmental performance. The task is to formulate and implement effective self-monitoring procedures in order to promote greener, more environmentally compatible forms of tourism, and to avoid the imposition of mandatory compliance measures.

Originally, facilitating travel was the primary focus in tourism planning, with the focus largely on tourism promotion. 'Subsequently, policies broadened to include spatial planning, but the emphasis remained on maximising economic development' (Getz 1986, in Godfrey 1996: 59). Since the publication of the World Conservation Strategy by the International Union for the Conservation of Nature, many countries and regions have begun working towards the goal of sustainable resource development (World Commission on Environment and Development 1987). Sustainable resource management is now widely accepted as the logical way to match the needs of conservation and development.

The era of environmental concern, ushered in by the World Conservation Strategy, is of immediate relevance to tourism. The environment represents not merely a constraint for tourism development, but a resource and an opportunity. Ideally, satisfying tourism settings grow out of complementary natural features and compatible social processes. At the same time, modern tourism amply demonstrates the capacity of human beings to manipulate the environment for better or for worse. Yet the consequences are not easily predictable. Tourism can certainly contribute to environmental degradation and be self-destructive; it also has the potential to bring about significant enhancement of the environment. With tourism-induced change, an important issue is irreversibility, which, in turn, is a function of factors outlined above.

Much attention is given by government and industry to the development of sustainable tourism (e.g. Butler 1990, 1991; Pigram 1990; Inskeep 1991; Bramwell and Lane 1993; Gunn 1994; Godfrey 1996; Dredge and Jenkins forthcoming).

> Sustainable development is positive socioeconomic change that does not undermine the ecological and social systems upon which communities and society are dependent. Its successful implementation requires integrated policy, planning and social learning processes; its political viability depends on the full support of the people it affects through their governments, their social institutions, and their private activities.
>
> (Rees 1989, in Gunn 1994: 85)

Sustainable development stresses that economic development is dependent upon the continued well-being of the physical and social environment (Dasmann 1985; Barbier 1987; Butler 1991). Dutton and Hall (1989) identified key mechanisms by which sustainable development could be achieved:

- developing cooperative and integrated control systems;
- developing mechanisms to coordinate the industry;
- raising consumer awareness;
- raising producer awareness;
- planning strategically to supersede conventional approaches.

While sustainability is an extremely influential concept in tourism planning, in practice it is fraught with problems (see Hall *et al.* 1997b). Perhaps there is some particular merit in Ashworth's (1992) rather cynical observation that the:

> tourism industry is tackling the criticisms being made of it, not the problems that cause the criticisms. If there is no resource or environmental problem, then it does not need to be defined nor do solutions need to be found. The problem is seen as one of promotion, and promotion is what the tourism industry is particularly good at. Buying off the grumblers with a few 'commitments' and 'mission statements'... is easier than the alternative [of sustainable tourism planning].
>
> (Ashworth 1992: 327)

These comments aside, and despite difficulties in achieving sustainable tourism, the integration of economic, sociocultural and environmental planning goals is increasingly being recognised as a vital component of longer-term tourism development that maintains cultural identity and biodiversity.

The realisation that more than one form or manifestation of tourism is possible, has prompted the development of alternative typologies seen as achievable and desirable, depending upon the circumstances. The term 'sustainable tourism' can be used to refer to tourist typologies, options or strategies preferable to mass tourism. This has led to some confusion as governments and industry attempt to avoid the mass tourism label. As Godfrey (1996: 60–1) pointed out:

> In trying to be different, common phrasing and synonyms such as soft (Kariel, 1989; Krippendorf, 1982), postindustrial (SEEDS, 1989), alternative (Gonsalves and Holden, 1985), responsible (WTO, 1990), appropriate (Singh *et al.* 1989), green (Bramwell, 1991), rural (Lane, 1989, 1990), low impact (Lillywhite and Lillywhite, 1991), eco- (Boo, 1990), and nature-based (Fennel and Eagles, 1990) have all been applied.
>
> (Godfrey 1996: 60–1)

Differences in conceptualisations and applications of sustainable tourism plans and policies are to be expected. Depending on a person's viewpoint, sustainable tourism may represent particular markets, may be about planning and policy processes, or it may be a governing principle (see Godfrey 1996). Whatever the case, positive elements in a strategy for sustainable tourism typically include:

- development within each locality of a special sense of place, reflected in architectural character and development style, sensitive to its unique heritage and environment;
- preservation, protection and enhancement of the quality of resources which are the basis of tourism;
- fostering development of additional visitor attractions with roots in their own locale and developing in ways which complement local attributes;
- development of visitor services which enhance awareness, understanding and development of local heritage and environment; and
- endorsement of growth when and where it improves things, not where it is destructive or exceeds natural and social carrying capacities, beyond which the quality of human life is adversely affected.

(Cox 1985: 6–7)

Summary

The concept of 'environment' is very broad, as is the concept of a 'tourist industry'. The relationship between tourism and the environment is extremely complex and dynamic, and not well understood. Research concerning the impacts of tourism on the economic, physical and social environments is lacking.

Integration of tourism and the environment is occurring at the global to site-specific levels, with benefits accruing to conservation of natural and cultural environments. Increasing attention is also being given to tourism's potential to contribute to regional economic development, for example, in national parks (see Chapter 10). Unfortunately, however, there are still numerous cases where the effects of tourism on the environment are negative, and unnecessarily so. The emergence of sustainable approaches to tourism development, encompassing notions of environmental compatibility, are laudable, but even then, the concept of a sustainable industry is open to challenge.

Travellers are becoming increasingly sophisticated and discerning. Many such travellers are looking to high quality, authentic, natural and cultural environments, where the likelihood of recreational satisfaction is high. A great responsibility rests with the industry and governments to develop sustainable industry practices which conserve the natural and cultural environments, and which, subsequently, will hold tourist appeal. Somewhat ironically, economic arguments relating to the generation of tourist revenue, often hold the key to the conservation of resources, whose 'real' values are intangible now and in the long term.

Guide to further reading

- Tourism landscapes: Urry (1995); Crouch (1999); Aitchison *et al.* (2000); Terkenli (2004).

- Seasonality and tourism: Baron (1975); Bonn *et al.* (1992); Calantone and Jotindar (1984); Chon (1989); Snepenger (1987); Uysal and Hagan (1994); Uysal *et al.* (1994); Butler and Mao (1997).
- *S*ustainable tourism: Barbier (1987); Pigram (1990); Bramwell and Lane (1993); Butler (1991); Butler *et al.* (1998); Hall (2003); Mercer (2000); Weaver and Lawton (2002); Holloway (2002); McCool and Moisey (2001); Lew *et al.* (2004); Theobald (2005).
- Nature-based tourism: Valentine (1992); Harper and Weiler (1992); McKercher (1998); Font and Tribe (2000); Newsome *et al.* (2002).
- Ecotourism: Boo (1990); Whelan (1991); Cater (1993); Wight (1993); Lindberg and Hawkins (1993); Richins *et al.* (1996); Farrell and Marion (2001); Weaver (2001a, 2001b); Fennell and Dowling (2003).
- Marine ecotourism: Orams (2002); Garrod and Wilson (2003).
- Impacts: Mathieson and Wall (1982); Mercer (2000); Weaver and Lawton (2002); Sonak (2004); Wong (2004); Journals such as *Journal of Sustainable Tourism* and *Annals of Tourism Research* contain many articles on the impacts of tourism.
- Tourism in developing countries: Lea (1988); Richter (1989); Opperman and Chon (1997); Mowforth and Munt (1998).
- Geographic Information Systems (GIS): Bahaire and Elliott-White (1999).

Review questions

1   Define sustainable tourism.
2   Compare and contrast definitions of 'tourist', 'tourism' and 'tourist industry'. Can you readily identify the 'tourist industry'? Examine contesting arguments about the existence of such an industry.
3   Critically examine the main forces affecting global tourism patterns and processes.
4   Conduct an inventory of the resource base for tourism and recreation in a place of your choice (perhaps your local area). In that inventory, identify the agencies responsible for the management of those resources. Have there been any recent, notable conflicts among those agencies with respect to recreational and tourist use of resources? Based on the resources you have identified, can you identify any potential recreational and tourist opportunities yet to be identified or explored by management agencies?
5   Discuss the relevance of sustainable development principles to tourism planning and management.
6   Discuss the relationship between tourism and the environment.
7   Present an overview of case studies or 'success stories' where tourism has contributed to conservation of the natural and built environments.
8   What approaches have been applied to study and manage seasonality in tourism?
9   Why have some less developed nations utilised tourism as a means of economic and social development? Overall, would you consider tourism has brought

many benefits to such countries? Explain your answer with reference to case studies.

10   What are the main factors affecting tourism's potential to impact on the physical environment?

11   What planning measures have been utilised to manage tourism's physical impacts in one or more natural areas in your country or local area? What have been the outcomes of these measures with respect to visitor management (e.g. satisfaction) and visitor impacts on the physical environment?

12 Planning for outdoor recreation in a changing world

At first sight, the notion of planning for outdoor recreation might seem a contradiction in terms, and likely to inhibit the spontaneity and freedom of choice associated with leisure activities. However, planning for outdoor recreation should be seen as being as essential as planning for other human needs such as health and welfare, transport and education.

By definition, planning should be proactive and forward looking, not relying merely, for example, on prohibitions and the *ad hoc* imposition of restrictions to people's leisure in reaction to problems as they arise. Emphasis in the planning process for outdoor recreation should be on the creation of physical and social settings in which people can exercise choice and satisfy their demands, within prevailing laws, economic limitations and resource constraints. It is in the expansion of choice by providing a broad range of opportunities for recreative use of leisure, and in the satisfaction of recreational participants, where the planning and management of recreation resources make an essential contribution.

In one sense, planning can be thought of as the ordering of space through time. In the planning of recreation space, the aim should be to provide a range of functional and aesthetically pleasing environments for outdoor recreation, which avoid the friction of unplanned development, without lapsing into uniformity and predictability. New spatial forms and settings need to be kept as open and flexible as possible, in keeping with an array of interests and dynamic physical, political, economic, social and technological circumstances. Recreation is generally marked by voluntary, discretionary behaviour. People choose to take part or not, and decide the location, timing, activities and costs to be incurred. Any one of these attributes can be modified or dispensed with by unforeseen or uncontrollable factors. Moreover, the process of choice is imperceptibly influenced by such factors as family relationships and personal characteristics, and pervasive adjustments to changes in income, education, lifestyle, social mores, traditions and culture.

Against such a background of change, planners seeking to cater for outdoor recreation demands must somehow anticipate a future influenced by a bewildering set of forces, many of which are difficult to predict. Given this uncertainty, planning initiatives become even more important to help underpin forms and patterns of outdoor recreation resilient and flexible enough to respond readily to environmental changes. This concluding chapter explores some of the implications of change for the planning and management of recreation resources, introduces a strategic planning approach as a means of adapting and coping with change, and looks towards what kind of future recreation planning should be directed.

Trends in leisure and outdoor recreation

The only things that are certain about the future are uncertainty and the inevitability of change and need for adjustment. Forecasts about possible leisure scenarios range from the fanciful prophecies of science fiction to more considered statistical predictions based on short-term projection or extrapolation of current trends. Such forecasts can only be expressed in terms of probability, and without the benefit of insight into innovations, changes in social circumstances and public policy, or technological breakthroughs. The demand dimension (e.g. population charact-eristics and recreation propensities) and the supply side of the equation (e.g. futuristic possibilities regarding the availability and use of recreation space) both lack clear definition. The ways in which demand and supply factors interact in terms of environmental impacts and recreation decision-making are not well understood, except in specific case studies. Moreover, any planning initiatives must be undertaken against a background of increasing environmental awareness and constraints on freedom of choice, because of concern for repercussions on nature and society. The travel industry, for example, is grappling with specific issues such as air and noise pollution, which have to be addressed regardless of cost in money or efficiency terms. 'Consumerism', too, is imposing greater demands on the recreation planner to provide quality products and experiences that do not always coincide with the earlier trends to mass participation and packaged tours. Currently overshadowing all these factors are economic considerations in public sector planning, particularly with respect to global financial markets.

Economic fluctuations

Globalisation has brought with it the realisation that no part of the world can be quarantined from the shock of economic reversals and the often painful restructuring which ensues for national, regional and local economies. The privatisation or out-sourcing of many government functions and the establishment of government business enterprises are accelerated by downturns in economic prospects, or by national and regional budget deficits. As one or a combination of factors such as inflation, unemployment, recession and outright poverty bite into scope for individual choice and relatively unfettered decision-making, many people's recreation opportunities inevitably contract. Curtailment of living standards, frustration of aspirations towards self-betterment, loss of self-esteem, destruction of long-held values, erosion of faith in 'the system' or government, and personal stress leading to emotional and behavioural trauma, can all be the outcomes of economic instability. Predictably, during such times, it is likely that recreation will decline in importance. This holds equally true for individuals, households and governments. Thus, pleasure travel, generally, is curtailed, purchases of recreation equipment are postponed, and participation in recreation, in so far as it involves spending, or even the use of resources (including time), which could be income-producing, is minimised. Moreover, governments and providers of recreation opportunities in the private sector, also experience difficulties in meeting their commitments during periods of inflation, economic recession, high interest rates, or any combination of these.

In times of adversity, the availability of recreation outlets takes on renewed urgency in helping to mitigate the effects of economic hardship. Recreation, in the sense of revitalisation, can act as a compensating mechanism in allowing people to forget their worries, or at least to cope better, and to grow as individuals. Fresh interests can be developed, and hitherto neglected or simpler pursuits rediscovered, in order to occupy an excess of 'leisure' time in a less cost-intensive manner. New skills and attitudes can be acquired which will enable disadvantaged sectors of the population to maintain their self-confidence, pride and hope.

At the same time, there is a brighter side to economic and social hardship (e.g. during periods of excessive inflation and during recession). When governments are forced to withdraw from, or reduce their involvement in, the field of recreation, communities have an opportunity to query the need for continued dependence on public funding, and are stimulated to examine the potential of self-help, cooperation and other means of economising. Thus, hard times can become a vehicle for bringing communities together. People can share frustrations and problems, substitute voluntary effort and talent for that previously provided and, in so doing, achieve a satisfying, cost-effective recreation programme at the neighbourhood or community level.

Of course, some governments and public agencies do not need the excuse of budgetary constraints to opt out of any responsibility for recreation. Even in 'normal' times, there are wide disparities at the national, state and local level in the commitment of funds and resources in this area. Some authorities maintain that recreation is not a legitimate field of interest, or priority, for publicly-elected bodies, and that private enterprise can best fill the gap. Others justify reductions in funding on the grounds of past excesses and waste. The notion that public provision for leisure and recreation is somehow dispensable, or at least low in priority, can only be overcome by a well-directed campaign from those affected (i.e. the community) to convince legislators that recreation is no longer a luxury or a privilege, but a right. In the meantime, competition for scarce public funds, overuse of available recreation resources, and intensified conflicts over shrinking recreation space, can only make the relevance of planning and management of outdoor recreation opportunities an even more urgent social issue.

Societal change

A second group of factors is that of changing lifestyles related to alterations in demographic patterns and social values. The changing age distribution of populations, the postponement of marriage and children (perhaps indefinitely), the liberation of women, the more frequent breakdown of family relationships, the proliferation of communication and other technological advances, and the change in societal attitudes to particular recreation forms and activities, are seen as some of the important influences on leisure and outdoor recreation in a changing world.

Societal changes normally do not exhibit the same cyclical characteristics as those of an economic nature. Rather, they are evolutionary and cumulative, and, sometimes, almost imperceptible. Ageing of Western populations has been accompanied by a measure of affluence, increased unobligated time, and a desire to remain active among older age groups. For example, it is no longer unusual to find 'elderly' people undertaking strenuous forms of outdoor recreation, uninhibited by misperceptions, restrictions and taboos of a time past.

The US, Canada, Britain and Australia share strong overtones of cultural pluralism in their population characteristics. Once again, a reaction can be discerned in leisure behaviour, as new groups of people are assimilated to a greater or lesser degree into the population as a whole. A heterogeneous society offers a richer spectrum of recreation opportunities, but at the same time generates difficulties for governments in providing a sufficiently diverse array of recreation experiences for a multicultural population.

Technological innovations

Outdoor recreation, as is the case for many other human activities, is feeling the shock of technological change. Improvements in transport and communications have decreased the friction of distance and made the greater part of the globe accessible. In tourism, for example, the advent of long-distance, large-capacity aircraft has made mass participation an international reality, while high-speed, computer-based communication facilities are now an integral part of the global tourism network. Not only do these facilities enable instantaneous links across the world, they have also added immeasurably to levels of awareness, both of tourists and those servicing the travelling public. With awareness comes stimulated demand for hitherto little-known sites and destinations. This, coupled with the ability to move vast numbers of people great distances in relatively short periods of time, means that few parts of the planet can be regarded any longer as out of reach for tourism and outdoor recreation.

Technological advances in motor vehicles, along with improvements to routeways and servicing, have increased the range and accessibility of places for recreation. The development of all-terrain vehicles, including the four-wheel drive, allows the recreating public to penetrate remote and possibly fragile environments. This brings with it potential problems of ecological disturbance, resource degradation, litter and overcrowding. At a larger scale, the prospect of wider introduction of high-speed rail transport is likely to impact on the recreation travel market and take business away from competitors. The Channel Tunnel is one example, and there are plans for fast trains to link airports to cities, and to cut travel time drastically between destinations. This could bring a whole new dimension to the limits on day and weekend recreation trips, and could reduce the obligated travel times associated with the journey to work. In the process, greater pressure is likely to be felt on recreation resources, both in near-urban areas and at more remote and sensitive sites, making the need for planning and resource management even more apparent.

New technology is also enabling people with disabilities to participate in outdoor recreation activities such as camping, sailing, fishing, rock climbing and snow skiing. At the other end of the spectrum, technology has opened up new opportunities in risk recreation. Examples include high-performance mountain bikes; lighter and more hydro-dynamic surf boards; stronger, more flexible climbing equipment; lighter, inflatable and more durable kayaks; and helicopter access to remote sites. Whereas advances in technological equipment and materials enhance recreation options, negative impacts such as loss of self-sufficiency, and minimising uncertainty and risk, must be addressed by recreation managers and planners (Hollenhorst 1995).

The internet has had an enormous impact on leisure and leisure industries. According to Buhalis (2003: 266), the internet has required businesses to alter their functions and operations. Leisure organisations are increasingly able to:

- accelerate knowledge and information distribution to prospective clientele and partners;
- reduce transportation, postage and communication costs;
- increase efficiency and productivity;
- support differentiation strategies by allowing better segmentation;
- enhance communication and co-ordination efficiency;
- improve and shorten the decision-making process; and
- support interactivity and interoperability with all stakeholders.

Interactive media, accessed through the internet, create new opportunities for recreation and tourism enterprises and services. It is estimated that more than 100 million people had been connected to the internet by the year 2000, with instant access to relevant, in-depth, up-to-date information about any country or recreation destination. Used interactively, the internet offers the facility to select sites and activities based on product information, images and the promise of a fulfilling leisure experience.

In the context of recreation planning, Geographic Information Systems (GIS) are a most exciting technological development. GIS use digital data collected from various sources that are processed and analysed by high-speed computers, and presented to planners and decision-makers for action. Some potential applications for GIS in outdoor recreation include: locating new trails with less potential for environmental damage; specifying fire hazard zones within a park; locating public facilities given constraints of access and proximity to park attractions; locating waste management facilities; and monitoring the environmental impacts of recreation use over time (see Box 12.1).

Given their multi-variable, multi-dimensional modelling capabilities, GIS appear to be valuable in site locational analysis for large recreation complexes and tourist

Box 12.1 Geographic Information Systems – applications to outdoor recreation

Geographic Information Systems (GIS) is a computerised mapping system that enables visualisation and analysis of spatial data and their non-spatial attributes (Nicholls 2003). This relatively new technology can help outdoor recreation managers to answer questions such as:

- Where is the best location to set a new recreation facility?
- Who and where are the potential users?
- Is the proposed facility accessible to residents in walking distance?
- What forms of public transport may be needed?

Lee and Graefe (2004) see GIS as a means of addressing these sorts of questions and as a helpful and effective tool for reducing some of the uncertainties in the decision-making process. As defined by Burrough (1986), GIS is an information system that can be used to capture, store, manipulate, integrate and display geographic information. Over some 40 years, GIS has evolved into a multidisciplinary instrument linking the physical and social sciences to provide a useful technology for the planning and management of outdoor recreation facilities.

Applications of GIS in outdoor recreation have focused on resource location, spatial patterns of distribution of phenomena and distance measurement. Lee and Graefe (2004) present examples of the use of GIS to depict spatial correlation between recreation sites and visitor occupations; boater attitudes, activity patterns and satisfaction levels with recreation sites and service provision; and terrain analysis as related to site preferences. Distance is an important factor influencing recreation behaviour and GIS has been used to predict participation rates and duration of stay according to distance travelled and to test the relevance of the distance-decay function.

Application of GIS in urban park and recreation research focuses on equitable allocation of facilities for users, services planning and issues of accessibility (Lee and Graefe 2004). Trends in the use of GIS for park and recreation management were identified by Wicks *et al.* (1983). Some prevalent applications included: displaying and analysing land use; selecting a location for new facilities; updating and displaying land records; documenting demographic patterns and trends; market area identification; and attitudes and interests surveys. Network analysis can also be used in conjunction with GIS to calculate travel distance to a recreation facility and to indicate an optimal route to minimise travel time between two locations (Bailey and Gattrell 1995).

A key feature of GIS is the ability to overlay different types of data for a specialised geographic area so that spatial relationships between them may

continued...

Box 12.1 continued

be identified and assessed. Nicholls (2003) demonstrates how GIS may be used by park and recreation managers to display service areas of their facilities on the basis of walking distance. Levels of accessibility can also be determined and access-deficit localities visually identified and mapped. This information, combined with population projections can be used to prioritise actions and guide recreation managers to areas where new facilities are most needed.

Lee and Graefe (2004) recognise the potential that GIS has in planning, managing, marketing and evaluating park and recreation facilities and service provision. However, they emphasise that its more extensive use will depend on the development of user-friendly technology and the willingness of park administrators to perceive it and adopt it as a key component of outdoor recreation management.

developments. Moreover, GIS can integrate data on spatial attributes and projected demand structures for a number of competing sites, for the purpose of analysis, evaluation and choice of optimum location. Coupled with a further advance in computer technology known as Digital Visualisation, the recreation planner is able to visualise a 'virtual landscape' undergoing a simulated change of use (e.g. a change from agricultural land to theme park) or degradation (e.g. the effect of traffic in a park). The planner can more easily assess the possible outcomes, according to estimated changes to different variables (e.g. number of users and patterns of use), before the event has happened, and thus formulate and implement appropriate management responses in a proactive manner.

Undoubtedly, the availability of technological advances is expanding recreation options and opening up access to more recreation opportunities. Whereas these developments may bring problems of conflict and resource degradation, the new tools, in the hands of planners and managers, will equip them to deal better with the uncertain dimensions of the outdoor recreation scene in the twenty-first century. The prospective changes and change agents, reviewed in this and previous chapters, should not be considered in negative terms when set against the promise of technology. Change is a powerful force when harnessed constructively. In the context of outdoor recreation, change will challenge both recreationists and planners to respond positively to the new opportunities it offers.

Planning for the future

> ... planning is a process, a process of human thought and action based upon that thought – in point of fact, thought for the future – nothing more or less than this is planning, which is a general human activity.
>
> (Chadwick 1971: 24).

The inevitability of change, and the need for flexibility mean that it is unwise and impractical to base planning for future patterns of recreation demand and resources on past experience. Many forms of outdoor recreation are responsive to variations in such factors as the cost of participation, the availability of transport, or even seasonal variations in opening times. As noted above, they are also likely to react to technological change, to rates of growth in population, to economic prospects, and to policies and priorities set by government and the community. Moreover, constraints on recreation planning can be imposed by unforeseen events, such as changes in international geopolitical or economic circumstances, or natural hazards (e.g. cyclones, earthquakes, tsunamis, floods or drought). None of these variables can be forecast with any real assurance (though they can be recognised as potential threats in contingency plans, in areas where they are likely to occur), nor can it be predicted how they might combine to influence opportunities for outdoor recreation.

Therefore, the preferred approach to meeting future demands for recreation is to plan strategically in terms of recognising a range of possible outcomes, in light of clear aims and objectives. This would encompass a limited number of flexible policy choices (options) and trade-offs, linked to a degree of acceptable risk in the assessment of those outcomes. The timeframe can also be important, given the cost and long lead time frequently involved in the acquisition of recreation space and the development of recreation facilities. In some circumstances, one year into the future might be too far away for planning purposes; for others, 2030 might be too close.

By adopting the alternative futures approach, the planner can develop combinations of strategies for meeting anticipated scenarios of future demand, relative to possible fluctuations in variables affecting recreation participation. Seeking technical solutions in 'knee-jerk' fashion to single-issue problems is not planning, particularly in the context of Chadwick's definition (see above). Increasingly, the recreation planning process will be expected to respond to a changing societal context, one that is marked by dynamic interaction between emerging technology and organisational change, political priorities and economic realities, as well as environmental constraints, to meet a complex array of recreation demands.

A strategic approach to outdoor recreation planning

Earlier in this chapter, it was suggested that planning could be thought of as the ordering of space. With recreation planning, that rather bland description needs further elaboration. Paraphrasing Getz (1987), recreation planning can be seen as a process, based on research and evaluation, which seeks to optimise the potential contribution of recreation to human welfare and environmental quality. Getz was actually focusing on tourism planning, but the message is the same. Rather than developing and promoting recreation for its own sake as a perceived desirable aspect of growth and change, the emphasis should be on socioeconomic and environmental enhancement, and the use of recreation to achieve those broad goals.

Getz (1987) advocates an integrative approach to tourism planning. Translating that approach to recreation, integrative recreation planning should be characterised as:

- *Goal-oriented*, emphasising the role to be played by recreation in achieving specified societal goals.
- *Democratic*, with meaningful input from the community level.
- *Integrative*, placing recreation planning issues within mainstream planning for other purposes.
- *Systematic*, based on research, prediction, evaluation and monitoring of outcomes.

Much earlier, Driver (1970a) pointed to key activities in a management process, in which 'planning' is but one activity:

- *The democratic process*: by which representation of interests and values are built into the political process of democracy.
- *The decision process*: of choosing among alternatives.
- *The administrative process*: whereby agencies created by decision-makers carry out the functions assigned to them.
- *The planning process*: accomplishing goals and providing information for decision-making, and the formulation, implementation and control of plans.

The conventional strategic management process (the terms 'strategic management' and 'strategic planning' will be used interchangeably) can be adopted by planners and resource managers. Such an approach accommodates Getz's integrative approach, Driver's concept of planning, and the concept of planning for alternative futures. By its very nature, strategic planning requires planners and their respective organisations to consider their existing operating environments (inside and outside of their organisations), and potential change in those environments.

Strategic outdoor recreation management is:

> The process of identifying, choosing and developing outdoor recreation resources activities that will enhance the long-term provision of satisfying recreational experiences by setting clear directions, and by creating ongoing compatibility between available skills and management resources, and the changing internal and external planning environments within which planners and managers operate.
>
> (adapted from Viljoen 1994: 4)

As noted above, the environment of recreation planning is inherently less predictable than in the past. Strategic management offers a means of dealing with change. The following discussion provides a brief introduction to the strategic management process. Readers are referred to the Guide to Further Reading at the end of this chapter for more references to comprehensive discussions of strategic management.

Strategic management is a process encompassing a range of interrelated activities, through which planners and managers move back and forth over time. At its broadest level, strategic management requires a clearly articulated mission statement. The 'mission' guides the organisation through several interrelated processes: strategy analysis, direction setting, strategy choice and strategy implementation, evaluation and control (see Figure 12.1) (Viljoen 1994: 40–3). The nature of the mission statement and the basic parameters of these four processes are discussed below.

Establishing a mission

The establishment of an organisation's mission is critical to its operations because:

> The mission statement, *inter alia*, articulates the overall purpose of the organisation and its distinctive characteristics ... This is important to guide the activities of strategic analysis (we should only analyse the environment and our resources in relation to our stated overall purpose), strategic choice (we should only choose strategies that are consistent with our overall purpose), and strategy implementation (we should implement strategies in a way that will help us better achieve our overall purpose). All aspects of strategic management, therefore, should be referenced to the mission of the organisation.
>
> (Viljoen 1994: 42)

Mission statements reflect the vision for an organisation, defining its purpose (or *raison d'être*) and outlining what it intends to accomplish in the larger environment (Rossman 1995). According to Drucker (1974: 94, in Hall and McArthur 1996), 'Defining the purpose and mission of the business is difficult, painful and risky. But it alone enables a business to set its objectives, to develop strategies, to concentrate on resources and go to work. It alone enables a business to be managed for performance' (see Box 12.2).

Strategy analysis

Strategy analysis involves the gathering and use of information to ascertain the strategic position of the organisation and the situations likely to be faced in the conduct of its activities. Strategy analysis requires managers to inventory all the major forces affecting their recreational product and to determine whether these represent opportunities or threats. These forces may include environmental forces (political, economic, social, technological and physical) influencing the availability of recreational opportunities, as well as the skills and resources (financial, human, physical and intangible) available to manage those opportunities. More specifically, strategic analysis involves several types of analyses such as those presented in Table 12.1.

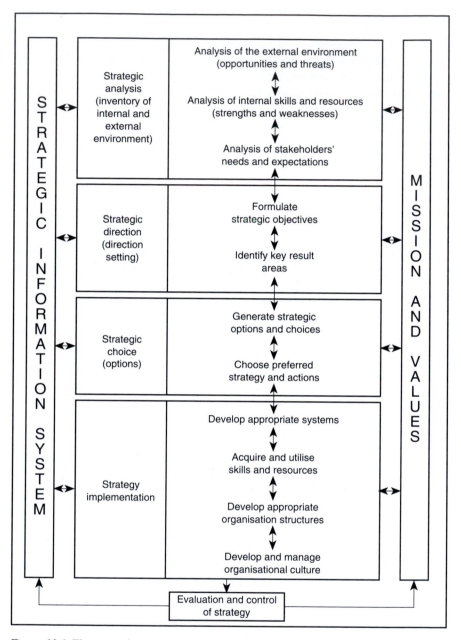

Figure 12.1 The strategic management process

Source: Adapted from Viljoen (1994: 43).

Box 12.2 Parks Agency Canada: Charter Mandate

On behalf of the people of Canada, we protect and present nationally significant examples of Canada's natural and cultural heritage, and foster public understanding, appreciation and enjoyment in ways that ensure the ecological and commemorative integrity of these places for present and future generations.

Our role

We are guardians of the national parks, the national historic sites and the national marine conservation areas of Canada.
We are guides to visitors from around the world, opening doors to places of discovery and learning, reflection and recreation.
We are partners, building on the rich traditions of our Aboriginal people, the strength of our diverse cultures and our commitments to the international community.
We are storytellers, recounting the history of our land and our people – the stories of Canada.

Our commitments

To protect, as a first priority, the natural and cultural heritage of our special places and ensure that they remain healthy and whole.
To present the beauty and significance of our natural world and to chronicle the human determination and ingenuity which have shaped our nation.
To celebrate the legacy of visionary Canadians whose passion and knowledge have inspired the character and values of our country.
To serve Canadians, working together to achieve excellence guided by values of competence, respect and fairness.

Parks Agency Canada has eight identified key service lines or result areas: establishment of national heritage places; heritage resource protection; *heritage preservation*; *visitor services*; town sites; through highways; management of Parks Agency Canada; People management. Under each of these areas, the agency has well targeted 'planned results' and 'performance expectations'. For example:

Service line 3: Heritage Preservation

Planned Result – Canadians, visitors and stakeholders appreciate and understand the significance of heritage places and support their protection.
Performance Expectations – 50% of national park visitors and 80% of national historic site visitors participate in a learning experience related to natural and/or cultural heritage; 85% of visitors are satisfied, 50% are very satisfied with onsite heritage presentations programming; 75% of visitors understand the significance of the heritage place; Canadians, visitors and stakeholders actively support the integrity of heritage places.

continued...

Box 12.2 continued

Service Line 4: Visitor Services

Planned Result – Visitors are welcomed, have safe visits, are satisfied with
service quality.
Performance Expectations – 10% increase in the number of visits to targeted
national historic sites by March 2008; 85% of visitors are satisfied and 50%
are very satisfied with their visit; minimize public safety incidents.

Source: Parks Agency Canada (2004).

Establishing strategic direction

Based on the strategy analysis, it becomes necessary to establish goals, objectives
and strategies that best suit the organisation. 'Strategic objectives should be derived
directly from an analysis of the internal and external environment and the
requirements of key stakeholders' (Viljoen 1994: 304). These objectives are tied
to Key Result Areas (KRAs) where actions are required. 'Each KRA becomes the
focus of the subsequent strategic management processes of strategy choice and
implementation' (Viljoen 1994: 41). If the organisation is a local government
agency, it may well be influenced more by social equity considerations – the
provision of a range of recreational activities accessible to a broad sector of the
community – than financial returns in recreation plans and programmes. Thus, an
equitable distribution of accessible recreation space may be the overriding strategic
objective.

Strategic choice

Strategic choice:

> involves the generation, evaluation and choice of a strategy that best suits the
> needs of the organisation. This process must be built on the previous phases
> of strategy analysis and direction setting. Generating strategic alternatives is
> essentially a brainstorming exercise where alternatives are identified and
> described without being evaluated.
>
> (Viljoen 1994: 41)

The SWOT (strengths, weaknesses, opportunities, threats) analysis (also see
Table 12.1) may be revisited at this or any other point. Utilising the outcomes of
the brainstorming session, and based on the SWOT analysis, choices may be made
about which recreational opportunities will be supplied, in what quantity they
will be supplied, and how and when they will be supplied.

Table 12.1 Strategic analysis

Type of analysis	Information to be gathered
Aspirations analysis	*Determine* stakeholders (i.e. those with an interest in the recreational activity or product, e.g. staff of the management agency, service providers, users), their power bases, and their possible reactions to particular strategies, the effects of their reactions, and means of shaping strategies to account for their desires.
Environment nature analysis	*Examine* the complexity of the planning environment (e.g. the number and nature of agencies involved in managing a multiple use resource or competing for use of a resource). *Examine* how the recreational planning environment interacts with other planning environments (e.g. recreation planning in environmentally sensitive areas or in urban environments). *Examine* flexibility in recreational choice or recreational supply (e.g. potentials for substitutability, degrees of recreational specialisation required).
Analysis of the structure of the environment	*Analyse* potential and actual markets (e.g. market segmentation analysis; assessment of latent demand for a recreational activity). *Analyse* environmental forces (see Chapter 11 and above) that might impinge on recreation supply and demand.
Resource analysis	*Develop* an inventory of resources (physical, human, systems and intangibles) in current areas of operation (finance, personnel, research and development, marketing, capital).
SWOT analysis	*Identify* significant opportunities for, and threats to, the supply of outdoor recreation opportunities (e.g. improved water treatment facilities for dams so that a wider range of recreational activities can be developed, or opposition from conservationists to recreation activities in sensitive/protected areas, respectively). *Identify* specific strengths (e.g. large land and water base, strong political support, sound financial backing, strong leadership and community backing) and weaknesses (e.g. very small land base, shrinking financial and other resources, weak community and political support) in the organisation's ability to supply recreational opportunities.

Source: Adapted from Viljoen (1994) and Hall and McArthur (1996).

Strategy implementation, monitoring and evaluation

Plans are useless by themselves. Once decisions have been reached on the elements to be included in the recreation plan or programme, and the strategies and actions to be pursued, the implementation stage has been reached. Strategy implementation concerns the operational strategies and systems that must be used to put the strategy into practice. It involves developing appropriate management, financial and communication systems; acquiring and utilising relevant human, physical, financial and other resources; developing an appropriate organisational structure; and ensuring that the organisational culture of the organisation is managed in a way that complements the organisation's tasks. These are interrelated activities. For instance, an organisation may acquire new people, whose ideas are innovative to the extent that new recreation supply opportunities or management techniques or pricing structures are identified. If these ideas are subjected to scrutiny in a strategic manner and then implemented, the nature of recreation demand may change, thereby perhaps bringing further alterations to existing structures, practices and strategies. In implementing plans and programmes, a number of techniques have been employed to keep the plan and its integral components (e.g. budget, resources and timing) on track. These techniques include: programme evaluation review technique (PERT), critical path analysis, milestone scheduling, and Goals Achievement Matrices.

Implementation of a plan is not an end in itself. If a plan or programme is to satisfy the needs of users for whom it was designed, it must be subjected to monitoring and evaluation. Monitoring is concerned with the collection of information about the developing state of a system to which any planning and management process is being addressed (see Haynes 1973). Evaluation is an activity designed to collect, analyse and interpret information, concerning the need for particular policies, plans or programmes, as well as the formulation, implementation, outcomes and impacts of policies, plans or programmes. The days of 'finishing the Master Plan', which then remained unaltered over time, have hopefully gone. The plan or programme should not be something set in concrete, but something to be worked and reviewed, and further refined and adapted, as needed, during implementation. The analysis of issues and best courses of action should not stop once the plan is in place, but remain a constant approach. In brief:

> … the effectiveness of a planning system must be judged by its continuing ability to influence change toward desired ends, and in its responsiveness to pressures to alter those ends in conformity with societal goals.
>
> (Haynes 1973: 5)

The general roles of, and reasons for, monitoring and evaluation in strategic planning include:

*   assessing the degree of need for particular plans;
*   continuous functioning of the planning process to enlighten, clarify and improve plans;

- conceptual and operational assistance to planners and decision-makers;
- specification of plan outcomes and impacts;
- assessing or measuring the efficiency and effectiveness of recreation plans in terms of resources;
- accountability reporting for resource allocation, distribution and redistribution;
- symbolic reasons – to demonstrate that something is being done;
- political reasons – the evaluation is directed and run in a way that will deliberately dictate the findings (e.g. evaluations may be motivated by self-service as well as public service, or by a desire to use analysis as ammunition for partisan political purposes) (see Hall and Jenkins 1995).

By way of example, Shivers (1967) identified several purposes of evaluation with respect to recreation plans and programmes (see Table 12.2).

Finally, two main forms of evaluation have been noted. *Formative evaluation*

Table 12.2 The purposes of evaluating a public recreation programme

1	To ensure that the recreation programme meets the stated needs and desires of the people in the community.
2	To promote professional growth and education among staff members of the recreation service system.
3	To ascertain the flexibility of policies within the system.
4	To appraise personnel quality and qualifications in relation to specific functions within the system.
5	To develop firmer grounds of agency philosophy so that a logical frame of reference is developed.
6	To effectively gauge public sector sentiment, attitudes and awareness of the recreation system.
7	To increase knowledge gained through practice and to additionally test current practice as to applicability in the public recreation setting.
8	To appraise existing facilities, physical property and plant as to their adequacy, accessibility, safety, attractiveness, appropriateness and utilisation.
9	To seek out and eliminate any detrimental features within the programme or agency.
10	To add any feasible and constructive devices, methods and experiences to the system in order to provide the most efficient and effective service to the community.
11	To promote recognition of the agency on the part of the community.
12	To replace outmoded concepts and invalid ideas which the public may have concerning the recreation agency.
13	As far as possible, to promote the professionalisation of agency personnel and the services provided.
14	To avoid unnecessary expenditure of public monies because of inadequate coordination in the provision of recreation services.
15	To ensure the agency and its personnel safeguard against political upheaval and partisan politics.
16	To ensure the adequate provision of spaces, areas and facilities will be safeguarded against any encroachment by establishing protection in perpetuity through dedication of all physical property for public recreation purposes only.

Source: Shivers (1967).

is periodic monitoring as the plan or programme is implemented, with a view to making necessary changes and to 'fine tuning' the plan. *Summative evaluation* is the assessment of the plan after it has been completed. Ideally, formative evaluations and, ultimately, summative evaluations, should be used.

The remainder of this chapter examines the principles and practices in setting goals and objectives, difficulties in implementing recreation and tourism plans and policies, and a number of broader recreation planning challenges.

Establishing goals and objectives: principles and practices

Conventionally, the planning process can be set out in terms of goals and objectives, which form the basis for programmes and courses of action to achieve the stated objectives, and move closer to attaining the goal(s) specified. As used in this discussion, the term 'goal' refers to a preferred state or condition towards which action is to be directed. An 'objective' is a specific, positive step, attainable as part of progress towards a particular goal. Goals are long-range targets, which give purpose to the planning process; objectives represent a yardstick by which achievement may be measured. The explicit identification of goals and objectives allows the definition of criteria to identify and evaluate alternative strategies, proposals and policies. It also helps determine the urgency of issues and problems, and the setting of priorities.

In the context of recreation planning, goals may be identified broadly with quantitative and qualitative advances in the availability of a range of opportunities for participation in recreation. The realisation of such goals should be in harmony with the maintenance, enhancement and, where necessary, restoration of desirable features of the physical and social environment. These goals may then be translated into attainable, quantifiable objectives, oriented towards specific aspects of recreation management and linked to designated policies, programmes and courses of action to achieve the stated objectives.

In the planning of urban recreation space, for example, a goal might be:

> to provide the community with the widest choice and maximum diversity of recreational opportunities consistent with economic feasibility, the expressed needs of the population, and societal goals.

Examples of objectives derived from this goal include:

- to provide a 'defined quantity' of open space of specified types per head of population;
- to provide a hierarchy of parks of varying size and distribution across a region;
- to provide for multiple use of specified areas of open space;
- to locate various quantities and types of open space in a given proximity to residential areas;
- to link open space into a linear network;
- to allow for conservation of specified areas of natural bushland.

An example of a coastal region's goals, objectives and strategies to address increasing environmental stress from tourism is provided in Table 12.3.

Table 12.3 Strategies for Germany's Baltic Coast (Rügen)

Tourism has been a major source of income for the north coast of Germany for many decades. The most important tourist resorts on the southern Baltic of Germany are Schleswig-Holstein, Fehmarn Island, the coast of Mecklenburg-Vorpommern and the island world off the shallow coast, especially Rügen.

Increasing visitor numbers have led to increasing environmental stress. According to Helfer (1993), tourism and leisure-time infrastructure on the Isle of Rügen have had a damaging impact particularly on the forests protecting the coast (namely, exposing, damaging and destroying tree roots; damage to trunks and branches; destruction of ground vegetation from trampling and driving on it; soil erosion and contamination; moderate to strong thinning out of the coastal-protective forest for car parks and paths, and clearings for activities).

A development plan at the county and regional level was developed in 1990 and revised in 1991. That plan includes guidelines for tourism and recreation development. In the area of tourism and recreation, the general goal is: to develop the area for the improvement of social, economic and environmental living conditions for the populace of and visitors to the county [of Rügen]. Of particular importance for this goal are the following points, which amount to strategic objectives for Key Result Areas:

- the securing of jobs;
- the preservation and development of the characteristic structure of settlements and the typical landscape; and
- the preservation and development of natural values and functions.

In particular, a considerable increase in the quality of the tourism sector is planned, in which the scenic and cultural aspects of Rügen and its environs will serve to develop a characteristic range of touristic and recreational attractions.

For the coastal area the 'zone ordinance' of the structural plan makes the following stipulations, some of which are more specific than strategic objectives and which amount to a combination of operational objectives, performance targets and strategy selection:

- in the entire shore zone, which is at least 100m wide, tourist use must take place only on the basis of the building plans;
- a strip of shoreline 100m wide is to be kept free of construction;
- on the long, drawn-out beaches of the Schaabe and between Gohren and Thiessow, the hinterland is to be kept free of construction and car parks as far as possible, to protect the landscape; access is to be secured by means of public transportation;
- the characteristic of the resorts as spas and medicinal baths should be built up again and developed in connection with improving the environmental situation;
- camping grounds are to be removed from the forests protecting the coasts;
- solid buildings which have been built without pertinent regulations should be examined as to their compatibility with the structure of settlements and landscape and removed if need be;
- aquatic sports, particularly pleasure boats, must be offered suitable facilities, without constructing unprofitable overcapacities, e.g. at moorings; in protected areas and shore zones aquatic sports are to be restricted;
- sports which are detrimental to nature-compatible recreation in the landscape are to be prohibited.

Source: German Federal Agency for Nature Conservation (1997: 184).

Based on the objectives that are set, policies and programmes can be formulated and courses of action specified, as essential inputs to the implementation phase of the planning process. These could include:

- inventory of existing supply of urban recreation space and its characteristics, distribution and use, relative to identified and predicted needs;
- identification of potential areas of open space to complement existing recreation space;
- determination of priorities for implementation;
- preparation of site development plans and management programmes for existing recreation space and areas to be acquired; and
- monitoring and review of the effectiveness of the planning process in addressing the changing recreation space needs of a dynamic urban system.

Strategic planning provides a framework for a very wide-ranging, yet integrated approach to recreation planning. The emphasis on provision of choice and diversity in recreation opportunities, tempered by relevance and realism, should ensure an effective contribution to the long-range goals of human welfare and environmental enhancement, with respect to recreation supply.

At the same time, specifying the goals and objectives for the planned use of resources is a demanding task. By definition, the goals adopted should be broad, comprehensive and long-range. However, care is needed to ensure that they do not become merely 'parenthood-type' statements, of little practical relevance or application. Nor should they be geared to an excessively narrow perception of the future, but be sufficiently flexible to accommodate new concepts and information as they emerge.

Objectives, as defined above, represent progressive steps towards goal attainment. Although the distinction is not always clear-cut, objectives may generally be expressed in more tangible terms. It is important, too, that the objectives delineated are not restricted only to recognised and familiar strategies. The kind of planning challenges identified in earlier chapters call for a strategic approach, incorporating a range of possible options, so that an appropriate response can be made as and when circumstances change.

It is important to note, also, that recreation planning, despite its importance, is only one component in an array of means of achieving the type of society and environment to which a nation and its people aspire. Clearly, the recreation planning process must be consistent with planning policies in associated areas of public and private sector responsibility. Therefore, goals and objectives related to recreation planning should take account of, and complement, those perceived to be worth pursuing by society as a whole. Taking those goals and objectives to fruition brings the focus to the implementation phase.

The implementation phase

Any number of examples exist of detailed plans for recreation, prepared by both public agencies and private interests, which have had no practical outcome. Apparently, a gulf can exist between plan formulation and implementation. Even when plans are approved and adopted, formidable barriers can arise which frustrate attempts to translate them into action. Elements of a plan may be discarded or amended, so that progress towards the achievement of objectives and, ultimately, goals, is interrupted and perhaps stalled altogether.

An implementation gap has been identified across a range of planning initiatives (Pigram 1992). The fact that it is often encountered in planning for recreation and tourism, probably reflects the relative lack of urgency and lower priority given to these activities by decision-makers, compared with those perceived as more fundamental to social and economic development. Impediments that stand in the way of progress towards realisation of planning outcomes determine the extent of the implementation gap. Bridging this gap is essentially a balancing exercise between political and social acceptability, economic and technological feasibility, and administrative reality (Pigram 1992).

Implementation of recreation plans and policies becomes increasingly more difficult when it crosses different tiers of responsibility, especially where both the public and private sector are involved. Studies examining the barriers that inhibit implementation of recreation plans at the local government level reveal the following constraints (Reid 1989):

* Inadequate funding;
* Lack of skilled personnel;
* Need for structural and operational change within city administration to accommodate planning for recreation;
* Environmental conditions – physical, socioeconomic and political;
* Complexity of the plan itself;
* Lack of opportunity for community endorsement and participation in the planning process and the implementation phase.

Overcoming these barriers is central to recreation planning, and to the successful culmination of the implementation phase.

In summary, strategic planning is a continuous process, whereby resource analysis, developing strategic directions, choosing between potential strategies, implementing strategies, and monitoring and evaluating the processes, outcomes and impacts of decisions and actions are ongoing. Planners and managers may even move backwards and forwards among these interrelated processes, because of changing circumstances. For example, the acceptance and facilitation of recreation activities in protected areas may be significantly influenced by political forces (e.g. changes in government), while in some urban areas waterfront development, shopping malls, gambling facilities, and festivals and events, appear to be the 'flavours of the month' in terms of the commitment of government resources.

The challenge of recreation planning

Most recreation experiences do not just happen; they have to be provided for in some way. Earlier, it was noted how the availability of recreation opportunities, services and facilities influences choice in outdoor recreation. It is in the expansion of choice, through provision of a diversity of outlets for leisure to meet the many aspirations of people and society, that planning plays an essential role. By providing a wider range of alternative recreation opportunities, the planner is contributing to the potential of leisure to stimulate and satisfy.

In short, planning for leisure environments of the future must progress beyond establishing a series of services or facilities, such as parks and playgrounds. The challenge is to create a physical and social environment in which individuals can satisfy their recreation interests within the economic limitations and resource constraints likely to be encountered. The emphasis is on recognising the multiplicity of individual and societal goals for leisure, and on the need for diversity, substitutability and choice, rather than uniformity, in addressing those goals. The recreation planner's concern, then, is with generating an appropriate array of leisure opportunities, rather than with provision of specific facilities alone. It is the interaction of people's values, needs and wants with those facilities and services, which generates leisure opportunities, and, ultimately, leads to participation and satisfaction – the end-products of the planning process.

Recreation planning is complex, partly because of the unstructured nature of recreation itself and the many conflicting interests and constituencies which have to be catered for, and also because of the rapidly changing context in which planning takes place (Mitchell 1983). In meeting the challenge of recreation planning, a number of aspects need to be considered – some limiting, and others with potential to contribute to positive outcomes. In particular:

1   The apparent inevitability of dwindling public sector support for provision of opportunities for recreation, and the consequent need to build partnerships with private enterprise, and to harness the promise of self-help schemes and voluntarism.
2   The need to plan within the capability of the resource base, the supporting infrastructure, and the thresholds of tolerance of affected communities, while, simultaneously, applying evolving technologies to expand these constraining horizons.
3   The need to recognise the plurality of the 'market' for recreation planning; to build in diversity and flexibility to accommodate change and compensate for equity deficiencies; and to use recreation opportunities, where possible, to offset negative social forces.
4   The adoption of an integrated perspective. Recreation planning would be one important component of overall planning for community welfare and environmental integrity, based on strategic management frameworks, and encompassing appropriate recreation planning frameworks such as the Recreation Opportunity Spectrum, the Limits of Acceptable Change, and the Visitor Impact Management framework and Visitor Activity Management Process.

5 The blending of 'bottom-up' responses from the participating public in the recreation planning process with balanced 'top-down' assessments from business interests and professional advisers and policy-makers.
6 Education about leisure to facilitate individual self-determination and freedom of choice.

Summary

There are considerable challenges to using leisure to find fulfilment and satisfaction in the true meaning of the term, 'recreation'. In an increasingly complex world, the task of creating and enhancing meaningful leisure environments and recreation opportunities becomes a most pressing issue. Failure to meet the challenge means that society must accept and tolerate existing constraints on leisure behaviour, and continue to condone the legacy of *ad hoc* development, out of balance with community needs and inevitably perpetuating the deficiencies of an inadequate system of provision for recreation (Pigram, 1983). Moreover, without planning, there is little prospect of recreation receiving proper priority in resource allocation, against the claims of competing uses.

Guide to further reading

- For detailed discussions of strategic planning and management: Viljoen (1994); Richardson and Richardson (1994); Mintzberg (1994).
- For detailed discussions of sustainable recreation and tourism planning and development: Pigram (1993); Hall (1995); Harrison and Husbands (1996); Butler *et al.* (1998); Mercer (2000); Broadhurst (2001); McCool and Moisey (2001); Weaver and Lawton (2002).
- For broad overviews of recreation and plan and programme evaluation: Theobald (1979); Hollick (1993); Howe (1980, 1993); Owen (1993); Rossman (1995). For triangulation in recreation evaluation, see Fuszek (1987, in Rossman 1995).
- Leisure policy and planning: Henry (1993); Veal (1994, 2002).
- Outdoor recreation policy and planning: Driver (1970b); LaPage (1970); Hutcheson *et al.* (1990); Mercer (1981b, 1994b); Hammitt and Cole (1998).
- Tourism policy: Pigram (1992); Hall and Jenkins (1995); Mercer (2000); Dredge and Jenkins (forthcoming, 2006).
- Internet: Laudon and Laudon (2002); Buhalis (2003).

Review questions

1 Why should we plan for outdoor recreation?
2 What are the main responsibilities of the public sector in outdoor recreation planning in your local area? Is the public sector fulfilling its responsibilities in that regard? Justify your answers using primary and secondary information sources.

3 Define planning. Define strategic planning. What are the main steps or features in strategic planning? Discuss the benefits and problems associated with strategic planning approaches generally, and with reference to recreation and tourism specifically.

4 Identify important factors that will affect future demand for outdoor recreation activities in your local community or region.

5 Identify important factors that will affect future supply for outdoor recreation activities in your local community or region.

6 How can planners best respond to uncertain futures in outdoor recreation demand and supply?

Bibliography

ABS *see* Australian Bureau of Statistics

Absher, J. and Brake, L. (1996) 'Aboriginal involvement in park management in Australia', *Trends*, 33, 4: 9–15.

Adelman, B.J. Eizen, Herberlein, T.A. and Bronnicksen, T.M. (1986) 'Social psychological explanations for the persistence of a conflict between paddling canoeists and motorcraft users in the Boundary Waters Canoe Area', *Leisure Sciences*, 5, 1: 45–61.

Aitchison, C. (2000) 'Young disabled people and everyday life: Reviewing conventional definitions for leisure studies', *Annals of Leisure Research*, 3: 1–20.

Aitchison, C., Macleod, N. and Shaw, S. (2000) *Leisure and Tourism Landscapes: Social and Cultural Geographies*, London: Routledge.

Aitken, S. (1991) 'Person-environment theories in contemporary perceptual and behavioural geography: personality, attitudinal and spatial choice theories', *Progress in Human Geography*, 15, 2: 179–93.

Alderson, W.T. and Low, S.P. (1996) *Interpretation of Historic Sites*, 2nd edn, Walnut Creek, CA: Sage Publications.

Allen, L. (1996) 'A primer: benefits-based management of recreation services', *Parks and Recreation*, March: 64–76.

Amato, P.R. (1997) 'Contact with non-custodial fathers and children's wellbeing', *Family Matters*, 36: 32–4.

Annals of Leisure Research (2004) 7: 1–2.

Archer, B. (1973) 'The uses and abuses of multipliers', *Tourist Research Paper TUR 1*, Bangor: University College of North Wales.

Armitage, J. (1977) *Man at Play*, London: Warne.

Ashley, R. (1990) 'The Visitor Activity Management Process and Canadian national historic parks and sites – a new commitment to the visitor', in Graham, R. and Lawrence, R. (eds) *Towards Serving Visitors and Managing Our Resources*, Proceedings of a North American Workshop on Visitor Management in Parks and Protected Areas, Waterloo: Tourism Research and Education Centre, University of Waterloo, pp. 249–56.

Ashworth, G.J. (1989) 'Urban tourism: an imbalance in attention', in Cooper, C.P. (ed.) *Progress in Tourism, Recreation and Hospitality Management*, 1: 33–54, London: Belhaven.

Ashworth, G.J. (1992) 'Planning for sustainable tourism: slogan or reality?', *Town Planning Review* 63, 3: 325–30.

Ashworth, G.J. and Tunbridge, J. (1990) *The Tourist-Historic City*, London: Belhaven.

Atkinson, J. (1991) *Recreation in the Aboriginal Community*, Canberra: Australian Government Publishing Service.

Australian Bureau of Statistics (ABS) (1997) *Family Characteristics*, Canberra: ABS.

Australian Bureau of Statistics (ABS) (2000) *Household Expenditure Survey*, Canberra: ABS,

Australian Bureau of Statistics (ABS) (2002) *Participation in Sport and Physical Activities*, Canberra: ABS (Cat. No. 4177.0).

Australian Bureau of Statistics (ABS) (2003) *Disability, Ageing and Carers: Summary of Findings*, Canberra: ABS (Cat. No. 4430).

Australian Bureau of Statistics (ABS) (2005a) *Year Book Australia*, Canberra: ABS (Cat. No. 1301.0).

Australian Bureau of Statistics (ABS) (2005b) *Australian National Accounts: Tourism Satellite Account* http://www.abs.gov.au/Ausstats/abs@.nsf/0/e2fcaf5e0fce6d12ca256d3b 0080a08c?OpenDocument (Cat. No. 52490), accessed 10 August 2005.

Bahaire, T. and Elliott-White, M. (1999) 'The application of geographical information systems (GIS) in sustainable tourism planning: a review', *Journal of Sustainable Tourism*, 7, 2: 159–74.

Bailey, P. (1978) *Leisure and Class in Victorian England: Rational Recreation and Contest for Control 1830–1885*, London: Routledge and Kegan Paul.

Bailey, T. and Gattrell, A. (1995) *Interactive Spatial Data Analysis*, London: Longman.

Ball, A. and Ball, B. (1996) *Basic Camp Management: An Introduction to Camp American Camping Association Administration*, Martinsville: American Camping Association.

Bammel, G. and Bammel, L. (1992) *Leisure and Human Behaviour*, Dubuque: Brown.

Bannon, J. (1976) *Leisure Resources*, Englewood Cliffs, NJ: Prentice Hall.

Barbier (1987) 'The concept of sustainable economic development', *Environmental Conservation*, 14, 2: 101–10.

Barker, R. (1968) *Ecological Psychology*, Stanford, CA: Stanford University Press.

Baron, R.V. (1975) *Seasonality in Tourism*, London: Economist Intelligence Unit.

Barrett, J. (1958) 'The seaside resort towns of England and Wales', unpublished PhD thesis, University of London, London.

Barrett, S. and Hough, M. (1989) 'Rethinking city landscapes', *Recreation Research Review*, 14, 3: 7–13.

Bateson, P., Wyman, S. and Sheppard, D. (eds) (1989) *National Parks and Tourism Seminar*, Sydney: NSW National Parks and Wildlife Service.

Bauer, M. (2004) 'Recreation conflict at six Boulder County Parks and open space properties: a baseline study', at http://www.co.boulder.co.us/openspace/recreating/public_parks/ user_study.htm, accessed 4 February 2005.

Beatty, J.E. and Torbert, W.R. (2003) 'The false duality of work and leisure', *Journal of Management Inquiry*, 12, 3: 239.

Beaulieu, J. and Schreyer, R. (1985) 'Choices of wilderness environments – differences between real and hypothetical choice situations', *Proceedings – Symposium on Recreation Choice Behaviour*, Ogden, UT: USDA Forest Service General Technical Report, INT184: 38–45.

Beckman, E.A. and Russell, R. (eds) (1995) 'Interpretation and the getting of wisdom', papers from the Fourth Annual Conference of the Interpretation Australia Association, Canberra: Australian Nature Conservation Agency and Australian Heritage Commission.

Bengtsson, A. (ed.) (1972) *Adventure Playgrounds*, New York: Praeger.

Bialeschki, M.D. and Hicks, H. (1998) 'I refuse to live in fear: the influence of violence on women's outdoor recreation activities', at http://www.unc.edu/depts/recreate/deb/ LSAfear.htm, accessed 25 February 2005.

Biddlecomb, C. (1981) *Pacific Tourism, Contrasts in Values and Expectations*, Suva: Pacific Conference of Churches.

Black, A. and Breckwoldt, R. (1977) 'Evolution of systems of national park policy-making in Australia', in Mercer, D.C. (ed.) *Leisure and Recreation in Australian*, Malvern: Sorrett, pp. 190–9.

Black. R. and Weiler, B. (2003) *Interpreting the Land Down Under: Australian Heritage Interpretation and Tour Guiding*, Golden, CO: Fulcrum.

Blair, A. and Hitchcock, D. (2001) *Environment and Business*, London: Routledge.

Blenkhorn, A. (1979) 'The attraction of water', *Parks and Recreation*, 44, 2: 17–23.

Blockley, M. (1996) 'Editorial: rights of access', *Interpretation*, 1, 2: 3–4.

Blood, R.O. and Wolfe, D.M. (1960) *Husbands and Wives*, Glencoe, IL: Free Press.

Boden, R. (1977) 'Ecological aspects of outdoor recreational planning', in Mercer, D. (ed.) *Leisure and Recreation in Australia*, Melbourne: Sorrett, pp. 222–31.

Boden, R. and Baines, G. (1981) 'National parks in Australia – origins and future trends', in Mercer, D. (ed.) *Outdoor Recreation: Australian Perspectives*, Malvern: Sorrett, pp. 148–55.

Bolman, L.G. and Deal, T.E. (1991) *Reframing Organisations: Artistry, Choice and Leadership*, San Francisco, CA: Jossey Bass.

Boniface, B. and Cooper, C. (1987) *The Geography of Travel and Tourism*, London: Heinemann.

Bonn, M.A., Furr, H.L. and Uysal, M. (1992) 'Seasonal variation of coastal resort visitors: Hilton Head Island', *Journal of Travel Research*, 31, 1: 50–6.

Boo, E. (1990) *Ecotourism: The Potentials and Pitfalls*, 2 vols, Washington, DC: World Wildlife Fund.

Borowski, A. (1990) *Australia's Population Trends and Prospects*, Canberra: Bureau of Immigration Research.

Bowler, I.R., Bryant, C.R. and Nellis, M.D. (eds) (1992) *Contemporary Rural Systems in Transition: Economy and Society, Vol. 1, Agriculture and Environment*, Wallingford: CABI.

Boyle, R. (1983) 'A survey of the use of small parks', *Australian Parks and Recreation*, November: 31–6.

Bradshaw, J. (1972) 'The concept of social need', *New Society*, 496, 30 March: 640–3.

Bramwell, B. and Lane, B. (1993) 'Sustainable tourism: an evolving global approach', *Journal of Sustainable Tourism*, 1, 1: 6–16.

Braverman, H. (1975) *Labor and Monopoly Capital; the Degradation of Work in the Twentieth Century*, New York: Monthly Review Press

Breckwoldt, R. (1983) *Wildlife in the Home Paddock*, North Ryde: Angus and Robertson.

Britton, N. (1991) 'The culture of emergency services: some relevant considerations', paper prepared for the Emergency Service Librarians Workshop, Australian Counter-Disaster College, Mt Macedon, Victoria, 8–13 September.

Britton, R. (1979) 'Some notes on the geography of tourism', *Canadian Geographer*, 23, 3: 276–82.

Britton, S. (1980) 'A conceptual model of tourism in a peripheral economy', in Pearce, D. (ed.) *Tourism in the Pacific*, Rarotonga: Proceedings of UNESCO Tourism Workshop, pp. 1–12.

Britton, S. (1982) 'The political economy of tourism in the Third World', *Annals of Tourism Research*, 9: 331–58.

Broadhurst, R. (2001) *Managing Environments for Leisure and Recreation*, London: Routledge.

Brotherton, D. (1973) 'The concept of carrying of countryside recreation areas', *Recreation News Supplement*, 9: 6–11.

Brown, K., Turner, R.K., Hameed, H. and Bateman, I. (1998) 'Comment – reply to Lindberg and McCool: "A critique of environmental carrying capacity as a means of managing the effects of tourism development"', *Environmental Conservation*, 25, 4: 293–4.

Brown, P. (1977) 'Information needs for river recreation planning and management', in *Proceedings, River Recreation Management Research Symposium*, USDA Forest Service General Technical Report NC-28, Minneapolis, MN: USDA Forest Service, pp. 193–201.

Brunson, M.W. (1997) 'Beyond wilderness: broadening the applicability of limits of acceptable change', in McCool, S.F. and Cole, D.N. (eds) *Proceedings – Limits of Acceptable Change and Related Planning Processes: Progress and Future Directions*, University of Montana's Lubrecht Experimental Forest, Missoula, MT, 20–22 May, Ogden, UT: USDA Forest Service, Rocky Mountain Research Station, pp. 44–8.

Buckley, R. (1996) 'Sustainable tourism: technical issues and information needs', *Annals of Tourism Research* 23: 925–8.

Buckley, R. (2000) 'Tourism in the most fragile environments', *Tourism Recreation Research*, 25, 1: 31–40.

Buckley, R. (2001) 'Environmental impacts', in Weaver, D. (ed.) *The Encyclopaedia of Ecotourism*, Wallingford: CABI.

Buckley, R. (2002) 'Tourism ecolabels', *Annals of Tourism Research*, 29, 1: 183–208.

Buckley, R. (2003a) 'Environmental assessment', in Jenkins, J.M. and Pigram, J.J. (eds) *Encyclopedia of Leisure and Outdoor Recreation,* London: Routledge, pp. 149–50.

Buckley, R. (2003b) 'Pay to play in parks: an Australian policy perspective on visitor fees in public protected areas', *Journal of Sustainable Tourism,* 11, 1: 56–73.

Buckley, R. (2003c) 'Trampling', in Jenkins, J.M. and Pigram, J.J. (eds) *Encyclopedia of Leisure and Outdoor Recreation,* London: Routledge, pp. 510–11.

Buckley, R. (2003d) *Case Studies in Ecotourism,* Wallingford: CABI.

Buckley, R. (2004) *Tourism in Parks: Australian Initiatives,* Gold Coast, QLD: Griffith University, International Centre for Ecotourism Research.

Buckley, R. (2004a) 'Environmental impacts of motorized off-highway vehicles', in Buckley, R. (ed) *Environmental Impacts of Ecotourism,* Wallingford: CABI, pp. 83–97.

Buckley, R. (2004b) 'Impacts of ecotourism on birds', in Buckley, R. (ed.) *Environmental Impacts of Ecotourism,* Wallingford: CABI, pp. 187–209.

Buckley, R. (2004c) 'Impacts of ecotourism on terrestrial wildlife', in Buckley, R. (ed) *Environmental Impacts of Ecotourism,* Wallingford: CABI, pp. 211–28.

Buckley, R. (2005) 'Recreation ecology research effort: an international comparison', *Tourism Recreation Research,* 30, 1: 99–101.

Buckley, R. and Pannell, J. (1990) 'Environmental impacts of tourism and recreation in national parks and conservation reserves', *Journal of Tourism Studies,* 1, 1: 24–32.

Budowski, G. (1976) 'Tourism and conservation: conflict, coexistence, or symbiosis?', *Environmental Conservation,* 3, 1: 27–31.

Buhalis, D. (2003) 'Internet', in Jenkins, J.M. and Pigram, J.J. (eds) *Encyclopedia of Leisure and Outdoor Recreation,* London: Routledge, pp. 264–6.

Bull, A. (1995) *The Economics of Travel and Tourism,* 2nd edn, Melbourne: Longman.

Burnett, J.J. and Bender Baker, H. (2001) 'Assessing the travel related behaviours of the mobility-disabled consumer', *Journal of Travel Research,* 40, 3: 4–11.

Burrough, P. (1986) *Principles of Geographic Information Systems for Land Resources Assessment,* Oxford: Clarendon Press.

Burton, J. (1975) *The Recreational Use of Malpas Reservoir,* Armidale, NSW: University of New England.

Bury, R. (1976) 'Recreation carrying capacity – hypothesis or reality', *Parks and Recreation,* 11: 22–5, 56–8.

Butler, R.W. (1975) 'Tourism as an agent of social change', in *Proceedings of IGU Working Group on Geography of Tourism and Recreation,* Peterborough: University of Trent, pp. 85–90.

Butler, R.W. (1980) 'The concept of a tourist area cycle of evolution: implications for management of resources', *Canadian Geographer,* 24, 1: 5–12.

Butler, R.W. (1984) 'The impact of informal recreation on rural Canada', in Bunce, M.F. and Troughton, M. J. (eds) *The Pressure of Change in Rural Canada,* Geographical Monograph 14, Downsview, Ontario: York University:.

Butler, R.W. (1990) 'Alternative tourism: pious hope or Trojan horse', *Journal of Travel Research,* 28, 3: 40–5.

Butler, R.W. (1991) 'Tourism, environment, and sustainable development', *Environmental Conservation,* 18, 3: 201–9.

Butler, R.W. (1995) 'Introduction', in Butler, R.W. and Pearce, D.G. (eds) *Change in Tourism: People, Places, Processes,* London: Routledge, pp. 1–11.

Butler, R.W. (1996) 'The concept of carrying capacity for tourist destinations: dead or merely buried?', *Progress in Tourism and Hospitality Research,* 2, 3: 283–92.

Butler, R.W. and Boyd, S.W. (eds) (2000) *Tourism and National Parks: Issues and Implications,* Chichester: John Wiley and Sons.

Butler, R.W. and Clark, G. (1992) 'Tourism in rural areas: Canada and the United Kingdom', in Bowler, I.R., Bryant, C.R. and Nellis, M.D. (eds) *Contemporary Rural Systems in Transition: Economy and Society,* Wallingford: CABI, pp. 161–85.

Butler, R.W. and Mao, B. (1997) 'Seasonality in tourism: problems and measurement', in Murphy, P. (ed.) *Quality Management in Urban Tourism*, Chichester: Wiley, pp. 9–24.

Butler, R.W. and Waldbrook, L.A. (1991) 'A new planning tool: the tourism opportunity spectrum', *Journal of Tourism Studies*, 2, 1: 2–14.

Butler, R.W., Hall, C.M. and Jenkins, J.M. (eds) (1998) *Tourism and Recreation in Rural Areas*, Chichester: John Wiley and Sons.

Calantone, R.J. and Jotindar, J.S. (1984) 'Seasonal segmentation of the tourism market using a benefit segmentation framework', *Journal of Travel Research*, 23, Fall: 14–24.

Calder, J. (1974) 'Recreational accessibility for the handicapped', in Australian Department of Tourism and Recreation, *Leisure A New Perspective*, 7.1–7.4.1, Canberra: Australian Department of Tourism and Recreation.

Caldwell, L.L., Smith, E.A. and Weissinger, E. (1992) 'Development of a leisure experience battery for adolescents: parsimony, stability and validity', *Journal of Leisure Research*, 24, 4: 361–76.

Cambridge City Council, Environment and Planning (2004) 'Cambridge City Council Open Space Standards: guidance for interpretation and implementation', at http://www2.cambridge.gov.uk/planning/reptdocs/open_space_standards_July04.pdf, accessed 6 February 2005.

Campaign to Protect Rural England (CPRE) (2003) *Lie of the Land*, London: CPRE.

Canadian Environmental Advisory Council (1991) *A Protected Areas Vision for Canada*, Ottawa: Environment Canada.

Carmichael, B., McBoyle, G. and Wall, G. (1995) 'Responding to environmental change and variability', in Thomson, J.L., Lime, D.W., Gartner, B. and Sames, W.M. (eds) *Proceedings of the Fourth International Outdoor Recreation and Tourism Trends Symposium and the 1995 National Recreation Resources Planning Conference*, 14–17 May, Minneapolis, MN: University of Minnesota, pp. 513–20.

Carroll, J. (1990) 'Foreword', in Hutcheson, J.D., Noe, F.P. and Snow, R.E. (eds) *Outdoor Recreation Policy: Pleasure and Preservation*, New York: Greenwood Press, pp. xiii–xviii.

Cater, C. and Cater, E. (2001) 'Marine environments', in Weaver, D.B. (ed.) *The Encyclopedia of Ecotourism*, Wallingford: CABI, pp. 265–85.

Cater, E. (1993) 'Ecotourism in the third world: problems for sustainable development', *Tourism Management*, 14, 2: 85–90.

Cecil, J. (2005) 'New AONB Boards', *Countryside Focus*, January: 1.

Chadwick, G. (1971) *A Systems View of Planning*, Oxford: Pergamon Press.

Chang, T.C., Milne, S., Fallon, D. and Pohlmann, C. (1996) 'Urban heritage tourism: the global–local nexus', *Annals of Tourism Research*, 23, 2: 284–305.

Chape, S., Blyth, S., Fish, L., Fox, P. and Spalding, M. (2003) *United National List of Protected Areas,*, Gland and Cambridge: IUCN and WCMC, UNEP.

Charters, T., Gabriel, M. and Prasser, S. (eds) (1996) *National Parks. Private Sector's Role*, Toowoomba: University of Southern Queensland Press.

Cherry, G.E. (1993) 'Changing social attitudes towards leisure and the countryside 1890–1990', in Glyptis, S. (ed.) *Leisure and the Environment*, London: Belhaven Press, pp. 22–33.

Chon, K. (1989) 'Understanding recreational traveler's motivation, attitude and satisfaction', *Tourist Review*, 44, 1: 3–7.

Christaller, W. (1963) 'Some considerations of tourism location in Europe', *Regional Science Association Papers,* XII: 95–105, Lund Congress.

Christiansen, M. (1977) *Park Planning Handbook*, New York: Wiley.

Chubb, M. and Chubb, H. (1981) *One Third of Our Time. An Introduction to Recreation Behavior and Resources*, New York: John Wiley and Sons.

City of Westminster (2005) *Facts and Figures 2004/05*, London: City of Westminster.

Clark, J. and Crichter, C. (1985) *The Devil Makes Work: Leisure in Capitalist Britain*, London: Macmillan.

Clark, R. (1976) *Control of Vandalism in Recreation Areas – Fact, Fiction or Folklore*, USDA Forest Service General Technical Report PSW-17, Portland, OR: USDA Forest Service.

Clark, R. and Stankey, G. (1979) *The Recreation Opportunity Spectrum: A Framework for Planning, Management and Research*, General Technical Report, PNW-98, Seattle, WA: USDA Forest Service.

Clawson, M. and Knetsch, J. (1963) 'Recreation research: some basic analytical concepts and suggested frameworks for recreation research', in *Proceedings of the National Conference on Outdoor Recreation Research*, Ann Arbor, MI: University of Michigan, School of Natural Resources and the US Bureau of Outdoor Recreation, pp. 9–32.

Clawson, M. and Knetsch, J. (1966) *Economics of Outdoor Recreation*, Baltimore, MD: Johns Hopkins Press.

Clawson, M., Held, R. and Stoddard, C. (1960) *Land for the Future*, Baltimore, MD: Johns Hopkins Press.

Cloke, P. (1993) 'The countryside as commodity: new spaces for rural leisure', in Glyptis, S. (ed.) *Leisure and the Environment*, London: Belhaven, pp. 53–67.

Cloke, P. and Goodwin, M. (1992) 'Conceptualising countryside change: from post-Fordism to rural structured coherence', *Transactions of Institute of British Geographers*, 17: 321–36.

Cloke, P. and Goodwin, M. (1993) 'Rural change: structured coherence or unstructured coherence', *Terra*, 105: 166–74.

Cloke, P. and Park, C. (1985) *Rural Resource Management*, London: Croom Helm.

Coalter, F. (1998) 'Leisure studies, leisure policy and social citizenship: the failure of welfare or the limits of welfare?', *Leisure Studies*, 17, 4: 21–36.

Cochrane, J. (1996) 'The sustainability of ecotourism in Indonesia', in Parnwell, M. and Bryant, R. (eds) *Environmental Change in Southeast Asia: People, Politics and Sustainable Development*, London: Routledge, pp. 237–59.

Cohen, E. (1978) 'The impact of tourism on the physical environment', *Annals of Tourism Research*, 2: 215–37.

Colby, K. (1988) 'Public access to private land: allemansrett in Sweden', *Landscape and Urban Planning*, 15: 253–64.

Cole, A. (1977) 'Perception and use of urban parks', in Mercer, D. (ed.) *Leisure and Recreation in Australia*, Melbourne: Sorrett, pp. 89–100.

Cole, D.N. (1995) 'Experimental trampling of vegetation. Relationship between trampling intensity and vegetation response', *Journal of Applied Ecology*, 32: 203–14.

Cole, D.N. (2003) 'Impacts of hiking and camping on soils and vegetation', in Buckley, R.C. (ed) *Environmental Impacts of Ecotourism*, Wallingford: CABI.

Cole, D.N. (2004) 'Impacts of hiking and camping on soils and vegetation: a review', in Buckley, R.C. (ed.) *Environmental Impacts of Ecotourism*, Wallingford: CABI.

Cole, D.N., Petersen, M.E. and Lucas, R.C. (1987) *Managing Wilderness Recreation Use: Common Problems and Potential Solutions*, General Technical Report INT-230, Ogden, UT: USDA Forest Service.

Coleman, D. (1993) 'Leisure based social support, leisure dispositions and health', *Journal of Leisure Research*, 25, 4: 350–61.

Coleman, D. and Iso-Ahola, S.E. (1993) 'Leisure and health: the role of social support and self determination', *Journal of Leisure Research*, 25, 2: 111–28.

Commonwealth Department of the Environment and Heritage http://www.deh.gov.au/parks/uluru/, accessed 2 April 2005.

Connecticut River Joint Commissions (1997) *Connecticut River Corridor Management Plan Volume 1: Riverwide Overview Upper Connecticut River in New Hampshire and Vermont* (original paper version Francis, S.F. and Mulligan, A.D., eds), Connecticut River Joint Commissions, Charlestown, NH, http://www.crjc.org/corridor-plan/plan-TOC.html, accessed 24 January 2005.

Conway, H. (1991) *People's Parks: the Design and Development of Victorian Parks in Britain*, Cambridge: Cambridge University Press.

Cooper, I. and Collins, M.F. (1998) *Leisure Management: Issues and Applications*, Wallingford: CABI.

Coppock, T., Duffield, B. and Sewell, D. (1974) 'Classification and analysis of recreation resources', in Lavery, P. (ed.) *Recreational Geography*, London: David and Charles, pp. 231–58.

Cordell, H.K., McDonald, B.L., Teasley, R.J., Bergstrom, J.C., Martin, J., Bason, J. and Leeworthy, V.R. (1999) 'Outdoor recreation participation trends: national survey on recreation and the environment', in Cordell, H.K. (ed.) *Outdoor Recreation in American Life: A National Assessment of Demand and Supply Trends*, Champaign, IL: Sagamore Publishing, pp. 219–321.

Countryside Agency (2002) *Landscape Character Assessment: Guidance for England and Scotland*, Cheltenham: Countryside Agency.

Countryside Agency (2004) *Countryside Focus*, Countryside Agency, also see http://www.countryside.gov.uk/LAR/Recreation/OpenAccess/index.asp, accessed 1 August 2005.

Countryside Commission (1970) *Countryside Recreation Glossary*, London: Countryside Commission.

Countryside Commission (1979) *Leisure and the Countryside*, Cheltenham: Countryside Commission.

Countryside Commission (1988) *Landscape Assessment of Farmland*, Cheltenham: Countryside Commission.

Countryside Commission (1989) *Forests for the Community*, Cheltenham: Countryside Commission.

Countryside Commission (1991) *Countryside Commission News*, July/August, Cheltenham: Countryside Commission.

Countryside Commission (1992) *City Links with National Parks*, Cheltenham: Countryside Commission.

Countryside Commission (1995) *London's Green Corridors*, Cheltenham: Countryside Commission.

Countryside Commission (1998) *Countryside Research*, January, Cheltenham: Countryside Commission.

Countryside Recreation Network (CRN) (1994) '1993 UK day visits survey: summary', *Countryside Recreation Network News*, 2, 1: 7–12.

Countryside Stewardship Scheme (2005) http://www.defra.gov.uk/erdp/schemes/css/default.htm, accessed 12 February 2005.

Cox, P. (1985) 'The architecture and non architecture of tourism developments', in Dean, J. and Judd, B. (eds) *Tourist Developments in Australia*, Canberra: Royal Australian Institute of Architects Education Division, pp. 46–51.

Craig-Smith, S. and Fagence, M. (eds) (1995) *Recreation and Tourism as a Catalyst for Urban Waterfront Redevelopment*, Westport: Praeger.

Crane, R. (2003) 'The South Downs: portrait of a landscape', *Countryside Voice*, Autumn: 21–4.

Crawford, D.W. and Godbey, G. (1987) 'Reconceptualizing barriers to family leisure', *Leisure Sciences*, 9: 119–27.

Crawford, D.W., Jackson, E.L. and Godbey, G. (1991) 'A hierarchical model of leisure constraints', *Leisure Sciences*, 13: 309–20.

Cross, G. (1990) *A Social History of Leisure Since 1600*, State College, PA: Venture Publishing.

Crouch, D. (ed.) (1999) *Leisure/Tourism Geographies: Practices and Geographical Knowledge*, London: Routledge.

Cullington, J.M. (1981) 'The public use of private land for recreation', unpublished MA thesis, Department of Geography, University of Waterloo, Waterloo, Ontario.

Cunningham, C. and Jones, M. (1987) 'Play needs of pre-adolescent children', paper presented to National Seminar of the Child Accident Prevention Foundation of Australia, Melbourne.

Cunningham, C. and Jones, M. (1994) 'The child-friendly neighbourhood', *International Play Journal*, 2: 79–95.

Cunningham, H. (1980) *Leisure in the Industrial Revolution*, London: Croom Helm.

Curry, N. (1996) 'Access: policy directions for the late 1990s', in Watkins, C. (ed.) *Rights of Way: Policy Culture and Management*, London: Pinter, pp. 24–34.

Curry, N. (2000) 'Community participation in outdoor recreation and the development of Millennium Greens in England', *Leisure Studies*, 19: 1–19.

Curry, N. (2001) 'Access rights for outdoor recreation in New Zealand: some lessons for open country in England and Wales', *Journal of Environmental Management*, 64, 110–23.

Curry, N. and Ravenscroft, N. (2000) 'Assessing the demand for countryside recreation: a case study in the County of Surrey', unpublished report to the Countryside Agency, South East and London Region, and Surrey County Council. Cheltenham: Countryside and Community Research Unit, University of Gloucestershire.

Curry, N. and Ravenscroft, N. (2001) 'Countryside recreation provision in England: exploring a demand-led approach', *Land Use Policy*, 18, 3: 281–91.

Cushman, G. and Hamilton-Smith, E. (1980) 'Equity issues in urban recreation services', in Mercer, D. and Hamilton-Smith, E. (eds) *Recreation Planning and Social Change in Urban Australia*, Melbourne: Sorrett, pp. 167–79.

Cushman, G. and Laidler, A. (1990) *Recreation, Leisure and Social Policy*, Occasional Paper No. 4, Canterbury, New Zealand: Department of Recreation and Tourism, Lincoln University.

Cushman, G., Veal, A.J. and Zuzanek, J. (eds) (1996a) 'Cross-national leisure participation research: a future', in Cushman, G., Veal, A.J. and Zuzanek, J. (eds) *World Leisure Participation: Free Time in the Global Village*, Wallingford: CABI, pp. 237–58.

Cushman, G., Veal, A.J. and Zuzanek, J. (eds) (1996b) *World Leisure Participation: Free Time in the Global Village*, Wallingford: CABI.

Cutter, S.L. and Renwick, W.H. (1999) *Exploitation, Conservation, Preservation: A Geographic Perspective on Natural Resource Use*, 3rd edn, New York: John Wiley and Sons.

Dales, J. (1972) 'The property interface', in Dorfman, R. and Dorfman, N. (eds) *Economies of the Environment*, New York: Norton, pp. 308–22.

Daly, J. (1987) *Decisions and Disasters: Alienation of the Adelaide Parklands*, Adelaide: Bland House.

Dann, G. (1976) 'Anomie, ego-enhancement and tourism', *Annals of Tourism Research*, 4, 4: 184–94.

Dann, G. (1981) 'Tourist motivation: an appraisal', *Annals of Tourism Research*, 8: 187–219.

Dann, G. (2000) 'Motivation', in Jafari, J. (ed.) *Encyclopedia of Tourism*, London: Routledge, pp. 393–5.

Darcy, S. (2003) 'Disability', in Jenkins, J.M. and Pigram, J.J. (eds) *Encyclopedia of Leisure and Outdoor Recreation*, London: Routledge, pp. 114–18.

Darling, F.F. and Eichhorn, N.D. (1967) 'The ecological implications of tourism in national parks', in *Ecological Impact of Recreation and Tourism Upon Temperate Environments*, IUCN Proceedings and Papers, New Series 7: 98–101, Morges, Switzerland.

Dasmann, R.F. (1985) 'Achieving the sustainable use of species and ecosystems', *Landscape Planning*, 12: 211–19.

Davidson, J. and Wibberley, G. (1977) *Planning and the Rural Environment*, Oxford: Pergamon.

Davidson, P. (1996) 'The holiday and work experiences of women with young children', *Leisure Studies*, 15: 89–103.

Davis, D., Banks, S., Birtles, A., Valentine, P. and Cuthill, M. (1997) 'Whale sharks in Ningaloo Marine Park: managing tourism in an Australian marine protected area', *Tourism Management*, 18, 5: 259–71.

de Kadt, E. (1979) *Tourism. Passport to Development*, New York: Oxford University Press.

Dearden, P. and Rollins, R. (eds) (1993) *Parks and Protected Areas in Canada: Planning and Management*, Toronto: Oxford University Press.

Deem, R. (1982) 'Women, leisure and inequality', *Leisure Studies*, 1, 1: 29–46.

DeGrazia, S. (1962) *Of Time, Work and Leisure*, New York: Twentieth Century Fund.

Delin, C.R. and Patrickson, M. (1994) 'An investigation into aspects of leisure among busy people', *Australian Journal of Leisure and Recreation*, 4: 5–12.

DeLuca, T.H., Patterson, W.A., Freimund, W.A. and Cole, D.N. (1998) 'Influence of llamas, horses, and hikers on soil erosion from established recreation trails in western Montana, USA', *Environmental Management*, 22, 2: 255–62.

Department of Environment, Food and Rural Affairs (2001) *Countryside Stewardship Scheme: Traditional Farming in the Modern Environment, Department of Environment, Food and Rural Affairs*, at http://www.defra.gov.uk/erdp/pdfs/cssnews/060CSSIntro.pdf, accessed 24 February 2005.

Department of Environment, Housing and Community Development (1977) *Leisure Planning Guide for Local Government*, Canberra: Australian Government Publishing Service.

Desbarats, J. (1983) 'Spatial choice and constraints on behavior', *Annals of the Association of American Geographers*, 73, 3: 340–57.

Devall, B. and Harry, J. (1981) 'Who hates whom in the great outdoors: the impact of recreational specialization and technologies of play', *Leisure Sciences*, 4, 4: 399–418.

Dewar, K. (2000) 'An incomplete history of interpretation from the big bang', *International Journal of Heritage Studies*, 6, 2: 174–80.

Dewar, K. (2003) 'Interpretation', in Jenkins, J.M. and Pigram, J.J. (eds) *Encyclopedia of Leisure and Outdoor Recreation*, London: Routledge, pp. 266–7.

Dibb, J.A. (1980) 'Coastal recreation for the disabled', *Australian Parks and Recreation*, May: 24–33.

Dickie, P. (1995) 'Money squeeze on parks', *Brisbane Sunday Mail*, 7 May.

Dillon, P. (1993) 'Irish in a stew over battle of the Burren', *National Parks Today*, 36: 5.

Dionigi, R. (2004) 'Competing for life: older people and competitive sport', unpublished PhD thesis, School of Social Sciences, The University of Newcastle, Callaghan.

Dionigi, R. (forthcoming 2005) 'A leisure pursuit that "goes against the grain": older people and competitive sport', *Annals of Leisure Research*, 8, 1.

Ditwiler, C. (1979) 'Can technology decrease natural resource use conflicts in recreation?', *Search*, 10, 12: 439–41.

Don, A. (1997) 'National parks, nature conservation and heritage in Ireland', *Australian Parks and Recreation*, 33, 2: 26–9.

Dower, J. (1965) *The Fourth Wave: The Challenge of Leisure: A Civic Trust Survey*, London: Civic Trust.

Doxey, G.V. (1974) 'A causation theory of visitor-resident irritants: methodology and research inferences', in *Proceedings of the Travel Research Association 6th Annual Conference*, San Diego: Travel Research Association, pp. 195–8.

Dredge, D. and Jenkins, J.M. (forthcoming, 2006) *Tourism Policy and Planning*, New York: John Wiley and Sons.

Dredge, S. and Moore, S. (1992) 'A methodology for the integration of tourism in town planning', *Journal of Tourism Studies*, 3, 1: 8–22.

Driml, S. and Common, M. (1995) 'Economic and financial benefits of tourism in major protected areas', *Australian Journal of Environmental Management*, 2, 2: 19–39.

Driver, B. (1970a) 'Some thoughts on planning, the planning process and related decision processes', in Driver, B. (ed.) *Elements of Outdoor Recreation Planning*, Ann Arbor, MI: University of Michigan Press, pp. 195–21.

Driver, B. (ed.) (1970b) *Elements of Outdoor Recreation Planning*, Ann Arbor, MI: University of Michigan Press.

Driver, B. (2003) 'Benefits', in Jenkins, J.M. and Pigram, J.J. (eds) *Encyclopedia of Leisure and Outdoor Recreation*, London: Routledge, pp. 31–6.

Driver, B. and Basssett, J. (1975) 'Defining conflicts among river users: a case study of Michigan's Au Sable River', *Naturalist*, 26: 19–23.

Driver, B. and Brown, P. (1978) 'A social-physiological definition of recreation demand, with implications for recreation resource planning', Appendix A, *Assessing Demand for Outdoor Recreation*, Washington, DC: US Bureau of Outdoor Recreation.

Driver, B. and Bruns, D. (1999) 'Concepts and uses of the benefits approach to leisure', in Jackson, E. and Burton, E. (eds) *Leisure Studies: Prospects for the Twenty-First Century*,

State College, PA: Venture Publishing, pp. 349–68.

Driver, B. and Tocher, S.R. (1974) 'Toward a behavioural interpretation of recreational engagements, with implications for planning', in Driver, B. (ed.) *Elements of Outdoor Recreation Planning*, Ann Arbor, MI: University of Michigan Press, pp. 9–31.

Driver, B., Brown, P. and Peterson, G. (eds) (1991) *The Benefits of Leisure*, State College, PA: Venture Publishing.

Driver, B., Brown, P., Stankey, G. and Gregoire, T. (1987) 'The ROS planning system: evolution, basic concepts and research needed', *Leisure Sciences*, 9: 201–12.

Driver, B., Bruns, D. and Booth, K. (2001) 'Status and common misunderstandings of the Net Benefits Approach to Leisure', in proceedings *Trends 2000*, Lansing, MI: Michigan State University, Department of Park, Recreation, and Tourism Resources, pp. 245–63.

Driver, B., Dustin, D., Baltic, T., Elsner, G. and Peterson, G. (eds) (1996) *Nature and the Human Spirit: Toward an Expanded Land Management Ethic*, State College, PA: Venture Publishing.

Duffield, B. and Owen, M. (1970) *Leisure and Countryside: A Geographical Appraisal of Countryside Recreation in Lanarkshire*, Edinburgh: University of Edinburgh Press.

Durham, M. (2004) 'The Autumn years', *Countryside Voice*, Autumn: 20–5.

Dutton, I. and Hall, C.M. (1989) 'Making tourism sustainable: the policy/practice conundrum', *Proceedings of the Environment Institute of Australia Second National Conference*, Melbourne, 9–11 October.

Dwyer, J.F. and Stewart, S.I. (1995) 'Restoring urban recreation opportunities: an overview with illustrations', in *Proceedings of the Fourth International Outdoor Recreation and Trends Symposium and the 1995 National Recreation Resource Planning Conference*, St Paul, MN: University of Minnesota Press, pp. 606–9.

Eagles, P.F.J. (1996) 'Issues in tourism management in parks: the experience in Australia', *Australian Leisure*, June: 29–37.

Eagles, P.F.J. (2003) 'National parks', in Jenkins, J.M. and Pigram, J.J. (eds) *Encyclopedia of Leisure and Outdoor Recreation*, London: Routledge, pp. 322–4.

Eagles, P.F.J. and McCool, S.F. (2002) *Tourism in National Parks and Protected Areas: Planning and Management*, Wallingford: CABI.

Earley, D. (1989) *Recommendations on Outdoor Playing Space: The NPFA Standard*, London: National Playing Fields Association (NPFA).

Earth Sanctuaries Limited (2000) http://esl.com.au/index.htm, accessed 22 February 2001.

Eckbo, G. (1967) 'The landscape of tourism', *Landscape*, 18, 2: 29–31.

Ecotourism Australia (EA) (2005) http://www.ecotourism.org.au/, accessed 15 August 2005.

Edington, J.M. and Edington, M.A. (1986) *Ecology, Recreation and Tourism*, Cambridge: Cambridge University Press.

Egyptian Environmental Affairs Agency (undated) *Ras Mohammed National Park Sector*, Cairo: Department of Natural Protectorates.

Eliot-Hurst, M. (1972) *A Geography of Economic Behavior*, North Scituate, MA: Duxbury.

Elson, J.B. (1978) 'Recreation demand forecasting: a misleading tradition', in *Planning for Leisure*, Seminar Proceedings, 7–8 July, Warwick: University of Warwick.

Environment Canada (1978) *Canada Land Inventory Report No. 14, Land Capability for Recreation: Summary Report*, Ottawa: Environment Canada.

Environment Canada (1988a) *Making a Difference: The Canada Land Inventory,* Fact Sheet 88-5, Ottawa: Environment Canada.

Environment Canada (1988b) *Canada Land Inventory Report No. 14, Land Capability for Recreation: Summary Report*, Ottawa: Environment Canada.

Environment Canada (2001) *Recreation: Importance of Water for Tourism and Recreation in Canada*, Ottawa: Environment Canada.

Evans, S. (2001) 'Community forestry: countering excess visitor demands in England's national parks', in McCool, S. and Moisey, R.N. (eds) *Tourism, Recreation, and Sustainability: Linking Culture and the Environment*, Wallingford: CABI, pp. 77–90.

Fagan, R.H. and Webber, M. (1994) *Global Restructuring: The Australian Experience*, Melbourne: Oxford University Press.

Farina, J. (1980) 'Perceptions of time', in Goodale, T. and Witt, P. (eds) *Recreation and Leisure*, State College, PA: Venture Publishing, pp. 19–29.

Farrell, B.H. and Marion, J.L. (2001) 'Identifying and assessing ecotourism visitor impacts in eight protected areas in Costa Rica and Belize', *Environmental Conservation*, 28: 215–25.

Federation of National and Nature Parks of Europe (FNNPE) (1993) *Loving Them to Death? Sustainable Tourism in Europe's Nature and National Parks*, Grafenau: Federation of National and Nature Parks of Europe.

Fedler, A.J. (1987) 'Introduction: are leisure, recreation and tourism interrelated?', *Annals of Tourism Research*, 14, 3: 311–13.

Fennell, D. (1999) *Ecotourism: An Introduction*, Wallingford: CABI.

Fennell. D. (2002) *Ecotourism Programme Planning*, Wallingford: CABI.

Fennell, D. and Dowling, R. (2003) *Ecotourism Policy and Planning*, Wallingford: CABI.

Ferrario, F. (1988) 'Emerging leisure market among the South African black population', *Tourism Management*, March: 23–38.

Ferris, A.L. (1962) *National Recreation Survey*, Study Report Number 19, Washington, DC: Outdoor Recreation Resources Review Commission.

Field, D. (1997) 'Parks and neighbouring communities: a symbiotic relationship', in Pigram, J. and Sundell, R. (eds) *National Parks and Protected Areas: Selection, Delimitation and Management*, Armidale: Centre for Water Policy Research, pp. 419–28.

Figgis, P. (2000) 'The double-edged sword: tourism and national parks', *Habitat Australia*, 28, 5: 24.

Firestone, J. and Shelton, B.A. (1994) 'A comparison of women's and men's leisure time: subtle effects of the double day', *Leisure Sciences*, 16: 45–60.

Fitzsimmons, A. (1976) 'National parks: the dilemma of development', *Science*, 191: 440–4.

Fletcher, R., Fairbairn, H. and Pascoe, S. (eds) (2004) *Fatherhood Research in Australia, Research Report*, March, Callaghan: The Family Action Centre and The University of Newcastle.

Font, X. and Tribe, J. (eds) (2000) *Forest Tourism and Recreation: Case Studies in Environmental Management*, Wallingford: CABI.

Forster, J. (1973) *Planning for Man and Nature in National Parks*, Morges: IUCN.

Forsyth, P. and Dwyer, L. (1996) 'Tourism in the Asia-Pacific region', *Asia-Pacific Literature*, 10, 1: 13–22.

Foster, J. (1979) 'A park system and scenic conservation in Scotland', *Parks*, 4, 2: 1–4.

Fraser, R. and Spencer, G. (1998) 'The value of an ocean view', *Australian Geographical Studies*, 36, 1: 94–8.

Freeman, P. (1992) *The Sydney Morning Herald*, 20 March: 10.

Galbraith, J.K. (1972) *The New Industrial State*, 2nd edn, Harmondsworth: Penguin.

Garling, T. and Golledge, R. (eds) (1993) *Behavior and Environment*, Amsterdam: North Holland.

Garrod, B. and Wilson, J.C. (eds) (2003) *Marine Ecotourism: Issues and Experiences*, Clevedon: Channel View.

Gartner, W. (1996) *Tourism Development: Principles, Processes and Policies*, New York: Van Nostrand Reinhold.

Geering, D. (1989) *Managing the Full Range of River Resources*, Sydney: New South Wales Department of Water Resources (unpublished).

Geoscience Australia (2002) *Australia's Identified Mineral Resources 2002*, Canberra: Geoscience Australia.

German Federal Agency for Nature Conservation (ed.) (1997) *Biodiversity and Tourism: Conflicts on the World's Seacoasts and Strategies for Their Solution*, Berlin: Springer-Verlag.

Getz, D. (1987) 'Tourism planning and research: traditions, models and futures', paper presented to the Australian Travel Research Workshop, Bunbury, Western Australia, November.

Gittins, J. (1973) 'Conservation and capacity: a case study of Snowdonia National Park', *The Geographical Journal*, 139, 3: 482–6.

Gittins, R. (2004) 'Ageing: what the bureaucrats can't see or won't admit', *Sydney Morning Herald*, 6 December: 36.

Glyptis, S. (1981) 'People at play in the countryside', *Geography*, 66, 4: 277–85.

Glyptis, S. (1991) *Countryside Recreation*, Harlow: Longman.

Glyptis, S. (1992) 'The changing demand for countryside recreation', in Bowler, I.R., Bryant, C.R. and Nellis, M.D. (eds) *Contemporary Rural Systems in Transition: Economy and Society*, Wallingford: CABI.

Glyptis, S. (ed.) (1993) *Leisure and the Environment*, London: Belhaven Press.

Godbey, G. (1981) *Leisure in Your Life*, Philadelphia, PA: Saunders.

Godbey, G. (1985) 'Non-use of public leisure services: a model', *Journal of Recreation and Park Administration*, 3: 1–12.

Godbey, G. and Parker, S. (1976) *Leisure Studies and Services*, Philadelphia, PA: W.B. Saunders.

Godfrey, K.B. (1996) 'Towards sustainability: tourism in the Republic of Cyprus', in Harrison, L.C. and Husbands, W. (eds) *Practicing Responsible Tourism: International Case Studies in Tourism Planning, Policy, and Development*, New York: Wiley and Sons.

Godin, V. and Leonard, R. (1977) 'Design capacity for backcountry recreation management planning', *Journal of Soil and Water Conservation*, 32: 161–4.

Gold, S. (1972) 'Non-use of neighbourhood parks', *Journal of the American Institute of Planners*, XXXVIII, 6: 369–78.

Gold, S. (1973) *Urban Recreation Planning*, Philadelphia, PA: Lea and Febiger.

Gold, S. (1974) 'Deviant behaviour in urban parks', *Journal of Health, Physical Education, Recreation*, Nov–Dec: 18–20.

Gold, S. (1980) *Recreation Planning and Design*, New York: McGraw-Hill.

Gold, S. (1988) 'Urban open space preservation: the American experience', *Australian Parks and Recreation*, 25, 1: 21–8.

Goldsmith, F. and Manton, R. (1974) 'The ecological effects of recreation', in Lavery, P. (ed.) *Recreational Geography*, London: David and Charles, pp. 259–69.

Goodale, T.L. and Godbey, G.C. (1988) *The Evolution of Leisure: Historical and Philosophical Perspectives*, State College, PA: Venture Publishing.

Goodall, B. and Whittow, J. (1975) *Recreation Requirements and Forest Opportunities*, Geographical Papers No. 37, Reading: Department of Geography, University of Reading.

Gordon, W.R. and Caltabiano, M.L. (1996) 'Youth leisure experiences in rural and urban North Queensland', *Australian Leisure*, 7, 2: 37–41.

Graefe, A.R. (1990) 'Visitor impact management', in Graham, R. and Lawrence, R. (eds) *Towards Serving Visitors and Managing Our Resources*, Proceedings of a North American Workshop on Visitor Management in Parks and Protected Areas, Waterloo: Tourism Research and Education Centre, University of Waterloo, pp. 213–34.

Graefe, A.R. (1991) 'Visitor impact management: an integrated approach to assessing the impacts of tourism in national parks and protected areas', in Veal, A.J., Jonson, P. and Cushman, G. (eds) *Leisure and Tourism: Social and Environmental Change*, Papers from the World Leisure and Recreation Association Congress, Sydney, Australia: University of Technology, pp. 74–83.

Graefe, A.R., Vaske, J.J. and Kuss, F.R. (1984) 'Social carrying capacity: an integration and synthesis of twenty years of research', *Leisure Sciences*, 6, 4: 395–431.

Graham, R. (1990) 'Visitor management in Canada's national parks', in Graham, R. and Lawrence, R. (eds) *Towards Serving Visitors and Managing Our Resources*, Proceedings of a North American Workshop on Visitor Management in Parks and Protected Areas, Waterloo: Tourism Research and Education Centre, University of Waterloo, pp. 271–96.

Graham, R. and Lawrence, R. (eds) (1990) *Towards Serving Visitors and Managing Our Resources, Proceedings of a North American Workshop on Visitor Management in Parks and Protected Areas*, Waterloo: Tourism Research and Education Centre, University of Waterloo.

Graham, R., Payne, R.J. and Nilsen, P. (1988) 'Visitor activity planning and management in Canadian national parks', *Tourism Management*, 9, 1: 44–62.

Gramann, J.H. and Burdge, R.J. (1981) 'The effect of recreation goals on conflict perception: the case of water skiers and fishermen', *Journal of Leisure Research*, 13, 1: 15–27.

Grandage, J. and Rodd, R. (1981) 'The rationing of recreational land use', in Mercer, D. (ed.) *Outdoor Recreation: Australian Perspectives*, Melbourne: Sorrett, pp. 76–91.

Grant, B. (2001). '"You're never too old": beliefs about physical activity and playing sport in later life', *Ageing and Society*, 21, 6: 777–98.

Grant, B. (2002) 'Over 65 and ready to play', *Australian Leisure Management*, October/November: 36, 38.

Gray, D. and Pelegrino, D. (1973) *Reflections on the Recreation and Park Movement*, Dubuque, IA: Brown.

Gray, F. (1975) 'Non-explanation in urban geography', *Area*, 7, 4: 228–35.

Gray, H.P. (1970) *International Travel – International Trade*, Lexington, MA: Heath Lexington.

Graziers' Association of New South Wales (1975) *Submission to the Select Committee of the Legislative Assembly upon the Fishing Industry*, Sydney: Graziers' Centre.

Green, B. (1977) 'Countryside planning: compromise or conflict?', *The Planner*, 63: 67–9.

Green, P. (1992) 'Parks management in New Zealand', *UNEP Industry and Environment*, 15, 3–4: 16–21.

Groome, D. (1993) *Planning and Rural Recreation in Britain*, Aldershot: Avebury.

Gunn, C. (1972) *Vacationscape, Designing Tourist Regions*, Austin, TX: University of Texas, Bureau of Business Research.

Gunn, C. (1979) *Tourism Planning*, New York: Crane Russack.

Gunn, C. (1988) *Vacationscape: Designing Tourist Regions*, 2nd edn, New York: Van Nostrand Reinhold.

Gunn, C. (1994) *Tourism Planning*, 3rd edn, Washington, DC: Taylor and Francis.

Gunn, S.L. and Peterson, C.A. (1978) *Therapeutic Recreation Program Design*, Englewood Cliffs, NJ: Prentice Hall.

Gutteridge, Haskins and Davey Pty Ltd (1988) *Draft Plan of Management: Wallis Island Crown Reserve*, Pyrmont: Industrial Publishing.

Hall, C.M. (1992) *Wasteland to World Heritage: Preserving Australia's Wilderness*, Melbourne: Melbourne University Press.

Hall, C.M. (1994) *Tourism in the Pacific Rim: Development, Impacts and Markets*, Melbourne: Longman.

Hall, C.M. (1995) *Introduction to Tourism in Australia: Impacts, Planning and Development*, Melbourne: Longman.

Hall, C.M. (1997) *Tourism in the Pacific Rim*, 2nd edn, Melbourne: Longman.

Hall, C.M. (1998) *Introduction to Tourism: Development, Dimensions and Issues*, 3rd edn, Melbourne: Longman.

Hall, C.M. (2003) *Introduction to Tourism: Dimensions and Issues*, French's Forest, NSW: Hospitality Press.

Hall, C.M. and Higham, J. (eds) (2005) *Tourism, Recreation and Climate Change*, Clevedon: Channel View.

Hall, C.M. and Jenkins, J.M. (1995) *Tourism and Public Policy*, London: Routledge.

Hall, C.M. and McArthur, S. (eds) (1993) *Heritage Management in New Zealand and Australia*, Oxford: Oxford University Press.

Hall, C.M. and McArthur, S. (eds) (1996) *Heritage Management in Australia and New Zealand: The Human Dimension*, Melbourne: Oxford University Press.

Hall, C.M. and Muller, D.K. (eds) (2004) *Tourism, Mobility, and Second Homes: Between Elite Landscape and Common Ground*, Clevedon: Channel View.

Hall, C.M. and Page, S.J. (eds) (1996) *Tourism in the Pacific: Issues and Cases*, London: International Thomson Business Press.

Hall, C.M. and Page, S.J. (2002) *The Geography of Tourism and Recreation*, 2nd edn, London: Routledge.

Hall, C.M., Jenkins, J.M. and Kearsley, G. (1997a) 'Tourism planning and policy in urban areas: introductory comments', in Hall, C.M., Jenkins, J.M. and Kearsley, G. (eds) *Tourism, Planning and Policy in Australia and New Zealand: Cases, Issues and Practice*, Sydney: Irwin.

Hall, C.M., Jenkins, J.M. and Kearsley, G. (1997b) *Tourism Planning and Policy in Australia and New Zealand: Cases, Issues and Practice,* Sydney: Irwin.

Hamilton-Smith, E. (1975) 'Issues in the measurement of community need', *Australian Journal of Social Issues*, 10, 1.

Hamilton-Smith, E. (2003) 'Site evaluation', in Jenkins, J.M. and Pigram, J.J. (eds), *Encyclopedia of Leisure and Outdoor Recreation*, London: Routledge, pp. 459–61.

Hamilton-Smith, E. and Mercer, D.C. (1991) *Urban Parks and their Visitors*, Melbourne: Melbourne and Metropolitan Board of Public Works.

Hammitt, W. (1990) 'Wildland recreation and resource impacts: a pleasure-policy dilemma', in Hutcheson, J.D., Noe, F.P. and Snow, R.E. (eds) *Outdoor Recreation Policy: Pleasure and Preservation*, 17–30, New York: Greenwood Press.

Hammitt, W. and Cole, D. (1987) *Wildland Recreation Ecology and Management*, New York: Wiley.

Hammitt, W. and Cole, D.N. (1998) *Wildland Recreation: Ecology and Management*, 2nd edn, New York: John Wiley and Sons.

Hampton, D. (undated) 'Review of multiple use options for the management of travelling stock reserves in New South Wales', discussion paper, Land Resources and Environment Branch, New South Wales Department of Lands, Sydney.

Hand, V. and Lewis, V. (2002) 'Fathers' views on family life and paid work', *Australian Institute of Family Studies*, 61: 26–9.

Hanley, N., Alvarez-Farizo, B. and Shaw, W.D. (2003b) 'Using economic instruments to manage access to rock-climbing sites in the Scottish Highlands', in Hanley, N., Shaw, W.D. and Wright, R.E. (eds) *The New Economics of Outdoor Recreation*, Cheltenham: Edward Elgar, pp. 40–58.

Hanley, N., Shaw, W.D. and Wright, R.E. (2003a) 'Introduction', in Hanley, N., Shaw, W.D. and Wright, R.E. (eds) *The New Economics of Outdoor Recreation*, Cheltenham: Edward Elgar, pp. 1–20.

Hanley, N., Shaw, W.D. and Wright, R.E. (eds) (2003c) *The New Economics of Outdoor Recreation*, Cheltenham: Edward Elgar.

Harmon, D. and Putney, A.D. (eds) (2003) *The Full Value of Parks: From Economic to the Intangible*, Lanham, MD: Rowman and Littlefield.

Harper, G. and Weiler, B. (eds) (1992) *Ecotourism*, Canberra: Bureau of Tourism Research.

Harrington, M., Dawson, D. and Bolla, P. (1992) 'Objective and subjective constraints on women's enjoyment of leisure', *Loisir et Société, Society and Leisure*, 15, 1: 203–21.

Harrington, R. (1975) 'Liability exposure in the operation of recreation facilities', *Outdoor Recreation Action*, 35: 22–5.

Harris, R. and Jago, L. (2001) Professional accreditation in the Australian tourism industry: an uncertain future, *Tourism Management*, 22, 4: 383–90.

Harrison, A. (1977) 'Getting your story across – interpreting the river resource', in proceedings, *River Recreation Management Research Symposium*, Minneapolis, MN: USDA Forest Service, General Technical Report NC-28, pp. 125–38.

Harrison, J. (1992) 'Protected area management guidelines', *Parks*, 3, 2: 22–5.

Harrison, L.C. and Husbands, W. (eds) (1996) *Practicing Responsible Tourism: International Case Studies in Tourism Planning, Policy, and Development*, New York: Wiley and Sons.

Hart, W. (1966) *A Systems Approach to Park Planning*, Morges: International Union for the Conservation of Nature.

Hartley, E. (1999) 'Visitor impacts at Logan Pass, Glacier National Park: a thirty year vegetation study', in Harmon, D. (ed.) *On the Frontiers of Conservation: Proceedings of the 10th Conference on Research and Management in National Parks and on Public Lands*, Asheville, NC and Hancock, MI: George Wright Society, pp. 297–305.

Haulot, A. (1978) 'Cultural protection policy in the field of tourism', *Parks*, 3, 3: 6–8.

Havighurst, R.J. (1961) 'The nature and values of meaningful free-time activity', in Kleemeier, R.W. (ed.) *Aging and Leisure*, New York: Oxford University Press.

Hawkins, D. (1997) 'The virtual tourism environment', in Cooper, C. and Wanhill, S. (eds) *Tourism Development,* pp. 43–58, Chichester: Wiley.

Hawkins, J.P., Van 't Hof, T., De Meyer, K., Tratalos, J. and Aldam C. (1999) 'Effects of recreational scuba diving on Caribbean coral and fish communities', *Conservation Biology*, 13: 888–97.

Hawks, S.R. (1991) 'Recreation in the family', in Bahr, S.J. (ed.) *Family Research: A Sixty Year Review, 1930–1990*, New York: Lexington Books, pp. 387–433.

Haynes, P. (1973) 'Towards a concept of monitoring', *Town Planning Review*, 45, 1

Hayslip, B. and Panek, P.E. (1989) *Adult Development and Aging*, New York: Harper & Row.

HaySmith, L. and Hunt, J.D. (1995) 'Nature tourism: impacts and management', in Knight, R.L. and Gutzwiller, K.J. (eds) *Wildlife and Recreationists: Coexistence through Management and Research*, Washington, DC: Island Press, pp. 203–19.

Haywood, K.M. (1989) 'Responsible and responsive approaches to tourism planning in the community', *Tourism Management*, 9, 2: 105–18.

Hedstrom, E. (1992) 'Preservation or profit?', *National Parks*, 66, 102: 18–20.

Heit, M. and Malpass, D. (undated) *Do Women Have Equal Play?*, Ottawa: Ministry of Culture and Recreation.

Helman, P., Jones, A., Pigram, J.J. and Smith, J. (1976) *Wilderness in Australia*, Armidale: Department of Geography, University of New England.

Hendee, J., Stankey, G. and Lucas, R. (1978) *Wilderness Management*, Washington: USDA Forest Service, Miscellaneous Publication 1365.

Henderson, K. (2003) 'Men', in Jenkins, J.M. and Pigram, J.J. (eds) *Encyclopedia of Leisure and Outdoor Recreation*, London: Routledge, pp. 303–7.

Henderson, K.A. (1991) 'The contribution of feminism to an understanding of leisure constraints', *Journal of Leisure Research*, 23: 363–77

Henderson, K.A. (1994) 'Broadening an understanding of women, gender and leisure', *Journal of Leisure Research*, 23: 1–7.

Henderson, K.A., Bedini, L.A., Hecht, L. and Shuler, R. (1993) 'The negotiation of leisure constraints by women with disabilities', in Fox, K. (ed.) *Proceedings of the 7th Canadian Congress on Leisure Research*, pp. 235–41, Winnipeg, Manitoba: Faculty of Physical Education and Recreation, University of Manitoba.

Henderson, K.A., Bedini, L.A., Hecht, L. and Shuler, R. (1995) 'Women with physical disabilities and the negotiation of leisure constraints', *Leisure Studies*, 14: 17–31.

Henderson, K.A., Bialeschki, M.D., Shaw, S.M. and Freysinger, V.J. (1989) *A Leisure of One's Own*, State College, PA: Venture Publishing.

Henderson, K.A., Bialeschki, M.D., Shaw, S.M. and Freysinger, V.J. (1996) *Both Gains and Gaps: Feminist Perspectives on Women's Leisure*, State College, PA: Venture Publishing.

Hendricks, W.W. (1995) 'A resurgence in recreation conflict research: introduction to the special issue', *Leisure Sciences*, 17, 3: 157.

Hendry, L.B. (1983) *Growing Up and Growing Out: Adolescents and Leisure*, Aberdeen: Aberdeen University Press.

Henning, D.H. (1974) *Environmental Policy and Administration*, New York: American Elsevier Publishing.

Henry, I. (1993) *The Politics of Leisure Policy*, Basingstoke: Macmillan.

Hersch, G. (1991) 'Leisure and aging, physical and occupational therapy', *Geriatrics*, 9, 2: 55–72.

Heywood, J. (1989) 'Recreation opportunity: the social setting', *Australian Parks and Recreation*, 25, 2: 18–20.

Hibberd, J. (ed.) (1978) *The Future of the Long Paddock: A Study of Travelling Stock Reserves, Routes and Roadside Verges in the Southern Tablelands of NSW*, Sydney: Nature Conservation Council of NSW, Eco-Press.

Higham, J.E. (1996) 'Wilderness perceptions of international visitors to New Zealand. The perceptual approach to the management of international tourists visiting wilderness areas within New Zealand's conservation estate', unpublished PhD thesis, University of Otago, Dunedin.

Higham, J.E. (1998) 'Tourists and albatrosses: the dynamics of tourism at the Northern Royal Albatross Colony', *Tourism Management*, 19: 521–31.

Hilary, E. (2003) 'My story: 50 years', *National Geographic*, May: 38–41.

Hogg, D. (1977) 'The evaluation of recreational resources', in Mercer, D. (ed.) *Leisure and Recreation in Australia*, Melbourne: Sorrett.

Holden, A. (2000) *Environment and Tourism*, London: Routledge.

Holland, S., Pybas, D. and Sanders, A. (1992) 'Personal watercraft: fun, speed – and conflict?' *Parks and Recreation*, 27, 11: 52.

Hollenhorst, S. (1995) 'Risk, technology driven, and other new activity trends', in *Proceedings of the Fourth International Outdoor Recreation Trends Symposium*, pp. 65–7, Minneapolis: University of Minnesota.

Hollenhorst, S. and Gardner, L. (1994) 'The indicator performance estimate approach to determining acceptable wilderness conditions', *Environmental Management*. 18, 6: 901–6.

Hollick, M. (1993) *An Introduction to Project Evaluation*, Melbourne: Longman.

Hollings, C. (1978) *Adaptive Environmental Assessment and Management*, Chichester: Wiley.

Holloway, J.C. (2002) *The Business of Tourism*, 6th edn, Harlow: Pearson.

Horna, J.L.A. (1991) 'The family and leisure domains: women's involvement and perceptions', *World Leisure and Recreation*, 33, 3: 11–14.

Horna, J.L.A. (1993) 'Married life and leisure: a multidimensional study of couples', *World Leisure and Recreation*, 35, 3: 17–21.

Horton, F.E. and Reynolds, D.R. (1971) 'Effects of urban spatial structure on individual behaviour', *Economic Geography*, 47: 36–48.

Hovinen, G. (1981) 'A tourist cycle in Lancaster County, Pennsylvannia', *Canadian Geographer*, 25, 3: 283–6.

Hovinen, G. (1982) 'Visitor cycles outlook for tourism in Lancaster County, Pennsylvannia', *Annals of Tourism Research*, 9: 563–83.

Howard, A. (1997) 'Conservation reserve boundaries and management implications in New South Wales', in Pigram, J. and Sundell, R. (eds) *National Parks and Protected Areas: Selection, Delimitation and Management*, Armidale: Centre for Water Policy Research, University of New England, pp. 387–401.

Howe, C.Z. (1980) 'Models for evaluating public recreation programs: what the literature shows', *Journal of Physical Education and Recreation*, 51, 8: 36–8.

Howe, C.Z. (1993) 'The evaluation of leisure programs: applying qualitative methods', *Journal of Physical Education, Recreation, and Dance*, 64, 8: 43–7.

Hultsman, W.Z. (1992) 'Constraints to activity participation in early adolescence', *Journal of Early Adolescence*, 12: 280–99.

Hultsman, W.Z. (1993) 'The influence of others as a barrier to recreation participation among early adolescents', *Journal of Leisure Research*, 25: 150–64.

Hultsman, W.Z. and Kaufman, J.E. (1990) 'The experience of leisure by youth in a therapeutic milieu', *Youth and Society*, 21: 496–510.

Hunt, G. (2000) 'Writing an environmental plan for the Community Forest of Mercia, England', in Font, X. and Tribe, J. (eds) *Forest Tourism and Recreation*, Wallingford: CABI, pp. 217–24.

Hurtes, K. (2003) 'Adolescent', in Jenkins, J.M. and Pigram, J.J. (eds) *Encyclopedia of Leisure and Outdoor Recreation*, London: Routledge, p. 8.

Hutcheson, J.D., Noe, F.P. and Snow, R.E. (eds) (1990) *Outdoor Recreation Policy: Pleasure and Preservation*, New York: Greenwood Press.

Hutchison, R. (1994) 'Women and the elderly in Chicago's public parks', *Leisure Sciences*, 16: 229–47.

Hutchison, S.L., Loy, D.P., Kleiber, D.A. and Dattilo, J. (2003) 'Leisure as a coping resource: variations in coping with traumatic injury and illness', *Leisure Sciences*, 25: 143–61.

Ibrahim, I. (1991) *Leisure and Society*, Dubuque, IA: Brown.

Ibrahim, H. and Cordes, K. (1993) *Outdoor Recreation*, Dubuque, IA: Brown and Benchmark.

Ilbery, R. (ed.) (1997) *The Geography of Rural Change*, London: Longman.

Information and Research Services (Department of the Parliamentary Library, Australia) (2000) *Living Standards, Current Issues Brief No. 4 2000–01*, Canberra: Department of the Parliamentary Library.

Inskeep, E. (1991) *Tourism Planning: An Integrated and Sustainable Development Approach*, New York: Van Nostrand Reinhold.

International Food Policy Research Unit (2004) *IFPRI Forum, International Food Policy Research Institute and its 2020 Vision Initiative*, October, Washington, DC.

International Hotel Association (1995) *Environmental Action Pack for Hotels*, Paris: UNEP.

Iso-Ahola, S.E. (1980) *The Social Psychology of Leisure and Recreation*, Dubuque, IA: William C. Brown.

Iso-Ahola, S.E. (1988) 'The social psychology of leisure: past, present and future research', in Barnett, L.A. (ed.) *Research About Leisure: Past, Present and Future*, Champaign, IL: Sagamore.

Iso-Ahola, S.E. and Crowley, E.D. (1991) 'Adolescent substance abuse and leisure boredom', *Journal of Leisure Research*, 23, 3: 260–71.

Iso-Ahola, S.E. and Weissinger, E. (1990) 'Perceptions of boredom in leisure: conceptualization, reliability and validity of the Leisure Boredom Scale', *Journal of Leisure Research*, 22, 1: 1–17.

Issaverdis, J.-P. (2001) 'The pursuit of excellence: benchmarking, accreditation, best practice and auditing', in Weaver, D. (ed.) *The Encyclopedia of Ecotourism*, Wallingford: CABI, pp. 579–94.

Ite, U. and Adams, W. (2000) 'Expectations, impacts and attitudes: conservation and development in Cross River National Park', *Journal of International Development*, 12: 325–42.

Ittelson, W., Franck, K. and O'Hanlon, T. (1976) 'The nature of environmental experience', in Wapner, S., Cohen, S. and B. Kaplan (eds), *Experiencing the Environment*, New York: Plenum Press.

IUCN (1994) *Guidelines for Protected Area Management Categories*, Gland: IUCN.

Ivy, M.I., Stewart, W.P. and Lue, Chi-Chuan (1992) 'Exploring the role of tolerance in recreational conflict', *Journal of Leisure Research*, 24, 4: 348–60.

Iwasaki, Y. and Mannell, R.C. (2000) 'The effects of leisure beliefs and coping strategies on stress, health relationships', *Leisure/Loisir*, 24, 1–2: 3–57.

Jackson, E.L. (1988) 'Leisure constraints: a survey of past research', *Leisure Sciences*, 10: 203–15.

Jackson, E.L. (1990) 'Variations in the desire to begin a leisure activity: evidence of antecedent constraints?', *Journal of Leisure Research*, 22: 150–64.

Jackson, E.L. (1991) 'Leisure constraints/constrained leisure: special issue introduction', *Journal of Leisure Research*, 22: 55–70.

Jackson, E.L. (1993) 'Recognising patterns of leisure constraints: results from alternative analyses', *Journal of Leisure Research*, 25: 129–49.

Jackson, E.L. (1994) 'Geographical aspects of constraints on leisure and recreation', *The Canadian Geographer*, 38: 110–21.

Jackson, E.L. and Burton, T.L. (eds) (1989) *Understanding Leisure and Recreation: Mapping the Past, Charting the Future*, State College, PA: Venture Publishing.

Jackson, E.L. and Henderson, K.A. (1995) 'Gender-based analysis of leisure constraints', *Leisure Sciences*, 17, 1: 31–51.

Jackson, E.L. and Rucks, V.C. (1995) 'Negotiation of leisure constraints by junior-high and high-school students: an exploratory study', *Journal of Leisure Research*, 27: 85–105.

Jackson, E.L. and Scott, D. (1999) 'Constraints to leisure', in Jackson, E.L. and Burton, T.L. (eds) *Leisure Studies: Prospects for the Twenty-First Century*, State College, PA: Venture Publishing, pp. 299–321.

Jackson, E.L. and Wong, R.A. (1982) 'Perceived conflict between urban cross-country skiers and snowmobilers in Alberta', *Journal of Leisure Research*, 14, 2: 47–62.

Jackson, E.L., Crawford, D.W. and Godbey, G. (1993) 'Negotiation of leisure constraints', *Leisure Sciences*, 15, 1: 1–11.

Jacob, G. and Schreyer, R. (1980) 'Conflict in outdoor recreation: a theoretical perspective', *Journal of Leisure Research*, 12, 4: 368–80.

Jafari, J. (1977) 'Editor's page', *Annals of Tourism Research*, 5: 6–11.

James, K. and Embrey, L. (2001) 'Anyone could be lurking around! Constraints on adolescent girls' recreational activities after dark', *World Leisure*, 43, 4: 44–52.

Jamrozik, A. (1986) 'Leisure as social consumption: some equity considerations for social policy', in Castle, R., Lewis, D. and Mangan, J. (eds) *Work, Leisure and Technology*, Melbourne: Longman, pp. 184–209.

Janiskee, R. (1976) 'On the recreation appeals of extra-urban environments', Mimeograph, 72nd Annual Meeting of Association of American Geographers, New York.

Jans, N.A. and Frazer-Jans, J.M. (1991) 'Organisational culture and organisational effectiveness', *Australian Journal of Public Administration*, 50, 3: 333–46.

Jansen-Verbeke, M. (1992) 'Urban recreation and tourism: physical planning issues', *Tourism Recreation Research*, 17, 2: 33–45.

Jansen-Verbeke, M. and van de Wiel, E. (1995) 'Tourism planning in urban revitalization projects: lessons from the Amsterdam waterfront development', in Ashworth, G.J. and Dietvorst, A. (eds) *Tourism and Spatial Transformations: Implications for Policy and Planning*, Wallingford: CABI, pp. 129–45

Jefferies, B.E. (1982) 'Sagamartha National Park: the impact of tourism in the Himalayas', *Ambio*, 11, 5: 274–82.

Jenkins, J.M. (1998) *Crown Lands Policy-Making in New South Wales, 1856–1991: The Life and Death of an Organisation, its Culture and a Project*, Canberra: Centre for Public Sector Management.

Jenkins, J.M. (2003) 'Action space', in Jenkins, J.M. and Pigram, J.J. (eds) *Encyclopedia of Leisure and Outdoor Recreation*, London: Routledge, p. 5.

Jenkins, J.M. and Prin, E. (1998) 'Rural landholder attitudes: the case of public recreational access to private rural lands', in Butler, R.W., Hall, C.M. and Jenkins, J.M. (eds) *Tourism and Recreation in Rural Areas*, Chichester: John Wiley and Sons.

Jenkins, J.M. and Wearing, S. (2003) 'Ecotourism and protected areas in Australia', in Fennell, D.A. and Dowling, R.K. (eds) *Ecotourism Policy and Planning*, Wallingford: CABI, pp. 205–33.

Jensen, C.R. (1977) *Outdoor Recreation in America: Trends, Problems and Opportunities*, 3rd edn, Minneapolis, MN: Burgess Publishing.

Johnston, R.J., Gregory, D. and Smith, D.M. (eds) (1986) *The Dictionary of Human Geography*, 2nd edn, Oxford: Basil Blackwell.

Jubenville, A. (1976) *Outdoor Recreation Planning*, Philadelphia, PA: Saunders.

Jubenville, A. (1978) *Outdoor Recreation Management*, Philadelphia, PA: Saunders.

Just, D. (1987) 'Appropriate amounts and design of open spaces', *Australian Parks and Recreation*, 25, 2: 32–9.

Kaiser, C. and Helber, L. (1978) *Tourism Planning and Development*, Boston, MA: CBI Publishing.

Kando, T.M. (1975) *Leisure and Popular Culture*, St Louis, MO: C.V. Mosby.

Kane, P. (1981) 'Assessing landscape attractiveness', *Applied Geography*, 1: 77–96.

Kaplan, M. (1975) *Leisure: Theory and Practice*, New York: John Wiley.

Kaplan, R. and Kaplan, S. (1989) *The Experience of Nature: A Psychological Perspective*, Cambridge: Cambridge University Press.

Kariel, H.G. (1989) 'Tourism and development: perplexity or panacea?', *Journal of Travel Research*, 28, 1: 2–6.

Kates, Peat, Marwick and Co. (1970) *Tourism and Recreation in Ontario*, Ontario: Minister of Tourism and Information.

Kay, C.E. and White, C.A. (2001) 'Reintroduction of bison into the Rocky Mountain parks of Canada: historical and archaeological evidence', in Harmon, D. (ed.) *Crossing Boundaries in Park Management: Proceedings of the 11th Conference on Research and Resource Management in Parks and on Public Lands*, Hancock, MI: George Wright Society, pp. 143–51.

Kay, T. and Jackson, G. (1991) 'Leisure despite constraint: the impact of leisure constraints on leisure participation', *Journal of Leisure Research*, 23: 301–13.

Kearsley, G. (1997) 'Tourism planning and policy in New Zealand', in Hall, C.M., Jenkins, J.M. and Kearsley, G. (eds) *Tourism Planning and Policy in Australia and New Zealand: Cases, Issues and Practice*, pp. 49–60, Sydney: Irwin.

Kearsley, G. (2000) 'Balancing tourism and wilderness qualities in New Zealand's native forests', in Font, X. and Tribe, J. (eds) *Forest Tourism and Recreation: Case Studies in Environmental Management*, Wallingford: CABI, pp. 75–91.

Kearsley, G. and Higham, J. (1996) 'Wilderness and back country motivations and satisfaction in New Zealand's natural areas and conservation estate', *Australian Journal of Leisure and Recreation*, 8: 30–4.

Kelleher, G.G. and Kenchington, R.A. (1982) 'Australia's Great Barrier Reef Marine Park: making development compatible with conservation', *Ambio*, 11, 5: 262–7.

Kelly, J.R. (1974) 'Socialization toward leisure: a developmental approach', *Journal of Leisure Research*, 6: 181–93.

Kelly, J.R. (1990) 'Leisure and aging: a second agenda', *Society and Leisure*, 13, 1: 145–67.

Kelly, J.R., Steinkamp, M.W. and Kelly, J.R. (1987) 'Later life satisfaction: does leisure contribute?', *Leisure Sciences*, 8: 189–200.

Kibicho, W. (2004a) 'Tourism and sex trade: roles male sex workers play in Malindi, Kenya', *Tourism Review International*, 7, 3/4: 129–41.

Kibicho, W. (2004b) 'A critical evaluation of how tourism influences the commercial sex workers' operations in Malindi Area, Kenya', *Annals of Leisure Research*, 7, 3: 188–201.

Kimble, G. (1951) 'The inadequacy of the regional concept', in Stamp, L. and Wooldridge, S. (eds) *London Essays in Geography*, London: Longman.

King, B.E.M. (1997) *Creating Island Resorts*, London: Routledge.

King, C.J. (1957) *An Outline of Closer Settlement in New South Wales: Part 1*, Sydney: Division of Marketing and Agricultural Economics, Department of Agriculture, NSW.

Kirkby, S. (1996) 'The World Wide Web as a provider of multimedia information to the ecotourist', in *Proceedings of National Seminar on the Role of Technology in Parks and Recreation*, Adelaide, April.

Kleiber, D. (1999) *Leisure Experience and Human Development: A Dialectical Approach*, New York: Basic Books.

Klekowski, L. (undated) http://www.bio.umass.edu/biology/conn.river/history.html, accessed 24 January 2005.

Knetsch, J. (1972) 'Interpreting demands for outdoor recreation', *Economic Record*, 48, September: 429–32.

Knetsch, J. (1974) 'Outdoor recreation and water resources planning', Water Resources Monograph No. 3, Washington, DC: American Geophysical Union.

Knopf, R. (1990) 'The limits of acceptable change (LAC) planning process: potentials and limitations', in Graham, R. and Lawrence, R. (eds) *Towards Serving Visitors and Managing Our Resources*, Waterloo: University of Waterloo Press, pp. 201–12.

Knight, R. and Gutzwiller, K.J. (eds) (1995) *Wildlife and Recreationists: Coexistence Through Management and Research*, Washington, DC: Island Press.

Knudson, D.M., Cable, T.T. and Beck, L. (1995) *Interpretation of Cultural and Natural Resources*, State College, PA: Venture Publishing.

Komarovsky, M. (1967) *Blue-collar Marriage*, New York: Random House.

König, U. (1998) *Tourism in a Warmer World: Implications of Climate Change due to Enhanced Greenhouse Effect for the Ski Industry in the Australian Alps*, Vol. 28, Zürich: Universität Zürich-Irchel Geographisches Institut.

Kotler, P. (1971) *Marketing Decision-Making: A Model Building Approach*, New York: Holt, Rinehart and Winston.

Kozlowski, J., Rosier, J. and Hill, G. (1988) 'Ultimate Environmental Threshold (UET) method in a marine environment (Great Barrier Reef Marine Park in Australia)', *Landscape and Urban Planning*, 15: 327–36.

Kraus, R. (1984) *Recreation and Leisure in Modern Society*, 3rd edn, Glenview, IL: Scott Foresman and Co.

Kraus, R. (2001) *Recreation and Leisure in Modern Society*, 6th edn, Boston, MA: Jones and Bartlett.

Kreutzwiser, R. (1989) 'Supply', in Wall, G. (ed.) *Outdoor Recreation in Canada*, Toronto: Wiley, pp. 21–41.

Krippendorf, J. (1987) *The Holidaymakers: Understanding the Impact of Leisure and Travel*, Oxford: Heinemann.

Krumpe, E. (1988) 'The role of information in people's leisure decision making process', in Killin, N., Paradice, W. and Engel, M. (eds) *Information in Planning and Management of Recreation and Tourist Services*, Newcastle: Hunter Valley Research Foundation.

Kruss, F., Graefe, A. and Vaske, J. (1990) *Visitor Impact Management,* Vols I, II, Washington, DC: National Parks and Conservation Association.

Lacey, P. (1996) 'Current recreation trends', *Australian Leisure*, 7, 4: 15–18.

Lake Tahoe Environmental Education Coalition (2003) 'Drinking water – another reason to protect Lake Tahoe', http://www.lteec.org/news.php?newsID=83, accessed 5 August 2005.

Lambley, D. (1988) 'The economic significance of different types of leisure, recreation and tourism in national parks: the Myall Lakes National Park', unpublished paper, Sydney: National Parks and Tourism Seminar.

Lancaster, R.A. (ed.) (1990) *Recreation, Park, and Open Space Standards and Guidelines*, Ashburn, VA: NRPA.

Landals, A.G. (1986) 'The tourists are ruining the parks', in *Tourism and the Environment: Conflict or Harmony? Proceedings of a Symposium sponsored by the Canadian Society of Environmental Biologists*, pp. 89–99, Alberta Chapter, Calgary, Canada, 18–19 March, Alberta: CSEB.

Lane, B. (1994) 'What is rural tourism?', *Journal of Sustainable Tourism*, 2, 1–2: 7–21.

LaPage, W. (1967) *Some Observations on Campground Trampling and Ground Cover Response*, Washington, DC: USDA Forest Service Research Paper NE-68.

LaPage, W. (1970) 'The mythology of outdoor recreation planning', *Southern Lumberman*, December: 118–21.

Laudon, K. and Laudon, J. (2002) *Management Information Systems: Managing the Digital Firm*, 7th edn, Englewood Cliffs, NJ: Prentice Hall.

Lavery, P. (1974) *Recreational Geography*, London: David and Charles.

Law, C. (2002) *Urban Tourism: The Visitor Economy and the Growth of Large Cities*, London: Continuum.

Law, C.M. (1992) 'Urban tourism and its contribution to economic regeneration', *Urban Studies*, 29, 3/4: 599–618.

Law, C.M. (1993) *Urban Tourism: Attracting Visitors to Large Cities*, London: Mansell.

Lawrence, G. (1987) *Capitalism and the Countryside*, Sydney: Pluto Press.

Lawrence, R.L. and Daniels, S.E. (1996) *Public Involvement in Natural Resource Decision Making: Goals, Methodology, and Evaluation*, Corvallis, OR: Forest Research Laboratory, Oregon State University.

Lea, J. (1988) *Tourism and Development in the Third World*, London: Routledge.

Leatherbury, E. (1979) 'River amenity evaluation', *Water Resources Bulletin*, 15, 5: 1281–92.

Leave No Trace (undated) http://www.lnt.org/about/index.html, accessed 4 March 2005.

Lee, B. and Graefe, A. (2004) 'GIS: a tool to locate a new park and recreation services', *Parks and Recreation*, 39, 10: 34–42.

Lee-Gosselin, M. and Pas, E.I. (1997) 'The implications of emerging contexts for travel-behaviour research', in Stopher, P. and Lee-Gosselin, M. (eds) *Understanding Travel Behaviour in an Era of Change*, Oxford: Elsevier Science, pp. 1–28.

Leiper, N. (1979) 'The framework of tourism', *Annals of Tourism Research*, 6, 4: 390–407.

Leiper, N. (1981) 'Towards a cohesive curriculum in tourism', *Annals of Tourism Research*, 8, 2: 69–84.

Leiper, N. (1984) 'Tourism and leisure: the significance of tourism in the leisure spectrum', in *Proceedings 12th New Zealand Geography Conference*, Christchurch: New Zealand Geography Society.

Leiper, N. (1995) *Tourism Management*, Collingwood: TAFE Publications.

Leitner, M.J. and Leitner, S.F. (2004a) *Leisure Enhancement*, 3rd edn, New York: Haworth Press.

Leitner, M.J. and Leitner, S.F. (2004b) *Leisure in Later Life*, 3rd edn, New York: Haworth Press.

Leopold, A., Cain, S., Cottam, C., Gabrielson, I. and Kimball, T. (1963) *Report of the Advisory Committee to the National Park Service on Research*, Washington, DC: National Academy of Sciences.

Leung, Y.-F. and Marion, J.L. (2000) 'Recreation impacts and management in wilderness: a state-of-knowledge review', in Cole, D.N., McCool, S.F., Borrie, W.T. and O'Loughlin, J. (eds) *Wilderness Science in a Time of Change Conference, Volume 5: Wilderness Ecosystems, Threats, and Management*. RMRS-P-15–Vol. 5, Ogden, UT: USDA Forest Service, Intermountain Research Station, pp. 23–48.

Levine, J.A. (2001) 'Work burden of women', *Science*, 10: 812.

Levy, J. (1977) 'A paradigm for conceptualising leisure behaviour', *Journal of Leisure Research*, 11, 1: 48–60.

Lew, A.A. (1989) 'Authenticity and sense of place in the tourism development experience of older retail districts', *Journal of Travel Research*, 27, 4: 15–22.

Lew, A.A., Hall, C.M. and Williams, A.C. (eds) (2004) *A Companion to Tourism*, Maldem, MA: Blackwell.

Liddle. M.J. (1975) 'A selective review of the ecological effects of human trampling on natural ecosystems', *Biological Conservation*, 7: 17–36.

Liddle, M. (1997) *Recreation Ecology*, London: Chapman & Hall.

Lilieholm, R.J., and Romney, L.R. (2000) 'Tourism, national parks, and wildlife', in Butler, R.W. and Boyd, S.W. (eds) *Tourism and National Parks: Issues and Implications*, New York: John Wiley and Sons, pp. 137–51.

Lime, D. (1974) 'Locating and designing campgrounds to provide a full range of camping opportunities', *Outdoor Recreation Research: Applying the Results*, General Technical Report NC-9, St Paul, MN: USDA Forest Service, pp. 56–66.

Lime, D. and Stankey, G. (1971) 'Carrying capacity: maintaining outdoor recreation quality', in *Proceedings, Forest Recreation Symposium*, Syracuse: New York College of Forestry, pp. 174–84.

Lindberg, K. (1991) *Policies for Maximizing Nature Tourism's Ecological and Economic Benefits*, Washington, DC: World Resources Institute.

Lindberg, K. and Hawkins, D.E. (eds) (1993) *Ecotourism: A Guide for Planners and Managers*, North Bennington, VT: Ecotourism Society.

Lindberg, K. and McCool, S.F. (1998) 'Comment: a critique of environmental carrying capacity as a means of managing the effects of tourism development', *Environmental Conservation*, 25, 4: 291–2.

Lindsay, J. (1980) 'Trends in outdoor recreation activity conflicts', in *Proceedings 1980 National Outdoor Recreation Trends Symposium*, Vol. 1, Broomall, PA: USDA Forest Service, pp. 215–21.

Ling Wong, J. (2004) 'A place in the country', *Countryside Voice*, Autumn: 42.

Linz, W. and Linz, M. (1996) 'Letter to the editor', *National Parks Magazine*, 70, 1–2: 10.

Lipscombe, N. (1986) 'Supply and demand in outdoor recreation: which should concern us most?', *Australian Parks and Recreation*, 23, 1: 16–18.

Lipscombe, N. (1993) 'Recreation planning: where have all the frameworks gone?', in McIntyre, N. (ed.) *Proceedings, Track to the Future: Managing Change in Parks and Recreation*, Melbourne: Royal Australian Institute of Parks and Recreation.

Lipscombe, N. (2003) 'Demand', in Jenkins, J.M. and Pigram, J.J. (eds) *Encyclopedia of Leisure and Outdoor Recreation*, London: Routledge, pp. 106–9.

Little, D.E. (2002) 'Women and adventure education: reconstructing leisure constraints and experiences to negotiate continuing participation', *Journal of Leisure Research*, 34, 2: 157–77.

Lobo, F. (1995) 'Recreation for all: an inappropriate concept for the unemployed', *Australian Parks and Recreation*, Winter: 1–26.

London, M., Crandall, R. and Fitzgibbons, D. (1977) 'The psychological structure of leisure: activities, needs, people', *Journal of Leisure Research*, 9, 4: 252–63.

Loomis, J., Gonzalez-Caban, A. and Englin, J. (2001), 'Testing for different differential effects of forest fires on hiking, mountain biking demand and benefits', *Journal of Agricultural and Resource Economics*, 26, 2: 508–23.

Lopata, H.Z. (1972) 'The life cycle of the social role of the housewife', in Bryant, C.D. (ed.) *The Social Dimensions of Work*, Englewood Cliffs, NJ: Prentice Hall, pp. 128–44.

Losier, G.F., Bourque, P.E. and Vallerand, R.J. (1993) 'A motivational model of leisure participation in the elderly', *The Journal of Psychology*, 127: 153–70.

Louviere, J. and Timmermans, H. (1990) 'Stated preference and choice models applied to recreation research: a review', *Leisure Sciences*, 12, 1: 9–32.

Lovelock, B. (2001) 'Interorganisational relations in the protected area-tourism policy domain: the influence of macro-economic policy', *Current Issues in Tourism*, 4, 2–4: 253–74.

Lucas, B. (1992) 'The Caracas Declaration', *Parks*, 3, 2: 7–8.

Lucas, R. (1964) *The Recreation Capacity of the Quetico – Superior Area*, St Paul, MN: USDA Forest Research Paper, LS-15.

Luckenbach, R.A., and Bury, R.B. (1983) 'Effects of off-road vehicles on the biota of Algodones Dunes, Imperial County, California', *Journal of Applied Ecology*, 20: 265–86.

Lundberg, D.E., Stavenga, M.H. and Krishnamoorthy, M. (1995) *Tourism Economics*, New York: John Wiley and Sons.

Lunn, A. (1986) 'Military matters', *Tarn and Tor*, 9: 6.

Lynch, R. and Veal, A.J. (1996) *Australian Leisure*, Melbourne: Longman.

Lyons, K.D. (2003) 'Organized camps', in Jenkins, J.M. and Pigram, J.J. (eds) *Encyclopedia of Leisure and Outdoor Recreation*, London: Routledge, p. 349.

McArthur, S. (2000) 'Visitor management in action: an analysis of the development and implementation of visitor management models at Jenolan Caves and Kangaroo Island', unpublished PhD thesis, University of Canberra, ACT.

McCleave, J., Booth, K. and Espiner, S. (2004) 'Love thy neighbour? The relationship between Kahurangi National Park and the border communities of Karamea and Golden Bay, New Zealand', *Annals of Leisure Research*, 7, 3–4: 202–21.

McCool, S.F. (1990a) 'Limits of acceptable change: evolution and future', in Graham, R. and Lawrence, R. (eds) *Towards Serving Visitors and Managing Our Resources*, Waterloo: University of Waterloo Press, pp. 185–93.

McCool, S.F. (1990b) 'Limits of acceptable change: some principles', in Graham, R. and Lawrence, R. (eds) *Towards Serving Visitors and Managing Our Resources*, Waterloo: University of Waterloo Press, pp. 195–9.

McCool, S.F. (2003) 'Carrying capacity', in Jenkins, J.M. and Pigram, J.J. (eds) *Encyclopedia of Leisure and Outdoor Recreation*, London: Routledge, pp. 142–4.

McCool, S.F. and Cole, D.N. (eds) (1998) *Proceedings – Limits of Acceptable Change and Related Planning Processes: Progress and Future Directions*, University of Montana's Lubrecht Experimental Forest, Missoula, MT, 20–22 May, Ogden, UT: USDA Forest Service, Rocky Mountain Research Station.

McCool, S.F. and Lime, D.W. (2001) 'Tourism carrying capacity: tempting fantasy or useful reality?', *Journal of Sustainable Tourism*, 9, 5: 372–88.

McCool, S.F. and Moisey, N. (eds) (2001) *Tourism, Recreation and Sustainability*, Wallingford: CABI.

McCool, S.F. and Stankey, G.H. (1986) *Visitor Attitudes Toward Wilderness Fire Management Policy 1971–84*, Res. Pap. INT-357, Ogden, UT: USDA Forest Service, Intermountain Research Station.

McCosh, R. (1973) 'Recreation site selection', in Gray, D. and Pelegrino, D. (eds) *Reflections on the Recreation and Park Movement*, Dubuque, IA: Brown, pp. 290–5.

McDonald, G. and Wilks, L (1986b) 'The regional impact of tourism and recreation in national parks', *Environment and Planning B: Planning and Design*, 13: 349–66.

McDonald, G. and Wilks, L. (1986a) 'Economic and financial benefits of tourism in major protected areas', *Australian Journal of Environmental Management*, 2, 2: 19–39.

McIntyre, N. (1990) 'Recreation involvement: the personal meaning of participation', unpublished PhD thesis, University of New England, Armidale.

McIntyre, N. (1993) 'Recreation planning for sustainable use', *Australian Journal of Leisure and Recreation*, 3, 2: 31–7 and 49.

McIntyre, N. and Boag, A. (1995) 'The measurement of crowding in nature-based tourism venues: Uluru National Park', *Tourism Recreation Research*, 20, 1: 37–42.

McIntyre, N., Cuskelly, G. and Auld, C. (1991) 'The benefits of urban parks', *Australian Parks and Recreation*, 27, 4: 11–18.

McKercher, B. (1998) *The Business of Nature-Based Tourism*, Melbourne: Hospitality Press.

McKnight, T.L. (1977) *The Long Paddock, Australia's Travelling Stock Routes*, Armidale: Department of Geography and Planning, University of New England.

McLoughlin, L. (1997) 'Sydney and the bush … Sydney or the bush', *Australian Planner*, 34, 3: 165–70.

McMeeking, D. and Purkayastha, B. (1995) 'I can't have my mom running me everywhere: adolescents, leisure and accessibility', *Journal of Leisure Research*, 27, 4: 360–78.

McNeely, J.A. and Thorsell, J.W. (1989) 'Jungles, mountains and islands: how tourism can help conserve the natural heritage', *World Leisure and Recreation*, 31, 4: 29–39.

MacPherson, B.D. (1991) 'Aging and leisure benefits: a life cycle perspective', in Driver, B.L., Brown, P.J. and Peterson, G.L. (eds) *Benefits of Leisure*, State College, PA: Venture Publishing.

Mak, J. and White, K. (1992) 'Comparative tourism development in Asia and the Pacific', *Journal of Travel Research*, 31, 1: 14–23.

Manfredo, M. and Driver, B. (2002) 'Benefits: the basis of action' in Manfredo, M. (ed.) *Wildlife Viewing: A Management Handbook*, Corvallis, OR: Oregon State University Press.

Mannell, R.C. and Zuzanek, J. (1991) 'The nature and variability of leisure constraints in daily life: the case of physically active leisure of older adults', *Leisure Sciences*, 13: 337–51.

Manning, R. (1979) 'Strategies for managing recreational use of national parks', *Parks*, 4, 1: 13–15.

Manning, R., McCool, S. and Graefe, A. (1995) 'Trends in carrying capacity', in Thompson, J., Lime, D., Gartner, B. and James, W. (eds) *Proceedings of the Fourth International Outdoor Recreation and Tourism Trends Symposium*, St Paul, MN: University of Minnesota, pp. 334–41.

Manning, R.E. (1979) 'Impacts of recreation on riparian soils and vegetation', *Water Resources Bulletin*, 15: 30–43.

Marion, J.L. (1998) 'Recreation ecology research findings: Implications for wilderness and park managers', in Kirchner, H. (ed.) *Proceedings of the National Outdoor Ethics Conference*; St Louis, MO and Gaithersburg, MD: Izaak Walton League of America, pp. 188–96.

Marion, J.L. and Cole, D.N. (1996) 'Spatial and temporal variation in soil and vegetation impacts on campsites', *Ecological Applications*, 6, 2: 520–30.

Markwell, K. (1996) 'Towards a gay and lesbian leisure research agenda', *Australian Leisure*, 7, 2: 42–4 and 48.

Marriott, K. (1993) 'Pricing policy for user pays', *Australian Parks and Recreation*, 29, 3: 42–5.

Marsh, J. (ed.) (1994) *Rails to Greenways, the Proceedings of a Conference*, 13–15 August 1993, Canadian Rails to Greenways Network and The Frost Centre for Canadian Heritage Development Studies, Trent University, Ontario.

Marsh, J. (2000) 'Tourism and national parks in polar regions', in Butler, R.W. and Boyd, S.W. (eds) *Tourism and National Parks: Issues and Implications*, Chichester: Wiley and Sons, pp. 125–36.

Martin, L., Bennett, R. and Gregory, D. (1985) 'The thirsty Algarve', *Geographical Magazine*, 57: 321–4.

Martin, W. and Mason, S. (1976) 'Leisure 1980 and beyond', *Long Range*, 9, 2: 58–65.

Massie, B. (2004) 'Disabled people want to enjoy the countryside too', *Countryside Focus*, July/August: 7.

Mather, A.S. (1986) *Land Use*, Harlow: Longman Scientific and Technical.

Mathieson, A. and Wall, G. (1982) *Tourism: Economic, Physical and Social Impacts*, London: Longman.

Mattyasovsky, E. (1967) 'Recreation area planning: some physical and ecological requirements', *Plan*, 8, 3: 91–109.

Mayes, G., Dyer, P. and Richins, H. (2004) Dolphin-human interaction: changing pro-environmental attitudes, beliefs, behaviours and intended actions of participants through management and interpretation programs, *Annals of Leisure Research*, 7, 1: 34–53.

Mayo, E.J. and Jarvis, L.P. (eds) (1981) *The Psychology of Leisure Travel: Marketing and Selling of Travel Services*, Boston, MA: CBI Publishing.

Medlik, S. (1972) *Planning for Coastal Recreation*, Monograph 1, Melbourne: Universities Recreation Research Group.

Medlik, S. (1975) 'The concept of recreational need', *Journal of Leisure Research*, 5, 1: 37–51.

Medlik, S. (1977) 'The factors affecting recreational demand', in Mercer, D. (ed.) *Leisure and Recreation in Australia*, Melbourne: Sorrett, pp. 59–68.

Medlik, S. (1980a) *In Pursuit of Leisure*, Melbourne: Sorrett.

Medlik, S. (1980b) 'Themes in Australian urban leisure research', in Mercer, D. and Hamilton-Smith, E. (eds) *Recreation Planning and Social Change in Urban Australia*, Melbourne: Sorrett, pp. 1–25.

Medlik, S. (1981a) 'Trends in recreational participation', in Mercer, D. (ed.) *Outdoor Recreation: Australian Perspectives*, Melbourne: Sorrett, pp. 24–44.

Medlik, S. (1981b) *Outdoor Recreation: Australian Perspectives*, Melbourne: Sorrett.

Medlik, S. (1993) *Dictionary of Travel, Tourism and Hospitality*, Oxford: Heinemann.

Medlik, S. (1994a) 'Monitoring the spectator society: an overview of research and policy issues', in Mercer, D.C. (ed.) *New Viewpoints in Australian Outdoor Recreation Research and Planning*, Melbourne: Hepper Marriott and Associates, pp. 1–28.

Medlik, S. (ed.) (1994b) *New Viewpoints in Australian Outdoor Recreation Research and Planning*, Melbourne: Hepper Marriott and Associates.

Medlik, S. (1995) *A Question of Balance: Natural Resources Conflict Issues in Australia*, 2nd edn, Annandale: Federation Press.

Mercer, D.C. (1970) 'The geography of leisure – contemporary growth point', *Geography*, 55: 261–73.

Mercer, D.C. (1972) *Planning for Coastal Recreation*, Monograph 1, Melbourne: Universities Recreation Research Group.

Mercer, D.C. (1975) 'The concept of recreational need', *Journal of Leisure Research*, 5, 1: 37–51.

Mercer, D.C. (1977) 'The factors affecting recreational demand', in Mercer, D. (ed.) *Leisure and Recreation in Australia*, Melbourne, Sorrett, pp. 59–68.

Mercer, D.C. (1980a) *In Pursuit of Leisure*, Melbourne: Sorrett.

Mercer, D.C. (1980b) 'Themes in Australian urban leisure research', in Mercer, D. and Hamilton-Smith, E. (eds) *Recreation Planning and Social Change in Urban Australia*, Melbourne: Sorrett, pp. 1–25.

Mercer, D.C. (1981a) 'Trends in recreational participation', in Mercer, D. (ed.) *Outdoor Recreation: Australian Perspectives*, Melbourne: Sorrett, pp. 24–44.

Mercer, D.C. (1981b) *Outdoor Recreation: Australian Perspectives*, Melbourne: Sorrett.

Mercer, D.C. (1994a) 'Monitoring the spectator society: an overview of research and policy issues', in Mercer, D.C. (ed.) *New Viewpoints in Australian Outdoor Recreation Research and Planning*, Melbourne: Hepper Marriott and Associates, pp. 1–28.

Mercer, D.C. (ed.) (1994b) *New Viewpoints in Australian Outdoor Recreation Research and Planning*, Melbourne: Hepper Marriott and Associates.

Mercer, D.C. (1995) *A Question of Balance: Natural Resources Conflict Issues in Australia*, 2nd edn, Annandale: Federation Press.

Mercer, D.C. (2000) *A Question of Balance: Natural Resources Conflict Issues in Australia*, 3rd edn, Leichhardt: Federation Press.

Mercer, D.C. (2003) 'Recreation', in Jenkins, J.M. and Pigram, J.J. (eds) *Encyclopedia of Leisure and Outdoor Recreation*, London: Routledge, pp. 412–15.

Mercer, D.C. (2004) 'Tourism and resource management', in Lew, A.A., Hall, C.M. and Williams, A.C. (eds) *A Companion to Tourism*, Maldem, MA: Blackwell, pp. 462–72.

Mercer, D. and Hamilton-Smith, E. (eds) (1980) *Recreation Planning and Social Change in Urban Australia*, Melbourne: Sorrett.

Merigliano, L., Cole, D.N. and Parsons, D.J. (1998) 'Applications of LAC type processes and concepts to nonrecreation management issues in protected areas', in McCool, S.F. and Cole, D.N. (eds) *Proceedings – Limits of Acceptable Change and Related Planning Processes: Progress and Future Directions*, Ogden, UT: University of Montana's Lubrecht Experimental Forest, Missoula, Montana, 20–22 May, USDA Forest Service, Rocky Mountain Research Station, pp. 37–43.

Messenger, J.C. (2004) *Working Time and Workers' Preferences in Industrialized Countries: Finding the Balance*, London: Routledge.

Meyer, K. (1994) *How to Shit in the Woods: An Environmentally Sound Approach to a Lost Art*, 2nd rev. edn, Berkeley, CA: Ten Speed Press.

Meyer-Arendt, K.J. (1990) 'Recreational Business Districts in Gulf of Mexico seaside resorts', *Journal of Cultural Geography*, 11: 39–55.

Middleton, V.Y.C. (1982) 'Tourism in rural areas', *Tourism Management*, 3: 52–8.

Mieczkowski, Z. (1990) *World Trends in Tourism and Recreation*, New York: Peter Lang.

Mill, R.C. and Morrison, A.M. (1985) *The Tourism System: An Introductory Text*, Englewood Cliffs, NJ: Prentice Hall.

Ministry for Planning and Environment (1989) *Planning Guide for Urban Open Space*, Melbourne: Ministry for Planning and Environment.

Mintzberg, H. (1994) *The Rise and Fall of Strategic Planning*, New York: Prentice Hall.

Mitchell, L. (1968) 'An evaluation of central place theory in a recreation context', *Southeastern Geographer*, 8: 46–53.

Mitchell, L. (1969) 'Toward a theory of public urban recreation', in *Proceedings of Association of American Geographers*, 1: 103–8.

Mitchell, L. (1983) 'Future directions of recreation planning', in Lieber, S. and Fesenmaier, D. (eds) *Recreation Planning and Management*, State College, PA: Venture Publishing, pp. 323–38.

Mobily, K.E. and Bedford, R.L. (1993) 'Language, play and work among elderly persons', *Leisure Studies*, 12: 203–19.

Mobily, K.E., Leslie, D.K., Lemkie, J.H., Wallace, R.B. and Kohout, F.J. (1986) 'Leisure patterns and attitudes of the rural elderly', *Journal of Applied Gerontology*, 5: 201–14.

Moir, J. (1995) 'Regional parks in Perth, Western Australia', *Australian Planner*, 32, 2: 88–95.

Moisey, R.N. (2002) 'The economics of tourism in national parks and protected areas', in Eagles, P.F.J. and McCool, S.F. (eds) *Tourism in National Parks and Protected Areas*, Wallingford: CABI, pp. 235–54.

Montgomery, M. (2004) Opening address to the Regional Cooperation and Development forum, National General Assembly of Local Governments, Canberra, Sunday 7 November, http://nga.alga.asn.au/generalAssembly/2004/01.presentations/01.mikeMontgomery.php, accessed 2 March 2005.

More, T.A., Bulmer, S. Henzel, L. and Mates, A. (2003) *Extending the Recreation Opportunity Spectrum to Non-federal Lands in the Northeast: An Implementation Guide*, Gen. Tech. Rep. NE-309. USDA Forest Service, Northeastern Research Station.

Morgan, D.J. (2003) 'Resource base', in Jenkins, J.M. and Pigram, J.J. (eds) *Encyclopedia of Leisure and Outdoor Recreation*, London: Routledge, pp. 430–2.

Morris, S. (1975) 'Owner rights and co-operation', in *Landscape Conservation*, Papers of Australian Conservation Foundation Conference, Canberra.

Morrison, J. (1994) 'Men at leisure: the implications of masculinity and leisure for "househusbands"', unpublished MA thesis, School of Leisure and Tourism Studies, University of Technology, Sydney.

Moseley, M. (1979) *Accessibility: The Rural Challenge*, London: Methuen.

Mowforth, M. and Munt, I. (1998) *Tourism and Sustainability: New Tourism in the Third World*, London: Routledge.

Mueller, E., Gurin G. and Wood, M. (1962) *Participation in Outdoor Recreation: Factors Affecting Demand Among American Adults*, Outdoor Recreation Resources Review Commission, Study Report Number 20, Washington, DC: US Government Printing Service.

Mullins, P. (1991) 'Tourism urbanization', *International Journal of Urban and Regional Research*, 3, September: 326–42.

Murdock, S.H., Backman, K., Nazrul Hoque, M. and Ellis, D. (1991) 'The implications of change in population size and composition on future participation in outdoor recreational activities', *Journal of Leisure Research*, 23, 3: 238–59.

Murphy, P.E. (1985) *Tourism: A Community Approach*, New York: Methuen.

Murphy, P.E. (ed.) (1997) *Quality Management in Urban Tourism*, Chichester: Wiley.

Myers, N. (1972) 'National parks in savannah Africa', *Science*, 178: 1255–63.

National Capital Development Commission (1978) *Planning Concept Papers*, Canberra: Australian Government Publishing Services.

National Community Forest Partnership (undated) http://www.communityforest.org.uk/, accessed 2 February 2005.

National Ecotourism Accreditation Program (NEAP) (2003) http://www.ecotourism.org.au/ EcoCertification3.pdf, accessed 8 August 2005.

National Park Service Organic Act, 16 U.S.C.1., http://www.nps.gov/legacy/organic-act.htm, accessed 3 August 2005.

National Parks Review Panel (1991) *Fit for the Future: Report of the National Parks Review Panel* (CCP 335), Cheltenham: Countryside Commission.

Nelson, J. and Butler, R.W. (1974) 'Recreation and the environment', in Manners, I. and Mikesell, M. (eds) *Perspectives on Environment*, Washington, DC: Association of American Geographers, pp. 290–310.

Nelson, J.G. (2000) 'Tourism and national parks', in Butler, R.W. and Boyd, S.W. (eds) *Tourism and National Parks: Issues and Implications*, Chichester: Wiley and Sons, pp. 303–21.

Nepal, S.K. (2000) 'Tourism, national parks and local communities', in Butler, R.W. and Boyd, S.W. (eds) *Tourism and National Parks: Issues and Implications*, Chichester: Wiley and Sons, pp. 73–94.

Netherlands Ministry for Cultural Affairs (1976) *National Parks and Landscape Parks in The Netherlands*, The Hague: Netherlands Government Publishing House.

Neulinger, J. (1981) *To Leisure: An Introduction*, Boston, MA: Allyn and Bacon.

Neulinger, J. (1982) 'Leisure lack and the quality of life', *Leisure Studies*, 1, 1: 53–64.

New South Wales Department of Lands (1986) *Land Assessment and Planning Process for Crown Lands in NSW*, Sydney: New South Wales Department of Lands.

New South Wales Department of Planning (1989) *Tourism Development Near Natural Areas*, Sydney: New South Wales Department of Planning.

New South Wales National Parks and Wildlife Service (1997a) *Draft Nature Tourism and Recreation Strategy*, Sydney: NSWNPWS.

New South Wales National Parks and Wildlife Service (1997b) *Draft National Parks Public Access Strategy*, Sydney: NSWNPWS.

New South Wales National Parks and Wildlife Service (2003) *New South Wales National Parks and Wildlife Service: Annual Report 2002–2003*, Sydney: NSWNPWS.

New South Wales National Parks and Wildlife Service (2004) *Living Parks*, Sydney: NSWNPWS.

New South Wales National Parks Association (1999) http://www.speednet.com.au/~abarca/NPApol01VEHICLES.htm, accessed 26 January 2005.

New Zealand National Parks Authority (1978) *Guidelines for Interpretative Planning*, Wellington: National Parks Authority.

Newcomb, R. (1979) *Planning the Past*, Dawson: Folkestone.

Newsome, D., Moore, S.A. and Dowling, R.K. (2002) *Natural Area Tourism: Ecology, Impacts and Management*, Clevedon: Channel View.

Nicholls, S. (2001) 'Measuring accessibility and equity of public parks: a case study using GIS', *Managing Leisure*, 6, 4: 201–19.

Nicholls, S. (2003) 'Measures of success: measuring park accessibility using GIS', *Parks and Recreation*, August: 53–5.

Nichols, D. and Freestone, R. (2003) 'Community valuations of historic parks: a Melbourne study', *Annals of Leisure Research*, 6, 2: 114–33.

Nilsen, P. and Taylor, G. (1998) 'A comparative analysis of protected area planning and management frameworks', in McCool, S.F. and Cole, D.N. (eds) *Proceedings – Limits of Acceptable Change and Related Planning Processes: Progress and Future Directions*, University of Montana's Lubrecht Experimental Forest, Missoula, Montana, 20–22 May, Ogden, UT: USDA Forest Service, Rocky Mountain Research Station, pp. 49–57.

Noble, B.F. (2000) 'Institutional criteria for co-management', *Marine Policy*, 24: 69–70.

Nolte, C. (1995) 'Few favor parking lot in Yosemite Valley', *San Francisco Chronicle*, 22 June: A21.

Notzke, C. (1995) 'A new perspective in aboriginal natural resource management: co-management', *Geoforum*, 22: 187–209.

Novelli, M. (ed.) (2005) *Niche Tourism: Contemporary Issues, Trends and Cases*, New York: Elsevier.

NSW Department of Lands – see New South Wales Department of Lands.

NSW Department of Planning – see New South Wales Department of Planning.

NSW National Parks and Wildlife Service – see New South Wales National Parks and Wildlife Service.

O'Brien, D. (2004) 'The New Forest National Park', *Viewpoint*, 41: 6.

O'Loughlin, T. (1993) '"Walk softly": the effectiveness of the Tasmanian minimal impact bushwalking campaign', in Hall, C.M. and McArthur, S. (eds) *Heritage Management in New Zealand and Australia*, Auckland: Oxford University Press, pp. 82–91.

O'Riordan, T. (1971) *Perspectives on Resource Management*, London: Pion.

OECD (1990) *Partnerships for Rural Development*, Paris: OECD.

OECD (1993) *What Future for Our Countryside? A Rural Development Policy*, Paris: OECD.

Office of the Deputy Prime Minister, UK (2002) *Planning Policy Guidance 17: Planning for Open Space, Sport and Recreation*, http://www.odpm.gov.uk/stellent/groups/odpm_control/documents/contentservertemplate/odpm_index.hcst?n=3425&l=3, accessed 1 February 2005.

Ohmann, L. (1974) 'Ecological carrying capacity', in *Outdoor Recreation Research: Applying the Results*, St Paul, MN: USDA Forest Service General Technical Report, NC-9, pp. 24–8.

Okotai, T. (1980) 'Research requirements of tourism in the Cook Islands', in Pearce, D. (ed.) *Tourism in the South Pacific*, Proceedings of UNESCO Tourism Workshop, Rarotonga, pp. 169–76.

Oliver, M. (1996) *Understanding Disability: From Theory to Practice*, Basingstoke: Macmillan.

Olokesusi, F. (1990) 'Assessment of the Yankari Game Reserve, Nigeria: problems and prospects', *Tourism Management*, 11, 2: 153–63.

Opperman, M. and Chon, K.S. (1997) *Tourism in Developing Countries*, London: International Thomson Business Press.

Orams, M. (2002) *Marine Tourism: Development, Impacts and Management*, London: Routledge.

Orcutt, J.J. (1984) 'Contrasting effects of two kinds of boredom on alcohol use', *Journal of Drug Issues*, 14: 161–73.

Osborne, D. and Gaebler, T. (1992) *Reinventing Government: How the Entrepreneurial Spirit is Transforming the Public Sector*, New York: Plume.

Ovington, J., Groves, K., Stevens, P. and Tanton, M. (1972) *A Study of the Impact of Tourism at Ayers Rock – Mt Olga National Park*, Canberra: Department of Forestry.

Owen, J.M. (1993) *Program Evaluation: Forms and Approaches*, St Leonards: Allen and Unwin.

Owen, P.L. (1984) 'Rural leisure and recreation research: a retrospective evaluation', *Progress in Human Geography*, 8, 2: 157–87.

Page, S. (1994) *Transport for Tourism*, London: Routledge.

Page, S. (1995) *Urban Tourism*, London: Routledge

Page, S. and Getz, D. (1997) *The Business of Rural Tourism: International Perspectives*, London: International Thomson Business Press

Palmer, L. (2004) 'Fishing lifestyles: "territorians", traditional owners and the management of recreational fishing in Kakadu National Park', *Australian Geographical Studies*, 42, 1: 60–76.

Parker, S. (1971) *The Future of Work and Leisure*, London: Granada Publishing.

Parker, S. (1983) *Leisure and Work*, London: Allen & Unwin.

Parker, S. and Paddick, R. (1990) *Leisure in Australia*, Melbourne: Longman.

Parks Agency Canada (2004) *Parks Agency Canada Corporate Plan 2004/05–2008/09*, Parks Agency Canada, http://parkscanada.pch.gc.ca/docs/pc/plans/plan-2004-05-2008-09/index_e.asp, accessed 17 January 2005.

Parks Victoria (2002) *Park Notes: Victoria's Marine National Parks and Marine Sanctuaries*, Parks Victoria, http://parkweb.vic.gov.au/resources/05_0796.pdf, accessed 22 February 2005.

Patmore, A.J. (1973) *Land and Leisure*, Harmondsworth: Penguin.

Patmore, A.J. (1983) *Recreation and Resources: Leisure Patterns and Leisure Places*, Oxford: Blackwell.

Patterson, I. and Pegg, S. (1995) 'Leisure, community integration and people with disabilities', *Australian Leisure*, 5, 1: 32–8.

Patterson, I. and Taylor, T. (2001) *Celebrating Inclusion and Diversity in Leisure*, Melbourne: HM Leisure Publishing.

Payne, R. and Nilsen, P. (1997) 'The role of social science research in establishing national parks: a Canadian case study', in Pigram, J. and Sundell, R. (eds) *National Parks and Protected Areas: Selection, Delimitation and Management*, Armidale: Centre for Water Policy Research, University of New England, pp. 403–17.

Pearce, D. (1978) 'A case study of Queenstown', in *Tourism and the Environment*, Information Series No. 6, Wellington: Department of Lands and Survey.

Pearce, D. (1979) 'Towards a geography of tourism', *Annals of Tourism Research*, 6, 3: 245–72.

Pearce, D. (ed.) (1980) *Tourism in the South Pacific*, Proceedings of UNESCO Tourism Workshop, Rarotonga, Christchurch: University of Canterbury.

Pearce, D. (1987) *Tourism Today: A Geographical Analysis*, Harlow: Longman Scientific and Technical.

Pearce, D. (1989) *Tourist Development*, 2nd edn, Harlow: Longman Scientific and Technical.

Pearce, P.L. (1982) *The Social Psychology of Tourist Behaviour*, Sydney: Pergamon Press.

Pearce, P.L., Morrison, A.M. and Rutledge, J.L. (1998) *Tourism: Bridges Across Continents*, Sydney: McGraw-Hill.

Pearce, P.L., Moscardo, G. and Ross, G.F. (1996) *Tourism Community Relationships*, Oxford: Pergamon.

Pearson, K. (1977) 'Leisure in Australia', in Mercer, D. (ed.) *Leisure and Recreation in Australia*, Melbourne: Sorrett, pp. 25–34.

Penning-Rowsell, E. (1975) 'Constraints on the application of landscape evaluation', *Transactions, Institute of British Geographers*, 66: 49–55.

Perdue, R.R., Long, P.T. and Allen, L. (1987) 'Rural resident tourism perceptions and attitudes', *Annals of Tourism Research*, 14: 420–9.

Perez de Cuellar, J. (1987) 'Statement', *World Leisure and Recreation*, 29, 1: 3.

Peterson, G., Driver, B.L. and Gregory, R. (eds) (1988) *Amenity Resource Valuation: Integrating Economics with other Disciplines*, State College, PA: Venture:

Peterson, G., Stynes, D., Rosenthal, D. and Dwyer, J. (1985) 'Substitution in recreation choice behaviour', in *Proceedings – Symposium on Recreation Choice Behaviour*, General Technical Report, INT-184, St Paul, MN: USDA Forest Service, pp. 19–30.

Phillips, A. and Roberts, M. (1973) 'The recreation and amenity value of the countryside', *Journal of Agricultural Economics*, 24, 1: 85–102.

Phillips, K. (1977) 'Forests and recreation', unpublished guest lecture, Armidale: Department of Geography, University of New England.

Pieper, J. (1952) *Leisure the Basis of Culture*, London: Faber.

Pigram, J.J. (1977) 'Beach resort morphology', *Habitat International*, 2, 5–6: 525–41.

Pigram, J.J. (1980) 'Environmental implications of tourism development', *Annals of Tourism Research*, 7, 4: 554–83.

Pigram, J.J. (1981) 'Outdoor recreation and access to the countryside: focus on the Australian experience', *Natural Resources Journal*, 21, 1: 107–23.

Pigram, J.J. (1983) *Outdoor Recreation and Resource Management*, London: Croom Helm.

Pigram, J.J. (1986) *Issues in the Management of Australia's Water Resources*, Melbourne: Longman.

Pigram, J.J. (1990) 'Sustainable tourism: policy considerations', *Journal of Tourism Studies*, 1, 2: 2–9.

Pigram, J.J. (1992) 'Alternative tourism: tourism and sustainable resource management', in Smith, V. and Eadington, W. (eds) *Tourism Alternatives*, 76–87, Philadelphia, PA: Pennsylvannia Press.

Pigram, J.J. (1993) 'Planning for tourism in rural areas: bridging the policy implementation gap', in Pearce, D.G. and Butler, R.W. (eds) *Tourism Research: Critiques and Challenges*, London: Routledge, pp. 156–74.

Pigram, J.J. (1995) 'Resource constraints on tourism: water resources and sustainability', in Pearce, D.G. and Butler, R.W. (eds) *Change in Tourism: People, Places, Processes*, London: Routledge, pp. 208–28.

Pigram, J.J. (2003a) 'Countryside Agency/Countryside Commission', in Jenkins, J.M. and Pigram, J.J. (eds) *Encyclopedia of Leisure and Outdoor Recreation*, London: Routledge, p. 85.

Pigram, J.J. (2003b) 'Pollution', in Jenkins, J.M. and Pigram, J.J. (eds) *Encyclopedia of Leisure and Outdoor Recreation*, London: Routledge, pp. 379–380.

Pigram, J.J. and Hobbs, J. (1975) 'The weather, outdoor recreation, and tourism', *Journal of Physical Education and Recreation*, 46, 9: 12–13.

Pigram, J.J. and Jenkins J.M. (1994) 'The role of the public sector in the supply of rural recreation opportunities', in Mercer, D. (ed.) *New Viewpoints in Outdoor Recreation Research in Australia*, Williamstown: Hepper Marriott and Associates, pp. 119–28.

Pigram, J.J. and Wahab, S. (1997) 'Sustainable tourism in a changing world', in Wahab, S. and Pigram, J.J. (eds) *Tourism, Development and Growth*, London: Routledge, pp. 17–32.

Pigram, J.J., Nguyen, S. and Rugendyke, B. (1997) 'Tourism and national parks in emerging regions of the developing world: Cat Ba Island National Park, Vietnam', Paper presented to Meeting of the International Academy for the Study of Tourism, Malacca, June.

Plog, S. (1972) 'Why destination areas rise and fall in popularity', unpublished paper presented to Southern California Chapter, Travel Research Association, Los Angeles.

Plog, S. (1998) 'Why destination presentation makes economic sense', in Theobald, W. (ed.) *Global Tourism*, 2nd edn, Oxford: Butterworth-Heinemann.

Pocock, B. and Clarke, J. (2004) 'Can't buy me love? Young Australians' views on parental work, time, guilt and their own consumption', Discussion Paper No. 61, The Australia Institute, February.

Poole, M. (1986) 'Adolescent leisure activities: social class, sex and ethnic differences', *Australian Journal of Social Issues*, 21, 1: 42–56.

Poon, A. (1993) *Tourism Technology and Competitive Strategies*, Wallingford: CABI.

Prentice, R. (1993) 'Motivations of the heritage consumer in the leisure market: an application of the Manning-Haas demand hierarchy', *Leisure Sciences*, 15: 273–90.

Pressey, R., Bedward, M. and Nicholls, A. (1990) 'Reserve selection in mallee lands', in Noble, J., Joss, P. and Jones, G. (eds) *Proceedings of the National Mallee Conference*, Melbourne: CSIRO, pp. 167–78.

Price, R. (1980) 'Tourism and outdoor recreation – some geographical definitions', paper presented to Annual Meeting of Travel Research Association, Savannah.

Price, R. (1981) 'Tourist landscapes and tourist regions', paper presented to 77th Annual Meeting of Association of American Geographers, Los Angeles.

Priddle, G. (1975) 'Identifying scenic rural roads', paper presented to 77th Annual Meeting, Association of American Geographers, Los Angeles.

Prior, T. and Clark, R. (1984) 'Technique for identifying scenic/recreational roads in the Hunter Valley, NSW', *Australian Geographer*, 16, 1: 50–4.

Przeclawski, K. (1986) *Humanistic Foundations of Tourism*, Warsaw: Institute of Tourism.

Pugh, D.A. (1990) 'Decision frameworks and interpretation', in Graham, R. and Lawrence, R. (eds) *Towards Serving Visitors and Managing Our Resources*, Proceedings of a North American Workshop on Visitor Management in Parks and Protected Areas, Waterloo: Tourism Research and Education Centre, University of Waterloo, pp. 355–6.

Pullen, J. (1977) *Greenspace and the Cities*, Canberra: Australian Institute of Urban Studies.

Ramthun, R. (1995) 'Factors in user group conflict between hikers and mountain bikers, *Leisure Sciences*, 17: 159–69.

Rapoport, R. and Rapoport, R.N. (1975) *Leisure and the Family Life Cycle*, London: Routledge and Kegan Paul.

Ravenscroft, N. (1992) *Recreation Planning and Development*, Basingstoke: Macmillan.

Ravenscroft, N. (1996) 'New access initiatives: the extension of recreation opportunities or the diminution of citizen rights?', in Watkins, C. (ed.) *Rights of Way: Policy Culture and Management*, London: Pinter.

Ravenscroft, N. and Curry, N. (2004) 'Constraints to participation in countryside recreation in England', *Annals of Leisure Research*, 7, 3–4: 172–87.

Ravenscroft, N. and Rogers, G. (2003) 'A critical incident study of barriers to participation on the Cuckoo Trail, East Sussex', *Managing Leisure*, 8, 4: 184–97.

Ravenscroft, N., Uzzell, D. and Leach, R. (2002) 'Danger ahead? The impact of fear of crime on people's recreational use of nonmotorised shared-use routes'. *Environment and Planning C: Government and Policy*, 20, 5: 741–56.

Ravenscroft, N., Groeger, J., Uzzell, D. and Leach, R. (2003) 'Conflict', in Jenkins, J.M. and Pigram, J.J. (eds) *Encyclopedia of Leisure and Outdoor Recreation*, London: Routledge, pp. 68–70.

Redcliff Ascent – Wilderness Treatment Programme (undated) http://www.redcliffascent.com/ accessed 3 January 2005.

Reid, D. (1989) 'Implementing senior government policy at the local level', *Journal of Applied Recreation Research*, 15, 1: 3–13.

Relph, E. (1976) *Place and Placelessness*, London: Pion.

Renard, Y. and Hudson, L. (1992) 'Community-based management of national parks', paper presented to Fourth World Parks Congress, Caracas, February.

Reynolds, R.J. (1990) 'VAMP and its application to camping: the Glacier National Park example', in Graham, R. and Lawrence, R. (eds) *Towards Serving Visitors and Managing Our Resources*, Proceedings of a North American Workshop on Visitor Management in Parks and Protected Areas, Waterloo: Tourism Research and Education Centre, University of Waterloo, pp. 257–70.

Richardson, B. and Richardson, R. (1994) *Business Planning: An Approach to Strategic Management*, 2nd edn, London: Pitman.

Richins, H., Richardson, J. and Crabtree, A. (eds) (1996) *Ecotourism and Nature-Based Tourism: Taking the Next Steps*, The Ecotourism Association of Australia National Conference Proceedings, Alice Springs, Northern Territory, 18–23 November.

Richter, L.K. (1989) *The Politics of Tourism in Asia*, Honolulu: University of Hawaii Press.

Roberts, K. (1983) *Youth and Leisure*, London: Allen and Unwin.

Roberts, L. and Hall, D. (2001) *Rural Tourism and Recreation: Principles to Practice*, Wallingford: CABI.

Robertson, B.J. (1999) 'Leisure and family: perspectives of male adolescents who engage in delinquent activity', *Journal of Leisure Research*, 31, 4: 335–59.

Robinson, D., Laurie, J., Wagar, J. and Traill, A. (1976) *Landscape Evaluation*, Manchester: University of Manchester.

Robinson, G. (1990) *Conflict and Change in the Countryside*, London: Belhaven.

Roehl, W. (1987) 'An investigation of the perfect information assumption in recreation destination choice models', paper presented to the Annual Conference of the Association of American Geographers, Portland, April.

Rokeach, M. (1973) *The Nature of Human Values*, New York: The Free Press.

Rosenzweig, R. (1983) *Eight Hours for What We Will: Workers and Leisure in an Industrial City, 1870–1920*, New York: Cambridge University Press.

Ross, G. (1994) *The Psychology of Tourism*, Melbourne: Hospitality Press.

Ross, G. (1998) *The Psychology of Tourism*, 2nd edn, Melbourne: Hospitality Press.

Rossman, J.R. (1995) *Recreation Programming: Designing Leisure Experiences*, 2nd edn, Champaign, IL: Sagamore.

Rothman, R. (1978) 'Residents and transients: community reaction to seasonal visitors', *Journal of Travel Research*, 16, 3: 8–13.

Rowan-Robinson, J. (2003) 'Reform of the law relating to access to the countryside: realising expectations', *Journal of Planning and Environment Law*, November, pp. 1394–400.

Rutledge, A. (1971) *Anatomy of a Park*, New York: McGraw-Hill.

Ryan, C. (1995) *Researching Tourist Satisfaction: Issues, Concepts and Problems*, London: Routledge.

Ryan, C. (2003) *Recreational Tourism: Demand and Impacts*, Clevedon: Channel View.

Ryan, C. and Hall, M. (2001) *Sex Tourism: Marginal People and Liminalities*, London: Routledge.

Ryan, C. and Montgomery, D. (1994) 'The attitudes of Bakewell residents to tourism and issues in community responsive tourism', *Tourism Management*, 15, 5: 358–70.

Samdahl, D. and Jekubovich, N. (1997) 'A critique of leisure constraints: comparative analyses and understandings', *Journal of Leisure Research*, 29: 430–52.

Sandercock, L. (2003) *Cosmopolis II: Mongrel Cities of the 21st Century*, London: Continuum.

Sax, J. (1980) *Mountains without Handrails: Reflections on the National Parks*, Ann Arbor, MI: University of Michigan Press.

Scharff, R. (1972) *Canada's Mountain National Parks*, Banff: Lebow Books.

Schindler B, Cheek, A., and Stankey, G.H. (1999) *Monitoring and Evaluating Citizen–Agency Interactions: A Framework Developed for Adaptive Management*, Pacific Northwest General Technical Report no. PNW-GTR-452, Portland, OR: USDA Forest Service.

Schneider, D.M. and Smith, R.T. (1973) *Class Differences and Sex Roles in American Kinship and Family Structures*, Englewood Cliffs, NJ: Prentice Hall.

Schneider, I.E. and Hammitt, W.E. (1995a) 'Visitor response to outdoor recreation conflict: a conceptual approach', *Leisure Sciences*, 17: 223–34.

Schneider, I.E. and Hammitt, W.E. (1995b) 'Visitor responses to on-site recreation conflict', *Journal of Applied Recreation Research*, 20, 4: 249–68.

Schomburgk, C. (1985) 'Urban parkland – how much, where and what types?', *Australian Parks and Recreation*, 22, 1: 21–7.

Schreyer, R., Knopf, R. and Williams, D. (1985) 'Reconceptualizing the motive/environment link in recreation choice behavior', *Proceedings – Symposium on Recreation Choice Behavior*, Ogden, UT: USDA General Technical Report, INT 184, pp. 9–18.

Seabrooke, W. and Miles, C. (1993) *Recreational Land Management*, 2nd edn, London: Spon Press.

Searle, M.S. and Jackson, E.L. (1985) 'Socioeconomic variations in perceived barriers to recreation among would-be participants', *Leisure Sciences*, 7: 227–49.

Seedsman, T.A. (1995) 'More to life! The value of leisure, recreation and education for the aged', *Australian Parks and Recreation*, 31, 4: 31–6.

Selin, S. and Chavez, D. (1995) 'Developing a collaborative model for environmental planning and management', *Environmental Management*, 19, 2: 189–95.

Sellar, D. (2003) 'Community rights and access to land in Scotland', *Proceedings from a Workshop on Old and New Commons*, Centre for Advanced Study, Oslo, Norway, 11–13 March, http://www.caledonia.org.uk/land, accessed 1 March 2005.

Serageldin, I. and Steer, A. (1994) *Valuing the Environment*, Washington, DC: World Bank.

Settergren, C.D. and Cole, D.N. (1970) 'Recreation effects on soil and vegetation in the Missouri Ozarks', *Journal of Forestry*, 68: 231–3.

Shackley, M. (1996) *Wildlife Tourism*, London: International Thomson Business Press.

Shackley, M. (ed.) (1998) *Visitor Management: Case Studies from World Heritage Sites*, Oxford: Butterworth-Heinemann.

Shafritz, J.M. and Ott, J.S. (1992) *Classics of Organization Theory*, 3rd edn, Pacific Grove, CA: Brooks/Cole.

Sharpley, R. (1997) *Rural Tourism: An Introduction*, London: International Thomson Business Press.

Sharpley, R. (2003) *Tourism and Leisure in the Countryside*, Huntingdon: Elm Publications.

Sharpley, R. and Sharpley, J. (1997) *Rural Tourism: An Introduction*, London: International Thomson Business Press.

Shaw, S.M. (1985) 'Gender and leisure: inequality in the distribution of leisure time', *Journal of Leisure Research*, 17: 266–82.

Shaw, S.M. (1992) 'Dereifying family leisure: an examination of women's and men's everyday experiences and perceptions of family time', *Leisure Sciences*, 14: 271–86.

Shaw, S.M. (1994) 'Gender, leisure and constraint: towards a framework for the analysis of women's leisure', *Journal of Leisure Research*, 26, 1: 8–22.

Shaw, S.M., Bonen, A. and McCabe, J.F. (1991) 'Do more constraints mean less leisure? Examining the relationship between constraints and participation', *Journal of Leisure Research*, 23, 4: 286–300.

Shelby, B. and Heberlein, T.A. (1986) *Carrying Capacity in Recreation Settings*, Corvallis, OR: Oregon State University Press.

Shivers, J. (1967) *Principles and Practices of Recreational Service*, New York: Macmillan.

Shoard, M. (1996) 'Robbers v. revolutionaries: what the battle for access is really all about', in Watkins, C. (ed.) *Rights of Way: Policy, Culture and Management*, London: Pinter, pp. 11–23.

Shoard, M. (1999) *A Right To Roam*, Oxford: Oxford University Press.

Shogan, D. (2002) 'Characterizing constraints of leisure: a Foucaultian analysis', *Leisure Studies*, 21, 1: 27–38.

Siegenthaler, K.L. and O'Dell, I. (2000) 'Leisure attitude, leisure satisfaction, and perceived freedom in leisure within family dyads', *Leisure Sciences*, 22: 281–96.

Simmons, B.A. and Dempsey, I. (1996) 'National profile of away from home leisure activities of persons aged 60 and over', *Australian Leisure*, 7, 1: 41–6.

Simmons, D. (1994) 'Community participation in tourism planning', *Tourism Management*, 15, 2: 98–108.

Simmons, D. and Leiper, N. (1993) 'Tourism: a social scientific perspective', in Perkins, H.C. and Cushman, G. (eds) *Leisure Recreation and Tourism*, Auckland: Longman Paul.

Simmons, I.G. (1975) *Rural Recreation in the Industrial World*, London: Edward Arnold.

Simmons, R., Davis, B.W., Chapman, R.J.K. and Sager, D.D. (1974) 'Policy flow analysis: a conceptual model for comparative policy research', *Western Political Quarterly*, 27, 3: 457–68.

Simpson, R. (1995) 'Channel vision of the future', *National Parks Today*, 39: 5.

Singleton, J.F. (1985) 'Activity patterns of the elderly', *Society and Leisure*, 8, 2: 805–19.

Slater, T. (1984) *The Temporary Community: Organized Camping for Urban Society*, Sutherland: Albatross Books.

Slatter, R. (1978) 'Ecological effects of trampling on sand dune vegetation', *Journal of Biological Education*, 12, 2: 89–96.

Slee, W., Joseph, D.H. and Curry, N.R. (2001) *Social exclusion in countryside leisure in the UK*, Cardiff: Countryside Recreation Network.

Slocombe, D.S. and Dearden, P. (2002) 'Protected areas and ecosystem-based management', in Dearden, P. and Rollins, R. (eds) *Parks and Protected Areas in Canada: Planning and Management*, Toronto: Oxford University Press, pp. 295–320.

Smale, B. (1990) 'Spatial equity in the provision of urban recreation opportunities', *Proceedings of Sixth Canadian Congress on Leisure Research*, Waterloo: University of Waterloo.

Smale, B.J.A. and Dupuis, S.L. (1993) 'The relationship between leisure activity participation and psychological well being across the lifespan', *Journal of Applied Recreation Research*, 18, 4: 281–300.

Smith, R. (1991) 'Beach resorts: a model of development evolution', *Landscape and Urban Planning*, 21: 189–210.

Smith, R. (1992) 'Review of integrated beach resort development in Southeast Asia', *Land Use Policy*, 9: 209–17.

Smith, R., Austin, D. and Kennedy, D. (2001) *Inclusive and Special Recreation: Opportunities for People with Disabilities*, Boston, MA: McGraw Hill.

Smith, V. (ed.) (1978) *Hosts and Guests: The Anthropology of Tourism*, Oxford: Basil Blackwell.

Smith, V. (1994) 'Privatisation in the Third World: small scale tourism enterprises', in Theobald, W.F. (ed.) *Global Tourism: The Next Decade*, Oxford: Butterworth-Heinemann, pp. 163–73.

Smyth, B. (2004) 'Fathers and family separation', in Fletcher, R., Fairbairn, H. and Pascoe, S. (eds) *Fatherhood Research in Australia, Research Report*, March, Callaghan: The Family Action Centre and The University of Newcastle.

Snead, S. (2003) 'Therapeutic recreation', in Jenkins, J.M. and Pigram, J.J. (eds) *Encyclopedia of Leisure and Outdoor Recreation*, London: Routledge, pp. 499–502.

Snepenger, D.J. (1987) 'Segmenting the vacation market by novelty-seeking role', *Journal of Travel Research*, 26, 2: 8–14.

Snepenger, D.J. and Karahan, R.S. (1991) 'Visitation to Yellowstone National Park after the fires of 1988', *Annals of Tourism Research*, 18: 319–20.

Snyder, E. and Spreitzner, E. (1973) 'Correlates of sport participation among adolescent girls', *Research Quarterly*, 47: 804–9.

Social and Community Planning Research (1997) *United Kingdom Day Visits Survey 1996*, London: Social and Community Planning Research.

Social and Community Planning Research (1999) 'United Kingdom Leisure Day Visits Survey 1998', unpublished data printouts, London: Social and Community Planning Research.

Sonak, S. (2004) 'Ecological footprint of production', *Journal of Tourism Studies*, 15, 2: 2–12.

Sorensen, A.D. and Epps, R. (eds) (1993) *Prospects and Policies for Rural Australia*, Melbourne: Longman.

Spinew, K., Tucker, D. and Arnold, M. (1996) 'Free time as a workplace incentive: a comparison between women and men', *Journal of Applied Recreation Research*, 21, 3: 195–212.

Sports Council (1991) *A Countryside for Sport: Towards a Policy for Sport and Recreation in the Countryside*, London: Sports Council.

Spring Ridge Lodge Retreat in Western Montana (undated) http://www.familyfirstaid.org/spring_creek.htm, accessed 3 January 2005.

Stafford, F.P. (1980) 'Women's use of leisure time converging with men's', *Monthly Labor Review*, 103: 57–9.

Stanfield, C. (1969) 'Recreation land use patterns within an American seaside resort', *The Tourist Review*, 24: 128–36.

Stanfield, C. and Rickert, J. (1970) 'The recreational business district', *Journal of Leisure Research*, 4: 213–25.

Stankey, G. (1977) 'Some social concepts for outdoor recreation planning', in *Proceedings of Symposium on Outdoor Advances in the Application of Economics*, Washington, DC: USDA Forest Service, General Technical Report WO-2.

Stankey, G. (1982) 'Carrying capacity, impact management and the recreation opportunity spectrum', *Australian Parks and Recreation*, May: 24–30.

Stankey, G., McCool, S. and Stokes, G. (1984) 'Limits of acceptable change: a new framework for managing the Bob Marshall Wilderness Complex', *Western Wildlands*, 10, 3: 33–7.

Stankey, G., Cole, D., Lucas, R., Peterson, M. and Frissell, S. (1985) *The Limits of Acceptable Change (LAC) System for Wilderness Planning*, Ogden, UT: USDA Forest Service.

Stebbins, R.A. (1982) 'Serious leisure: a conceptual statement', *Pacific Sociological Review* 25: 251–72.

Stebbins, R.A. (1997) 'Casual leisure: a conceptual statement', *Leisure Studies*, 16: 17–25.

Stebbins, R.A. (2003a) 'Casual leisure', in Jenkins, J.M. and Pigram, J.J. (eds) *Encyclopedia of Leisure and Outdoor Recreation*, London: Routledge, pp. 44–6.

Stebbins, R.A. (2003b) 'Serious leisure', in Jenkins, J.M. and Pigram, J.J. (eds) *Encyclopedia of Leisure and Outdoor Recreation*, London: Routledge, pp. 452–5.

Stemerding, M., Oppewal, H. and Timmermans, H. (1999) 'A constraints-induced model of park choice', *Leisure Sciences*, 21, 2: 145–58.

Stephens, W.N. (1983) *Explanations for Failures of Youth Organizations*, ERIC Document No. ED 228 440.

Stohlgren, T.J. and Parsons, D.J. (1979) 'Vegetation and soil recovery in wilderness campsites closed to visitor use', *Environmental Management*, 10, 3: 375–80.

Stokowski, P. (1990) 'The social networks of spouses', in Smale, B. (ed.) *Proceedings of the Sixth Canadian Congress on Leisure Research*, Waterloo: University of Waterloo.

Stopher, P. and Lee-Gosselin, M. (eds) (1997) *Understanding Travel Behaviour in an Era of Change*, Oxford: Elsevier Science.

Stott, D. (1998) 'Korea crisis slashes tourist numbers', *The Sydney Morning Herald*, 7 January: 2.

Strauss, A. and Corbin, J. (1990) *Basics of Qualitative Research: Grounded Theory Procedures and Techniques*, Newburt Park: Sage.

Stumbo, N.J. (2002) *Client Assessment in Therapeutic Recreation Services*, State College, PA: Venture Publishing.

Stumbo, N.J. and Peterson, C.A. (2004) *Therapeutic Recreation Program Design: Principles and Procedures*, 4th edn, San Francisco: Benjamin Cummings.

Sullivan, R. (1993) *Recreation. A Healthy Alternative to Crime*, Woodville: Royal Australian Institute of Parks and Recreation.

Sundell, R. (1991) 'The use of spatial modelling to evaluate park boundaries and delineate critical resource areas', unpublished PhD thesis, Northwestern University, Evanston.

Swinglehurst, E. (1994) 'Face to face: the socio-cultural impacts of tourism', in Theobald, W.F. (ed.) *Global Tourism: The Next Decade*, Oxford: Butterworth-Heinemann, pp. 92–102.

Swinnerton, G. (1982) *Recreation on Agricultural Land in Alberta*, Edmonton: Environment Council of Alberta.

Sylvester, C., Voelkl, J.E. and Ellis, G.D. (2001) *Therapeutic Recreation: Theory and Practice*, State College, PA: Venture Publishing.

Szadkowski, J. (1995) 'Taking a stand for children', *The Washington Times*, 16 December, http://www.vachss.com/av_articles/wash_times_95.html, accessed 20 December 2004.

Taffel, J. (2004) 'Fixed up in a foreign land', *Sydney Morning Herald*, Health and Science Supplement, 9 December: 9.

Tangi, M. (1977) 'Tourism and the environment', *Ambio*, 6: 336–41.

Taylor, J.Y. (1997) 'Leave only footprints? How backcountry campsite use affects forest structure', *Yellowstone Science*, 5 (1–winter): 14–17.

Taylor, P.W. (1959) '"Need" statements', *Analysis* 19: 106–11.

Terkenli, T. (2004) 'Tourism and landscape', in Lew, A.A., Hall, C.M. and Williams, A.C. (eds) *A Companion to Tourism*, Malden, MA: Blackwell, pp. 339–48.

Terkenli, T.S. (2002) 'Landscapes of tourism: towards a global cultural economy of space?', *Tourism Geographies*, 4, 3: 227–54.

The Australian (1993) 10 November.

The Countryside Agency (2000) Board meeting minutes, http://www.countryside.gov.uk/ WhoWeAreAndWhatWeDo/boardMeetings/boardPapers/CA_AP00_30.asp? printable=true, accessed 9 August 2005.

The Macquarie Dictionary (1987), The Macquarie Library, Macquarie University.

The Ramblers (UK), http://www.ramblers.org.uk/info/contacts/govt.html.

Theberge, J. (1989) 'Guidelines to drawing ecologically sound boundaries for national parks and nature reserves', *Environmental Management*, 13, 6: 695–702.

Theberge, J. (1992) 'Concepts of conservation biology and boundary delineation in parks', in *Proceedings of a Seminar on Size and Integrity Standards for Natural Heritage Areas in Ontario*, Toronto: Ministry of Natural Resources, pp. 16–24.

Theobald, W. (1979) *Evaluation of Recreation and Park Programs*, New York: John Wiley and Sons.

Theobald, W. (ed.) (2005) *Global Tourism*, 3rd edn, New York: Elsevier.

Thomas, J. and Neill, K. (1993) 'Benchmarking industrial R&D', *Search*, 24, 6: 158–9.

Thomas, L. and Middleton, J. (2003) *Guidelines for Management Planning of Protected Areas*, World Commission on Protected Areas, Best Practice Protected Area Guidelines Series No. 10, Gland: The World Conservation Union (IUCN).

Thompson, P. (1992) ' "I don't feel old": subjective ageing and the search for meaning in later life', *Ageing and Society*, 12: 23–47.

Thomson, J., Lime, D., Gartner, W. and Sames, W. (1995) *Proceedings Fourth International Outdoor Recreation and Tourism Trends Symposium*, St Paul, MN: University of Minnesota Press.

Thomson, K. and Whitby, M. (1976) 'The economics of public access in the countryside', *Journal of Agricultural Economics*, 27, 3: 307–19.

Thorsell, J.W. (1984) 'National parks from the ground up: experience from Dominica, West Indies', in: McNeely, J.A. and Miller, K.R. (eds) *National Parks, Conservation and Development. The Role of Protected Areas in Sustaining Society*, Proceedings of the World Congress on National Parks, Bali, Indonesia, 11–22 October 1982. pp. 616–21.

Tilden, F. (1977) *Interpreting Our Heritage*, 3rd edn, Chapel Hill, NC: North Carolina University Press.

Tinsley, H.E.A., Colbs, S.L., Teaff, J.D. and Kauffman, N. (1987) 'The relationship of age, gender, health and economic status to the psychological benefits older adults report from participation in leisure activities', *Leisure Sciences*, 9: 53–65.

Towner, J. (1996) *An Historical Geography of Recreation and Tourism in the Western World 1540–1940*, Chichester: John Wiley.

Toyne, P. (1974) *Recreation and Environment*, London: Macmillan.

Trapp, S., Gross, M. and Zimmerman, R. (1994) *Signs, Trails, and Wayside Exhibits: Connecting People and Places*, Stevens Point, WI: University of Wisconsin, SP Foundation Press.

Tratalos, J.A. and Austin, T.J. (2001) 'Impacts of recreational SCUBA diving on coral communities of the Caribbean Island of Grand Cayman', *Biological Conservation*, 102: 67–75.

Tribe, J. (1995) *The Economics of Leisure and Tourism: Environments and Markets*, Oxford: Butterworth-Heinemann.

Tribe, J. (2003) 'Cost–benefit analysis', in Jenkins, J.M. and Pigram, J.J. (eds) *Encyclopedia of Leisure and Outdoor Recreation*, London: Routledge, pp. 83–4.

Tubb, K.N. (2003) 'An evaluation of the effectiveness of interpretation within Dartmoor National Parks in reaching the goals of sustainable tourism development', *Journal of Sustainable Tourism*, 11, 6: 476–98.

Turner, A. (1987) 'The management of impacts in recreational use of nature areas', paper presented to 22nd Conference of the Institute of Australian Geographers, Canberra, August.

Turner, A. (1994) 'Managing impacts: measurement and judgement', in Mercer, D. (ed.) *New Viewpoints in Outdoor Recreation Research and Planning*, Melbourne: Hepper Marriott and Associates, pp. 129–40.

Turner, L. (1976) 'The international division of leisure: tourism and the Third World', *Annals of Tourism Research*, 4, 1: 12–24.

UNESCO (1976) 'The effects of tourism on socio-cultural values', *Annals of Tourism Research*, 4, 2: 74–105.

United States Bureau of Outdoor Recreation (1973) *Outdoor Recreation. A Legacy for America*, Washington, DC: Department of the Interior.

United States Bureau of Outdoor Recreation (1975) *Assessing Demand for Outdoor Recreation*, Washington, DC: US Bureau of Outdoor Recreation.

United States Department of Agriculture (USDA) (1981) *National Agricultural Lands Study*, Washington, DC: USDA.

United States Department of Agriculture (USDA), Forest Service (1985) *Bob Marshall Great Bear Scapegoat Wilderness Action Plan for Managing Recreation (The Limits of Acceptable Change)*, Flathead National Forest: USDA Forest Service.

United States Department of Agriculture (USDA), Forest Service (2000) *National Survey on Recreation and the Environment*, Knoxville, TN: USDA Forest Service and University of Tennessee.

United States Department of Agriculture (USDA), Forest Service, http://www.fs.fed.us/.

United States Department of the Interior, National Parks Service (1997) *VERP – The Visitor Experience and Resource Protection Framework: A Handbook for Planners and Managers*, US Department of the Interior, National Parks Service, Denver, http://planning.nps.gov/document/verphandbook.pdf.

United States Department of the Interior (1978) *National Urban Recreation Study*, Washington DC: United States Department of the Interior.

United States Government (1996) *Multiple Use Sustained Yield Act of 1960* (as amended), http://www.fs.fed.us/emc/nfma/includes/musya60.pdf, accessed 20 January 2005.

University of Brighton (2001) *Water-Based Sport and Recreation: The Facts*, Report to Department of the Environment, Food and Rural Affairs, Countryside Division. School of the Environment, University of Brighton, http://www.defra.gov.uk/wildlife-countryside/resprog/findings/watersport.pdf.

Unwin, K. (1975) 'The relationship of observer and landscape in landscape evaluation', *Transactions, Institute of British Geographers*, 66: 130–4.

USA Today (1994), 11 April: 7.

Urry, J. (1995) *Consuming Places*, London: Routledge.

US Bureau of Outdoor Recreation – see United States Bureau of Outdoor Recreation.

US Department of Agriculture – see United States Department of Agriculture.

US Department of the Interior – see United States Department of the Interior.

Uysal, M. and Hagan, L.A.R. (1994) 'Motivation of pleasure travel and tourism', in Khan, M.A., Olsen, M.D. and Var, T. (eds) *VNR's Encyclopedia of Hospitality and Tourism*, New York: Van Nostrand Reinhold.

Uysal, M., Fesenmaier, D.R. and O'Leary, J.T. (1994) 'Geographic and seasonal variation in the concentration of travel in the United States', *Journal of Travel Research*, 32, 3: 61–4.

Valentine, P.S. (1991) 'Ecotourism and nature conservation: a definition with some recent development in Micronesia', in Weiler, B. (ed.) *Ecotourism: Incorporating the Global Classroom*, Canberra: Bureau of Tourism Research, pp. 4–10.

Valentine, P.S. (1992) 'Review: nature-based tourism', in Weiler, B. and Hall, C.M. (eds) *Special Interest Tourism*, London: Belhaven, pp. 105–28.

Vancouver-Clark Parks and Recreation, Washington, WA (undated), http://www.ci.vancouver.wa.us/parks-recreation/parks_trails/planning/standards.htm, accessed 31 January 2005

Vandalism in the Parks (undated), http://www.cityofboise.org/parks/caring/index.aspx?id= vandalism, accessed 2 February 2005.

Vanhove, N. (1997) 'Mass tourism: benefits and costs', in Wahab, S. and Pigram, J.J. (eds) *Tourism, Development and Growth*, London: Routledge, pp. 50–77.

Vanhove, N. (2005) *Economics of Tourism Destinations*, Oxford: Elsevier Butterworth-Heinemann.

Van Lier, H.N. and Taylor, P.D. (eds) (1993) *New Challenges in Recreation and Tourism Planning*, Amsterdam: Elsevier Science Publishers.

Vaske, J.J., Decker, D.J. and Manfredo, M.J. (1995) 'Human dimensions of wildlife management: an integrated framework for coexistence', in Knight, R.L. and Gutzwiller, K.J. (eds) *Wildlife and Recreationists: Coexistence Through Management and Research*, Washington, DC: Island Press, pp. 33–47.

Veal, A.J. (1987) *Leisure and the Future*, London: Allen and Unwin.

Veal, A.J. (1994) *Leisure Policy and Planning*, Harlow: Longman.

Veal, A.J. (1997) *Research Methods for Leisure and Tourism: A Practical Guide*, 2nd edn, London: Pitman in Association with the Institute of Leisure and Amenity Management.

Veal, A.J. (2002) *Leisure and Tourism Policy and Planning*, 2nd edn, Wallingford: CABI.

Veal, A.J. (2003) 'Contingency valuation', in Jenkins, J.M. and Pigram, J.J. (eds) *Encyclopedia of Leisure and Outdoor Recreation*, London: Routledge, pp. 81–2.

Veal, A.J. and Lynch, R. (2001) *Australian Leisure*, 2nd edn, Sydney: Pearson Education.

Veblen, T. (1899, republished 1970) *The Theory of the Leisure Class*, London: Allen and Unwin.

Vertinsky, P. (1995) 'Stereotypes of aging women and exercise: a historical perspective', *Journal of Aging and Physical Activity*, 3: 223–37.

Veverka, J.A. (1994) *Interpretive Master Planning*, Helena, MT: Falcon Press.

Viljoen, J. (1994) *Strategic Management: Planning and Implementing Successful Corporate Strategies*, Melbourne: Longman.

Vining, J. and Fishwick, L. (1991) 'An exploratory study of outdoor recreation site choices', *Journal of Leisure Research*, 23, 2: 114–32.

Wade, M.G. (ed.) (1985) *Constraints on Leisure*, Springield, IL: Charles C. Thomas.

Wahab, S. and Pigram, J.J. (eds) (1997) *Tourism, Development and Growth*, London: Routledge.

Walker, B. and Nix, H. (1993) 'Managing Australia's biological diversity', *Search*, 24, 5: 173–8.

Wall, G. (1989) *Outdoor Recreation in Canada*, New York: Wiley.

Wall, G. and Wright, C. (1977) *The Environmental Impact of Outdoor Recreation*, Department of Geography Publications Series No. 11, Waterloo: University of Waterloo.

Walmsley, D.J. (2003) 'Cognition', in Jenkins, J.M. and Pigram, J.J. (eds) *Encyclopedia of Leisure and Outdoor Recreation*, London: Routledge, p. 57.

Walmsley, D.J. and Jenkins, J.M. (1994) 'Evaluations of recreation opportunities: tourist images of the New South Wales North Coast', in Mercer, D.C. (ed.) *New Viewpoints in Australian Outdoor Recreation Research and Planning*, Melbourne: Hepper Marriott and Associates, pp. 89–98.

Walmsley, D.J. and Jenkins, J.M. (1999) 'Cognitive distance: a neglected issue in travel behaviour', in Pizam, A. and Mansfield, Y. (eds) *Consumer Behaviour in Travel and Tourism*, New York: Haworth Hospitality Press, pp. 287–304.

Walmsley, D.J. and Jenkins, J.M. (2003) 'Leisure', in Jenkins, J.M. and Pigram, J.J. (eds) *Encyclopedia of Leisure and Outdoor Recreation*, London: Routledge, 452–5.

Walmsley, D.J. and Lewis, G.J. (1984) *Human Geography: Behavioural Approaches*, London: Longman

Walmsley, D.J., Boskovic, R. and Pigram, J.J. (1981) *Tourism and Crime, Report to the Criminology Research Council*, Armidale: University of New England.

Walters, C. (1986) *Adaptive Management of Renewable Resources*, New York: Macmillan.

Washburne, R.F. (1982) 'Wilderness recreation carrying capacity: are numbers necessary?', *Journal of Forestry*, 80: 726–8.

Watkins, C. (ed.) (1996) *Rights of Way: Policy, Culture and Management*. London: Pinter.

Watson, A.E. (1995) 'An analysis of recent progress in recreation conflict research and perceptions of future challenges and opportunites', *Leisure Sciences*, 17, 3: 235.

Watson, A.E., Williams, D.R. and Daigle, J.J. (1991) 'Sources of conflict between hikers and mountain bike riders in the Rattlesnake NRA', *Journal of Park and Recreation Administration*, 9, 3: 59.

WCED – *see* World Commission on Environment and Development.

Wearing, S. and Huyskens, M. (2001) 'Moving on from joint management policy regimes in Australian national parks', *Current Issues in Tourism*, 4, 2–4: 182–209.

Weaver, D.B. (2001a) *Ecotourism*, Milton, QLD: John Wiley and Sons.

Weaver, D.B. (ed.) (2001b) *The Encyclopedia of Ecotourism*, Wallingford: CABI.

Weaver, D.B. (2002) 'Asian ecotourism: patterns and themes', *Tourism Geographies*, 4, 2: 153–72.

Weaver, D.B. and Lawton, L. (2002) *Tourism Management*, Milton, QLD: John Wiley and Sons.

Weinmayer, M. (1973) 'Vandalism by design: a critique', in Gray, D. and Pelegrino, D. (eds) *Reflections on the Recreation and Park Movement*, Dubuque, IA: Brown, pp. 246–8.

Wellman, J.D., Roggenbuck, J.W. and Smith, A.C. (1982) 'Recreational specialization and norms of depreciative behavior among canoeists', *Journal of Leisure Research*, 14, 4: 323–40.

Wethers, B. with Michaud, S.G. (2000) *Left for Dead: My Journey Home from Everest*, London: Warner Books.

Wharton, A. (2005) 'Online database tells tale of our landscape', *Countryside Focus*, 2.

Whelan, T. (1991) 'Ecotourism and its role in sustainable development', in Whelan, T. (ed.) *Nature Tourism*, Washington, DC: Island Press.

WHO – see World Health Organisation.

Wicks, B., Backman, K., Allen, J. and Blaricom, D. (1993) 'Geographic information systems: a tool for marketing, managing and planning municipal park systems', *Journal of Park and Recreation Administration*, 11, 1: 9–23.

Wight, P.A. (1993) 'Sustainable ecotourism: balancing economic, environmental and social goals within an ethical framework', *Journal of Tourism Studies*, 4, 2: 54–66.

Wilcox, A.T. (1969) 'Professional preparation for interpretive services', *Rocky Mountain-High Plains Parks and Recreation Journal*, 4, 1: 11–14.

Wilensky, H. (1961) 'The uneven distribution of leisure', *Social Problems*, 9, 1: 107–45.

Williams, D.R. (2003) 'Choice', in Jenkins, J.M. and Pigram, J.J. (eds) *Encyclopedia of Leisure and Outdoor Recreation,* London: Routledge, pp. 51–3.

Williams, P., Dosa, K. and Fulton, A. (1994) 'Tension on the slopes: managing conflict between skiers and snowboarders', *Journal of Applied Recreation Research*, 19, 3: 191.

Williams, S. (1995) *Outdoor Recreation and the Environment*, London: Routledge.

Williamson, P. (1995) 'Occupational therapy and "serious" leisure: promoting productive occupations through leisure', *Australian Journal of Leisure and Recreation*, 5, 1: 61–4.

Willits, W. and Willits, F. (1986) 'Adolescent participation in leisure activities: "the less the more" or "the more the more"?', *Leisure Sciences*, 8: 189–205.

Wilson, P. and Biberbach, P. (1994) 'Community forests – northern experience', *Countryside Campaigner*, Spring: 23.

Wilson, W. (1941) 'The study of administration', *Political Science Quarterly*, 55: 481–506.

Wingo, L. (1964) 'Recreation and urban development: a policy perspective', *Annals of the American Academy of Political Science*, 35: 129–40.

Wolfe, R. (1964) 'Perspectives on outdoor recreation', *Geographical Review*, 54: 203–38.

Wolfe, R. (1982) 'Recreational travel: the new migration, revisited', *Ontarion Geography*, 19: 103–24.

Wong, P.P. (2004) 'Environmental impacts of tourism', in Lew, A.A., Hall, C.M. and Williams, A.C. (eds) *A Companion to Tourism*, Malden, MA: Blackwell, pp. 450–61.

Woodford, J. (2005) 'Aborigines may get right to hunt in parks', *Sydney Morning Herald*, 4 January: 6.

Woodward, D. and Green, E. (1988) '"Not tonight, dear!" The social control of women's leisure', in Wimbush, E. and Talbot, M. (eds) *Relative Freedoms: Women and Leisure*, Trowbridge: Open University Press, pp. 131–46.

Worboys, G., Lockwood, M. and De Lacy, T. (2001) *Protected Area Management: Principles and Practice*, Melbourne: Oxford University Press.

World Commission on Environment and Development (WCED) (1987) *Our Common Future*, New York: Oxford University Press.

World Commission on Protected Areas (2003) http://www.iucn.org/themes/wcpa/wpc2003/english/outputs/intro.htm, accessed 8 August 2005.

World Health Organisation (WHO) (1980) *International Classification of Impairments, Disabilities and Handicaps*, Geneva: World Health Organisation.

World Tourism Organisation (WTO) (1977) *World Travel Statistics*, Madrid: WTO.

World Tourism Organisation (WTO) (1979) *World Travel Statistics*, Madrid: WTO.

World Tourism Organisation (WTO) (1991) *Resolutions of International Conference on Travel and Tourism (Recommendation No. 29)*, Ottawa: WTO.

World Tourism Organisation (WTO) (1992) *Guidelines: Protection of National Parks and Protected Areas for Tourism*, Madrid: WTO.

World Tourism Organisation (WTO) (1993a) *Recommendations on Tourism Statistics*, Madrid: WTO.

World Tourism Organisation (WTO) (1993b) *Global Tourism Forecasts to the Year 2000 and Beyond*, Madrid: WTO.

World Tourism Organisation (WTO) (1994) *Compendium of Tourism Statistics*, Madrid: WTO.

World Tourism Organisation (WTO) (1997a) *Tourism Highlights 1996*, Madrid: WTO.

World Tourism Organisation (WTO) (1997b) *WTO News*, March, Madrid: WTO.

World Tourism Organisation (WTO) (2000) *World Tourism Highlights 2000*, Madrid: WTO.

World Tourism Organisation (WTO) (2001) *The Tourism Satellite Accounts as an Ongoing Process: Past, Present, and Future*, WTO, Madrid.

World Tourism Organisation (WTO) (2003) *World Tourism Highlights 2002*, WTO, Madrid.

WTO – see World Tourism Organisation.

World Travel and Tourism Council (WTTC) (1995) *Travel and Tourism: A New Economic Perspective*, London: Pergamon.

Wunderlich, G. (1979) 'Land ownership: a status of facts', *Natural Resources Journal*, 19, 1: 97–118.

Yarwood, R. and Gardner, G. (2000) 'Fear of crime, cultural threat and countryside', *Area*, 32, 4: 403–11.

Yeo, E. and Yeo, S. (eds) (1981) *Popular Culture and Class Conflict 1590–1914: Explorations in the History of Labour and Leisure*, Brighton: Harvester.

Yin, Z., Katims, D. and Zapata, J. (1999) 'Participation leisure activities and involvement in delinquency by Mexican American adolescents', *Hispanic Journal of Behavioural Sciences*, 21, 2: 170–86.

Young, M. (1999) 'The social construction of tourist spaces', *Australian Geographer*, 30, 3: 373–89.

Young, N. (2000) *Surf Rage*, Angourie: Nymboida Press.

Young, S. (1973) *Tourism. Blessing or Blight?*, London: Penguin.

Youniss, J. (1980) *Parents and Peers in Social Development*, Chicago, IL: University of Chicago Press.

Zabinski, C., Wojtowicz, T. and Cole, D. (2000) 'The effects of recreation disturbance on subalpine seed banks in the Rocky Mountains of Montana', *Canadian Journal of Botany*, 78, 5: 577–82.

Zabriske, R. and McCormick, B. (2001) 'The influences of family leisure patterns on perceptions of family functioning', *Family Relations: Interdisciplinary Journal of Applied Family Studies*, 50, 3: 66–74.

Zbicz, D. (2000) *Transfrontier Ecosystems and Internationally Adjoining Protected Areas*, http://www.unep-wcmc.org/protected_areas/transboundary/adjoining.pdf.

Zimmerman, E. (1951) *World Resources and Industries*, New York: Harper.
Zube, E.H. and Busch, M. (1990) 'Park–people relationships: an international review', *Landscape and Urban Planning*, 19: 117–31.

Index

Note: numbers in italics refer to illustrations